# PHYSICS OF CONTINUOUS MEDIA

## PROBLEMS AND SOLUTIONS IN ELECTROMAGNETISM, FLUID MECHANICS AND MHD

### SECOND EDITION

# PHYSICS OF CONTINUOUS MEDIA

## PROBLEMS AND SOLUTIONS
## IN ELECTROMAGNETISM,
## FLUID MECHANICS AND MHD

### SECOND EDITION

## GRIGORY VEKSTEIN

**CRC Press**
Taylor & Francis Group
Boca Raton  London  New York

CRC Press is an imprint of the
Taylor & Francis Group, an **informa** business

First published 2013 by CRC Press

Published 2019 by CRC Press
Taylor & Francis Group
6000 Broken Sound Parkway NW, Suite 300
Boca Raton, FL 33487-2742

ISBN-13: 978-1-4665-1763-9 (pbk)

**Visit the Taylor & Francis Web site at**
**http://www.taylorandfrancis.com**

**and the CRC Press Web site at**
**http://www.crcpress.com**

# Contents

## Contents

# Preface

*Physics of Continuous Media: Problems and Solutions in Electromagnetism, Fluid Mechanics and MHD, Second Edition* is a revised and extended edition of the text, which was originally published by IOP (The Institute of Physics, UK) in 1992. It is based on lectures and tutorials given for many years, first at the Novosibirsk State University in Russia, and then at the University of Manchester in UK. Its composition, which is a set of problems followed with detailed solutions, is aimed to make this textbook useful as a complementary source to readily available systematic courses on the subject. Therefore, each chapter begins with a brief summary of the definitions and equations, which are necessary to understand and tackle the problems that follow. A number of specific references, mainly to the relevant paragraphs of the famous course of theoretical physics by L. D. Landau and E. M. Lifshitz, as well as to the original journal publications, are provided throughout the text. I am deeply grateful to my friends and former colleagues back in Novosibirsk: Boris Breizman, Dmitri Ryutov and Gennady Stupakov, for numerous critical comments and constructive suggestions. Many thanks also go to the younger colleagues in Manchester: Adam Stanier and, especially, Mykola Gordovskyy, whose help in preparing the present edition was invaluable.

Grigory Vekstein

# 1

## A bit of math: vectors, tensors, Fourier transform, etc.

### Vector operations in cylindrical polar coordinates $(r, \theta, z)$

$$\vec{\nabla} f = \frac{\partial f}{\partial r}\vec{e_r} + \frac{1}{r}\frac{\partial f}{\partial \theta}\vec{e_\theta} + \frac{\partial f}{\partial z}\vec{e_z}$$

$$\vec{\nabla} \cdot \vec{A} = \frac{1}{r}\frac{\partial (rA_r)}{\partial r} + \frac{1}{r}\frac{\partial A_\theta}{\partial \theta} + \frac{\partial A_z}{\partial z}$$

$$\vec{\nabla}^2 f = \frac{1}{r}\frac{\partial}{\partial r}\left(r\frac{\partial f}{\partial r}\right) + \frac{1}{r^2}\frac{\partial^2 f}{\partial^2 \theta} + \frac{\partial^2 f}{\partial^2 z}$$

$$\vec{\nabla} \times \vec{A} = \left(\frac{1}{r}\frac{\partial A_z}{\partial \theta} - \frac{\partial A_\theta}{\partial z}\right)\vec{e_r}$$

$$+ \left(\frac{\partial A_r}{\partial z} - \frac{\partial A_z}{\partial r}\right)\vec{e_\theta} + \left[\frac{1}{r}\frac{\partial (rA_\theta)}{\partial r} - \frac{1}{r}\frac{\partial A_r}{\partial \theta}\right]\vec{e_z}$$

$$\vec{\nabla}^2 \vec{A} = \left(\vec{\nabla}^2 A_r - \frac{A_r}{r^2} - \frac{2}{r^2}\frac{\partial A_\theta}{\partial \theta}\right)\vec{e_r}$$

$$+ \left(\vec{\nabla}^2 A_\theta - \frac{A_\theta}{r^2} + \frac{2}{r^2}\frac{\partial A_r}{\partial \theta}\right)\vec{e_\theta} + \vec{\nabla}^2 A_z\vec{e_z}$$

### Vector operations in spherical coordinates $(r, \theta, \phi)$

$$\vec{\nabla} f = \frac{\partial f}{\partial r}\vec{e_r} + \frac{1}{r}\frac{\partial f}{\partial \theta}\vec{e_\theta} + \frac{1}{r\sin\theta}\frac{\partial f}{\partial \phi}\vec{e_\phi}$$

$$\vec{\nabla} \cdot \vec{A} = \frac{1}{r^2}\frac{\partial (r^2 A_r)}{\partial r} + \frac{1}{r\sin\theta}\frac{\partial (\sin\theta A_\theta)}{\partial \theta} + \frac{1}{r\sin\theta}\frac{\partial A_\phi}{\partial \phi}$$

$$\vec{\nabla}^2 f = \frac{1}{r^2}\frac{\partial}{\partial r}\left(r^2\frac{\partial f}{\partial r}\right) + \frac{1}{r^2\sin\theta}\frac{\partial}{\partial \theta}\left(\sin\theta\frac{\partial f}{\partial \theta}\right) + \frac{1}{r^2\sin^2\theta}\frac{\partial^2 f}{\partial^2 \phi}$$

$$\vec{\nabla} \times \vec{A} = \frac{1}{r \sin \theta} \left( \frac{\partial (\sin \theta A_\phi)}{\partial \theta} - \frac{\partial A_\theta}{\partial \phi} \right) \vec{e_r}$$

$$+ \left[ \frac{1}{r \sin \theta} \frac{\partial A_r}{\partial \phi} - \frac{1}{r} \frac{\partial (r A_\phi)}{\partial r} \right] \vec{e_\theta} + \left[ \frac{1}{r} \frac{\partial (r A_\theta)}{\partial r} - \frac{1}{r} \frac{\partial A_r}{\partial \theta} \right] \vec{e_\phi}$$

$$\vec{\nabla}^2 \vec{A} = \left[ \vec{\nabla}^2 A_r - \frac{2}{r^2} A_r - \frac{2}{r^2 \sin \theta} \frac{\partial (\sin \theta A_\theta)}{\partial \theta} - \frac{2}{r^2 \sin \theta} \frac{\partial A_\phi}{\partial \phi} \right] \vec{e_r}$$

$$+ \left[ \vec{\nabla}^2 A_\theta + \frac{2}{r^2} \frac{\partial A_r}{\partial \theta} - \frac{1}{r^2 \sin^2 \theta} A_\theta - \frac{2 \cos \theta}{r^2 \sin^2 \theta} \frac{\partial A_\phi}{\partial \phi} \right] \vec{e_\theta}$$

$$+ \left[ \vec{\nabla}^2 A_\phi + \frac{2}{r^2 \sin \theta} \frac{\partial A_r}{\partial \phi} + \frac{2 \cot \theta}{r^2 \sin \theta} \frac{\partial A_\theta}{\partial \phi} - \frac{1}{r^2 \sin^2 \theta} A_\phi \right] \vec{e_\phi}$$

---

## Some useful identities of vector analysis

$$\vec{\nabla} \cdot (f \vec{A}) = f \vec{\nabla} \cdot \vec{A} + \vec{A} \cdot \vec{\nabla} f$$

$$\vec{\nabla} \times (f \vec{A}) = f \vec{\nabla} \times \vec{A} - \vec{A} \times \vec{\nabla} f$$

$$\vec{\nabla} \cdot (\vec{A} \times \vec{B}) = \vec{B} \cdot (\vec{\nabla} \times \vec{A}) - \vec{A} \cdot (\vec{\nabla} \times \vec{B})$$

$$\vec{\nabla} \times (\vec{A} \times \vec{B}) = \vec{A}(\vec{\nabla} \cdot \vec{B}) + (\vec{B} \cdot \vec{\nabla})\vec{A} - \vec{B}(\vec{\nabla} \cdot \vec{A}) - (\vec{A} \cdot \vec{\nabla})\vec{B}$$

$$\vec{\nabla}(\vec{A} \cdot \vec{B}) = \vec{A} \times (\vec{\nabla} \times \vec{B}) + (\vec{B} \cdot \vec{\nabla})\vec{A} + \vec{B} \times (\vec{\nabla} \times \vec{A}) + (\vec{A} \cdot \vec{\nabla})\vec{B}$$

---

## Hermitian and antihermitian tensors

A tensor of rank two $T_{ik}$ is a hermitian tensor, if $T_{ki} = T_{ik}^*$, and it is an antihermitian one, if $T_{ki} = -T_{ik}^*$. Any tensor $T_{ik}$ can be represented as a superposition of the hermitian tensor $T_{ik}^{(H)}$ and the antihermitian one $T_{ik}^{(A)}$ as

$$T_{ik} = T_{ik}^{(H)} + T_{ik}^{(A)}, \ T_{ik}^{(H)} = \frac{1}{2}(T_{ik} + T_{ki}^*), \ T_{ik}^{(A)} = \frac{1}{2}(T_{ik} - T_{ki}^*)$$

Vector $\vec{A}$ is called an **eigenvector** of a matrix $T_{ik}$, if $T_{ik} A_k = \lambda A_i$, where $\lambda$ is the respective **eigenvalue** of this matrix ( summation over "dumb" indices, in this case over the index $k$, is assumed). If matrix $T_{ik}$ is a hermitian one, then all its eigenvalues are real and, vice versa, all eigenvalues of an antihermitian matrix are imaginary.

## Problem 1.0.1

Derive the averaged values for the following expressions: $(\vec{a} \cdot \vec{n})\vec{n}$, $(\vec{a} \times \vec{n})^2$ and $(\vec{a} \times \vec{n})(\vec{b} \cdot \vec{n})$, if $\vec{a}$ and $\vec{b}$ are given vectors, and $\vec{n}$ is a unit vector whose orientation in space is random but isotropic.

It is convenient to use here tensor notation. Thus, $< (\vec{a} \cdot \vec{n})\vec{n} >=< a_k n_k n_i >= a_k < n_k n_i >$, where the symbol $<>$ means averaging over all possible orientations of vector $\vec{n}$. Since the latter is distributed isotropically, the tensor of rank two $< n_k n_i >$ should be the same in any coordinate system, i.e., it is an **invariant tensor**. Therefore, it is proportional to the unit tensor of rank two $\delta_{ik}$ : $< n_i n_k >= A\delta_{ik}$. The constant $A$ can be determined by comparing traces of these two tensors. Since $\vec{n}$ is a unit vector, the trace of $< n_i n_k >$ is equal to 1. On the other hand, the trace of $A\delta_{ik}$ is equal to $3A$. Thus, $A = 1/3$, and $a_k \langle n_k n_i \rangle = a_i/3$, or, in vector notations, $< (\vec{a} \cdot \vec{n})\vec{n} >= \vec{a}/3$.

Two other expressions can be dealt with similarly. Thus, $< (\vec{a} \times \vec{n})^2 >=< e_{ikl} a_k n_l e_{ipq} a_p n_q >= a_k a_p e_{ikl} e_{ipq} < n_l n_q >= \frac{1}{3}\delta_{lq} a_k a_p e_{ikl} e_{ipq} = \frac{1}{3} a_k a_p e_{ikl} e_{ipl}$, where $e_{ikl}$ is the unit antisymmetric pseudotensor of rank three. By taking into account that $e_{ikl} e_{ipl} = 2\delta_{kp}$, one gets $< (\vec{a} \times \vec{n})^2 >= \frac{2}{3} a^2$. Finally, $< (\vec{a} \times \vec{n})(\vec{b} \cdot \vec{n}) >=< e_{ikl} a_k n_l b_m n_m >= e_{ikl} a_k b_m < n_l n_m >= \frac{1}{3} e_{ikl} a_k b_m \delta_{lm} = \frac{1}{3} e_{ikl} a_k b_l = \frac{1}{3}(\vec{a} \times \vec{b})$.

## Problem 1.0.2

Derive the time-averaged expressions for tensors $E_\alpha E_\beta, B_\alpha B_\beta, E_\alpha B_\beta$, where vectors $\vec{E}$ and $\vec{B}$ are the electric and magnetic fields of a circularly polarized electromagnetic wave propagating in free space with the wave vector $\vec{k}$ and the electric field amplitude equal to $E_0$.

Consider first tensor $< E_\alpha E_\beta >$. By the very meaning of the time averaging involved here, it is clear that this tensor is invariant under any rotation of the coordinate system about the wave vector $\vec{k}$. Therefore, tensor $< E_\alpha E_\beta >$ can be constructed only from the following "building blocks:" the invariant unit tensor $\delta_{\alpha\beta}$ and tensor $k_\alpha k_\beta/k^2 = n_\alpha n_\beta$, where $\vec{n} = \vec{k}/k$:

$$< E_\alpha E_\beta >= a\delta_{\alpha\beta} + bn_\alpha n_\beta \tag{1.1}$$

In order to find constants $a$ and $b$ in (1.1), consider traces of the above-written tensors and their convolution with tensor $n_\alpha n_\beta$. Thus, $\mathrm{Tr} < E_\alpha E_\beta >=< E^2 >= E_0^2 = 3a + b$, while $< E_\alpha E_\beta > n_\alpha n_\beta =< E_\alpha E_\beta n_\alpha n_\beta >= 0 = a + b$ (since in such electromagnetic wave $\vec{E} \cdot \vec{k} = 0$). These two relations yield

$a = -b = E_0^2/2$, hence

$$< E_\alpha E_\beta > = \frac{E_0^2}{2}(\delta_{\alpha\beta} - n_\alpha n_\beta) \qquad (1.2)$$

Since magnetic field of the wave is equal to $\vec{B} = \vec{n} \times \vec{E}$, the magnetic field vector rotates in the same way as the electric field one, therefore $< B_\alpha B_\beta > = < E_\alpha E_\beta >$. Finally, one gets $< E_\alpha B_\beta > = < E_\alpha e_{\beta\gamma\delta} n_\gamma E_\delta >$, which, according to (1.2), is equal to $e_{\beta\gamma\delta} n_\gamma \frac{E_0^2}{2}(\delta_{\alpha\delta} - n_\alpha n_\delta) = \frac{E_0^2}{2} e_{\alpha\beta\gamma} n_\gamma$.

## Problem 1.0.3

Find the electromagnetic field of a point charge $q$ moving with a constant velocity $\vec{v}$ in free space by solving Maxwell's equations with the help of Fourier transformation.

Fourier transformation of the electromagnetic fields from variables $(\vec{r}, t)$ to variables $(\vec{k}, \omega)$ obeys the following rule:

$$[\vec{E}(\vec{r}, t), \vec{B}(\vec{r}, t)] = \frac{1}{(2\pi)^2} \int d\vec{k}d\omega [\vec{E}(\vec{k}, \omega), \vec{B}(\vec{k}, \omega)] \exp[i(\vec{k} \cdot \vec{r} - \omega t)], \quad (1.3)$$

where

$$[\vec{E}(\vec{k}, \omega), \vec{B}(\vec{k}, \omega)] = \frac{1}{(2\pi)^2} \int d\vec{r}dt [\vec{E}(\vec{r}, t), \vec{B}(\vec{r}, t)] \exp[-i(\vec{k} \cdot \vec{r} - \omega t)] \quad (1.4)$$

Such a procedure is valid for any function of $\vec{r}$ and $t$ that tends sufficiently rapidly to zero at $r \to \infty$, $t \to \pm\infty$. Then, Maxwell's equations,

$$\vec{\nabla} \times \vec{E} = -\frac{1}{c}\frac{\partial \vec{B}}{\partial t}, \quad \vec{\nabla} \times \vec{B} = \frac{1}{c}\frac{\partial \vec{E}}{\partial t} + \frac{4\pi}{c}\vec{j},$$

become greatly simplified in the $(\vec{k}, \omega)$ representation, where they take the form

$$\vec{B}(\vec{k}, \omega) = \frac{c}{\omega}[\vec{k} \times \vec{E}(\vec{k}, \omega)], \quad [\vec{k} \times \vec{B}(\vec{k}, \omega)] = -\frac{\omega}{c}\vec{E}(\vec{k}, \omega) - \frac{4\pi i}{c}\vec{j}(\vec{k}, \omega) \quad (1.5)$$

The Fourier transform of the electric current, obtained according to the rule (1.4), yields

$$\vec{j}(\vec{k}, \omega) = \frac{q\vec{v}}{(2\pi)^2} \int d\vec{r}dt\delta(\vec{r} - \vec{v}t) \exp[-i(\vec{k} \cdot \vec{r} - \omega t)]$$

$$= \frac{q\vec{v}}{(2\pi)^2} \int dt \exp[i(\omega - \vec{k} \cdot \vec{v})t] = \frac{q\vec{v}}{2\pi}\delta(\omega - \vec{k} \cdot \vec{v})$$

By eliminating $\vec{B}$ from (1.5), one gets

$$\vec{k}(\vec{k} \cdot \vec{E}) - k^2 \vec{E} = -\frac{\omega^2}{c^2} \vec{E} - \frac{4\pi i \omega}{c^2} \vec{j}$$

The required value of $(\vec{k} \cdot \vec{E})$ can be obtained by taking the scalar product of the second of the equations (1.5) with $\vec{k}$, which yields $(\vec{k} \cdot \vec{E}) = -\frac{4\pi i}{\omega}(\vec{k} \cdot \vec{j})$. Thus, the Fourier transform of the electric and magnetic fields reads:

$$\vec{E}(\vec{k}, \omega) = \frac{2iq\delta(\omega - \vec{k} \cdot \vec{v})}{[k^2 - \omega^2/c^2]}(-\vec{k} + \omega \vec{v}/c^2),$$

$$\vec{B}(\vec{k}, \omega) = \frac{c}{\omega}(\vec{k} \times \vec{E}) = \frac{1}{c}[\vec{v} \times \vec{E}(\vec{k}, \omega)] \qquad (1.6)$$

Then, according to (1.3),

$$
\begin{aligned}
\vec{E}(\vec{r}, t) &= \frac{2iq}{(2\pi)^2} \int d\vec{k} d\omega \frac{(-\vec{k} + \vec{v}\omega/c^2)}{(k^2 - \omega^2/c^2)} \exp[i(\vec{k} \cdot \vec{r} - \omega t)]\delta(\omega - \vec{k} \cdot \vec{v}) \\
&= \frac{2iq}{(2\pi)^2} \int d\vec{k} \frac{[-\vec{k} + \vec{v}(\vec{k} \cdot \vec{v})/c^2]}{[k^2 - (\vec{k} \cdot \vec{v})^2/c^2]} \exp[i(\vec{k} \cdot \vec{r} - \vec{k} \cdot \vec{v}t)] \qquad (1.7)
\end{aligned}
$$

In order to calculate the integral in (1.7), it is useful to note that in the particular case of $\vec{v} = 0$ the electric field should be reduced to the Coulomb one, thus

$$-\frac{2iq}{(2\pi)^2} \int d\vec{k} \frac{\vec{k}}{k^2} \exp(i\vec{k} \cdot \vec{r}) = \frac{q\vec{r}}{r^3} \qquad (1.8)$$

If the charge is moving along, say, the $x$-axis, then one can introduce vector $\vec{R} = [(x - vt), y, z]$ and rewrite the integral in (1.7) as

$$\int d\vec{k} \frac{[-\vec{k} + \vec{v}(\vec{k} \cdot \vec{v})/c^2]}{[k^2 - (\vec{k} \cdot \vec{v})^2/c^2]} \exp(i\vec{k} \cdot \vec{R}) = -\int \gamma d\vec{k}^* \frac{\exp(i\vec{k}^* \cdot \vec{R}^*)}{(k^*)^2} \left[ \frac{k_x^*}{\gamma}, \vec{k}_\perp^* \right], \quad (1.9)$$

where $\vec{R}^* = (\gamma R_x, \vec{R}_\perp)$, $\vec{k}^* = (k_x/\gamma, \vec{k}_\perp)$, and $\gamma = (1 - v^2/c^2)^{-1/2}$ is the relativistic factor. By comparing expressions (1.8) and (1.9), one now concludes that

$$\vec{E}(\vec{r}, t) = \frac{\gamma q}{(R^*)^3}[\gamma^{-1} R_x^*, \vec{R}_\perp^*] = \frac{\gamma q \vec{R}}{[\gamma^2(x - vt)^2 + y^2 + z^2]^{3/2}}$$

According to (1.6), the magnetic field is equal to $\vec{B}(\vec{r}, t) = \frac{1}{c}[\vec{v} \times \vec{E}(\vec{r}, t)]$.

## Problem 1.0.4

Derive the electromagnetic force exerted on a neutron moving with velocity $\vec{v}$ in an external electric, $\vec{E}$, and magnetic, $\vec{B}$, fields.

Electromagnetic properties of the neutron are characterized by its magnetic moment $\vec{m}^{(0)}$ measured in the neutron's rest frame $K_0$. In the laboratory frame, $K$, where the neutron is moving with the velocity $\vec{v}$, it acquires, in general case, the electric dipole moment $\vec{d} \neq 0$, and the magnetic moment $\vec{m} \neq \vec{m}^{(0)}$. The transformation rules for the electric and magnetic moments of any system can be obtained by introducing the polarization vector $\vec{P}$ and the magnetization vector $\vec{M}$, which are volumetric densities of, respectively, the electric and magnetic dipole moments, so that

$$\vec{d} = \int \vec{P} dV, \ \vec{m} = \int \vec{M} dV \tag{1.10}$$

Then, the electric charge density $\rho$ and the electric current density $\vec{j}$ inside the system can be expressed in terms of $\vec{P}$ and $\vec{M}$ by the well-known relations (see, e.g., W. K. H. Panofsky and M. Phillips, *Classical Electricity and Magnetism*, Addison-Wesley, 1962):

$$\rho = -\vec{\nabla} \cdot \vec{P}, \ \vec{j} = \frac{\partial \vec{P}}{\partial t} + c(\vec{\nabla} \times \vec{M}) \tag{1.11}$$

In terms of relativistic four-vectors of the coordinate $r_\alpha = (\vec{r}, ict)$ and the current $j_\alpha = (\vec{j}, ic\rho)$ (Greek indices are used for four-dimensional quantities), one can re-write (1.11) in the four-dimensional form as $j_\alpha = -c\partial M_{\alpha\beta}/\partial x_\beta$, where vectors $\vec{P}$ and $\vec{M}$ are components of the four-dimensional polarization-magnetization tensor $M_{\alpha\beta}$:

$$M_{\alpha\beta} = \begin{pmatrix} 0 & -M_z & M_y & -iP_x \\ M_z & 0 & -M_x & -iP_y \\ -M_y & M_x & 0 & -iP_z \\ iP_x & iP_y & iP_z & 0 \end{pmatrix}$$

By comparing this tensor with the field-stength tensor of electromagnetic field $F_{\alpha\beta}$ (see, e.g., J. D. Jackson, *Classical Electrodynamics*, Wiley, 1964):

$$F_{\alpha\beta} = \begin{pmatrix} 0 & B_z & -B_y & -iE_x \\ -B_z & 0 & B_x & -iE_y \\ B_y & -B_x & 0 & -iE_z \\ iE_x & iE_y & iE_z & 0 \end{pmatrix}$$

one can notice that $M_{\alpha\beta}$ can be obtained from $F_{\alpha\beta}$ by the substitution $\vec{E} \to \vec{P}$, $\vec{B} \to -\vec{M}$. Therefore, relativistic transformation rules for vectors $\vec{P}$ and $\vec{M}$ follow straightforwardly from those for vectors $\vec{E}$ and $\vec{B}$, yielding

$$P_\parallel = P_\parallel^{(0)}, \ \vec{P}_\perp = \gamma[\vec{P}_\perp^{(0)} + \frac{1}{c}(\vec{v} \times \vec{M}^{(0)})]$$

$$M_\parallel = M_\parallel^{(0)}, \ \vec{M}_\perp = \gamma[\vec{M}_\perp^{(0)} - \frac{1}{c}(\vec{v} \times \vec{P}^{(0)})],$$

where symbols ∥ and ⊥ mean the components along and across the velocity vector $\vec{v}$, and the index (0) denotes quantities in the rest frame $K_0$. By using now relations (1.10) and taking into account the Lorentz contraction in the laboratory frame $K$, one finally obtains the following expressions for the electric and magnetic dipole moments:

$$\vec{d} = \vec{d}^{(0)} - \frac{(\gamma - 1)\vec{v}}{\gamma v^2}(\vec{d}^{(0)} \cdot \vec{v}) + \frac{1}{c}(\vec{v} \times \vec{m}^{(0)}), \tag{1.12}$$

$$\vec{m} = \vec{m}^{(0)} - \frac{(\gamma - 1)\vec{v}}{\gamma v^2}(\vec{m}^{(0)} \cdot \vec{v}) - \frac{1}{c}(\vec{v} \times \vec{d}^{(0)}) \tag{1.13}$$

The electromagnetic force exerted on the system, which is equal to

$$\vec{F} = \int dV [\rho\vec{E} + \frac{1}{c}(\vec{j} \times \vec{B})],$$

can be represented, by using relations (1.11), as $\vec{F} = \vec{F}_1 + \vec{F}_2 + \vec{F}_3$, where $\vec{F}_1 = -\int dV (\vec{\nabla} \cdot \vec{P})\vec{E}$, $\vec{F}_2 = \int dV (\vec{\nabla} \times \vec{M}) \times \vec{B}$, $\vec{F}_3 = \frac{1}{c}\int dV(\frac{\partial \vec{P}}{\partial t} \times \vec{B})$. Consider now the contribution $\vec{F}_1$, which, in tensor notation, after integration by parts, takes the form

$$F_{1i} = -\int dV E_i \frac{\partial P_k}{\partial x_k} = -\int dV \frac{\partial (E_i P_k)}{\partial x_k} + \int dV P_k \frac{\partial E_i}{\partial x_k} \tag{1.14}$$

The first integral on the right-hand side of (1.14) is equal to zero because it can be transformed into the surface integral outside the system where $\vec{P} = 0$. In the second integral the spatial derivative of the electric field can be considered as a constant inside such a small system as the neutron, thus

$$F_{1i} \approx \frac{\partial E_i}{\partial x_k} \int dV P_k = d_k \frac{\partial E_i}{\partial x_k}$$

$$\vec{F}_1 = (\vec{d} \cdot \vec{\nabla})\vec{E},$$

which is a well-known electric force exerted on an electric dipole. Similarly,

$$F_{2i} = \int dV B_k \left( \frac{\partial M_i}{\partial x_k} - \frac{\partial M_k}{\partial x_i} \right)$$

$$= \int dV \left[ \frac{\partial (B_k M_i)}{\partial x_k} - M_i \frac{\partial B_k}{\partial x_k} - \frac{\partial (B_k M_k)}{\partial x_i} + M_k \frac{\partial B_k}{\partial x_i} \right]$$

Again, the divergence-type terms make no contribution since $\vec{M} = 0$ outside the system, $\partial B_k / \partial x_k = \vec{\nabla} \cdot \vec{B} = 0$, so that

$$F_{2i} \approx \frac{\partial B_k}{\partial x_i} \int dV M_k = m_k \frac{\partial B_k}{\partial x_i} = \frac{\partial}{\partial x_i}(m_k B_k), \tag{1.15}$$

a standard force on a magnetic dipole in a non-uniform magnetic field.

Consider now the force $\vec{F}_3$, which is caused by the polarization current $\partial \vec{P}/\partial t$. Although it is assumed that the distribution of $\vec{P}$ and $\vec{M}$ is steady in the rest frame of the system, it is not so in the laboratory frame, where the system is moving with velocity $\vec{v}$. Indeed, the former means that $d\vec{P}/dt = (\partial \vec{P}/\partial t) + (\vec{v} \cdot \vec{\nabla})\vec{P} = 0$, hence $\partial P/\partial t = -(\vec{v} \cdot \vec{\nabla})\vec{P}$, and

$$F_{3i} = -\frac{1}{c} \int dV \, e_{ikl} v_m \frac{\partial P_k}{\partial x_m} B_l = -\frac{1}{c} e_{ikl} v_m \int dV \left[ \frac{\partial (P_k B_l)}{\partial x_m} - P_k \frac{\partial B_l}{\partial x_m} \right]$$

where $e_{ikl}$ is the antisymmetric unit tensor of rank three. Similarly to previous calculations, the divergence-type term makes no contribution, and

$$F_{3i} \approx \frac{1}{c} e_{ikl} v_m \frac{\partial B_l}{\partial x_m} \int dV \, P_k = \frac{1}{c} e_{ikl} v_m d_k \frac{\partial B_l}{\partial x_m},$$

or, in vector form,

$$\vec{F}_3 = \frac{1}{c} \vec{d} \times (\vec{v} \cdot \vec{\nabla})\vec{B} \qquad (1.16)$$

Thus, motion of the neutron (or any other compact system of electric charges and currents) has the two-fold effect on the electromagnetic force: it not only changes the electric and magnetic dipole moments of the system, but also results in the additional force contribution given by equation (1.16).

Altogether, the electromagnetic force exerted on the moving neutron is equal to

$$\vec{F} = (\vec{d} \cdot \vec{\nabla})\vec{E} + \vec{\nabla}(\vec{m} \cdot \vec{B}) + \frac{1}{c} \vec{d} \times (\vec{v} \cdot \vec{\nabla})\vec{B},$$

where the dipole moments $\vec{d}$ and $\vec{m}$ are given by expressions (1.12) and (1.13) with $\vec{d}^{(0)} = 0$ (according to modern particle physics, the own electric dipole moment of neutrons, if a non-zero, is very small).

It is instructive to complement the above-given formal derivations with a simple particular example, which provides explicit demonstration of the physical origin of the transformations (1.12-1.13) and the extra force (1.16). Thus, consider a square current loop of side $a$, with the electric current $I$ in the rest frame $K_0$ as shown in Figure 1.1(a). This system has no electric dipole moment of its own, $\vec{d}^{(0)} = 0$, and its magnetic dipole moment, directed along the $z$-axis, is equal to $m_0 = Ia^2/c$. What will be observed in the laboratory frame $K$ where the current loop is moving with the velocity $v$ along the $x$-axis (Figure 1.1(b))? By applying Lorentz transformation rules for the current and charge densities: $j_x = \gamma(j_x^{(0)} + v\rho^{(0)}), j_y = j_y^{(0)}, j_z = j_z^{(0)}, \rho = \gamma(\rho^{(0)} + vj_x^{(0)}/c^2)$, one gets $j_1 = j_3 = \gamma I/S_0, j_2 = j_4 = I/S_0, \rho_1 = -\rho_3 = \gamma vI/c^2 S_0$, where $S_0$ is the leg cross section area in the rest frame $K_0$. Since Lorentz contraction results in the reduction of the cross section area of legs 2 and 4, as well as in the reduced length of legs 1 and 3, the net currents and electric charges observed in the laboratory frame $K$ will be as follows [see Figure 1.1(b)]:

$$I_1 = I_3 = \gamma I, \; I_2 = I_4 = I/\gamma, \; q_1 = -q_3 = \rho_1 S_0 a/\gamma = \frac{vaI}{c^2}$$

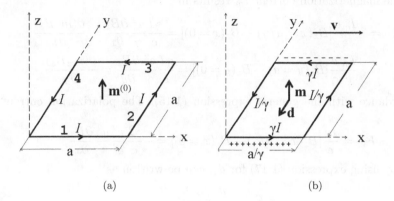

(a)                             (b)

**FIGURE 1.1**
A rectangular current loop observed in its rest frame (a) and in the laboratory
frame (b)

Therefore, the loop acquires the electric dipole moment

$$d_y = q_3 a = -\frac{v a_2 I}{c^2} = -\frac{v}{c} m_0, \qquad (1.17)$$

in agreement with the transformation rule (1.12). As far as the magnetic dipole
moment in the laboratory frame is concerned, its derivation requires a bit more
attention due to the unequal net currents in the loop legs: $I_1 = I_3 \neq I_2 = I_4$.
Thus, one needs to separate the closed magnetization current (that determines
the magnetic dipole moment), and the polarization current, which is now
also present due to a non-zero electric dipole moment of the moving loop:
$I_{total} = I_m + I_p$. To do so, one may note that polarization current should
be parallel to the electic dipole moment; therefore, in this case, when $\vec{d}$ is
directed along the $y$-axis, $I_p$ is present only in the legs 2 and 4 of the loop (see
Figure 1.1(b)). Hence, the magnetization current $I_m = I_1 = I_3 = \gamma I$, while
$I_2 = I_4 = I_m + I_p = I/\gamma$, which yields $I_p = -\gamma I v^2/c^2$. It follows then, that
the magnetic dipole moment of the loop remains equal to $m_0$, in accordance
with the transformation rule (1.13), since the increase in the magnetization
current by a factor of $\gamma$ is offset by the reduction in the loop surface area by
the same factor due to Lorentz contraction of legs 1 and 3 (see Figure 1.1(b)).

Consider now the force exerted on the loop, if the magnetic field $B_z(x, y)$
is present in the laboratory frame $K$. Due to simplicity of the system, this
force can be easily calculated by straightforward application of Ampere's law.

Thus, the magnetization current $I_m$ results in

$$
\begin{aligned}
F_{2x} &= \frac{I_m a}{c}[B_z(x=a/\gamma) - B_z(x=0)] = \frac{\gamma I}{c}\frac{a^2}{\gamma}\frac{\partial B_z}{\partial x} = \frac{\partial(m_z B_z)}{\partial x}, \\
F_{2y} &= \frac{I_m}{c}\frac{a}{\gamma}[B_z(y=a) - B_z(y=0)] = \frac{\gamma I}{c}\frac{a^2}{\gamma}\frac{\partial B_z}{\partial y} = \frac{\partial(m_z B_z)}{\partial y},
\end{aligned}
$$

in accordance with the general expression (1.15). The polarization current yields

$$
F_{3x} = \frac{Ia}{c}[B_z(x=a/\gamma) - B_z(x=0)] = -\frac{\gamma I}{c}\frac{v^2}{c^2}\frac{a^2}{\gamma}\frac{\partial B_z}{\partial x},
$$

which, by using expression (1.17) for $d_y$, can be written as

$$
F_{3x} = \frac{d_y}{c}v\frac{\partial B_z}{\partial x},
$$

in agreement with expression (1.16) for the contribution of the polarization current to the electromagnetic force.

## Problem 1.0.5

In linear electrodynamics the electric current in a medium, $\vec{j}$, is proportional to the electric field, $\vec{E}$, so that $j_\alpha = \hat{\sigma}_{\alpha\beta}E_\beta$, where $\hat{\sigma}$ is some linear operator. Prove, that in the general case, the relation between the vectors $\vec{j}$ and $\vec{E}$ can be written in the following form:

$$
j_\alpha(\vec{r},t) = \int d\vec{r}'\,dt'\,\sigma_{\alpha\beta}(\vec{r},\vec{r}',t,t')E_\beta(\vec{r}',t') \tag{1.18}
$$

By using Fourier transform, the electric field $\vec{E}(\vec{r}, t)$ can be represented as

$$
E_\beta(\vec{\xi}) = \int d\vec{q}\,A_\beta(\vec{q})\psi_q(\vec{\xi}),
$$

$$
\psi_q(\vec{\xi}) = \frac{1}{(2\pi)^2}\exp[i(\vec{k}\cdot\vec{r} - \omega t)],
$$

$$
\vec{\xi} = (\vec{r}, t),\ \vec{q} = (\vec{k}, \omega)
$$

The coefficients

$$
A_\beta(\vec{q}) = \int d\vec{\xi}'\,E_\beta(\vec{\xi}')\psi_q^*(\vec{\xi}'),
$$

hence

$$
j_\alpha(\vec{\xi}) = \hat{\sigma}_{\alpha\beta}E_\beta = \int d\vec{q}\,A_\beta(\vec{q})\hat{\sigma}_{\alpha\beta}\psi_q(\vec{\xi}) =
$$

$$
\int d\vec{q}\int d\vec{\xi}'\,E_\beta(\vec{\xi}')\psi_q^*(\vec{\xi}')\hat{\sigma}_{\alpha\beta}\psi_q(\vec{\xi}) \tag{1.19}
$$

By introducing the following definition:

$$\sigma_{\alpha\beta}(\vec{\xi}, \vec{\xi}') = \int d\vec{q}\psi_q^*(\vec{\xi}')\hat{\sigma}_{\alpha\beta}\psi_q(\vec{\xi}),$$

the expression (1.19) for $\vec{j}$ can be written as

$$j_\alpha = \int d\vec{\xi}' \sigma_{\alpha\beta}(\vec{\xi}, \vec{\xi}')E_\beta(\vec{\xi}'),$$

which is equivalent to expression (1.18).

## Problem 1.0.6

Derive the general expression for the total kinetic energy of an ideal incompressible fluid of density $\rho$ due to its potential (vortex-free) flow around a solid body, which is moving in the fluid with velocity $\vec{u}$.

Such a flow can be described by the flow potential $\phi(\vec{r})$, defined as $\vec{v} = -\vec{\nabla}\phi$, which, due to the fluid incompressibility ($\vec{\nabla} \cdot \vec{v} = 0$), should satisfy Laplace equation $\vec{\nabla}^2\phi = 0$. Its particular solution is specified by the following boundary conditions: i) $v_n|_S = (-\partial\phi/\partial n)|_S = \vec{u} \cdot \vec{n}$, where $\vec{n}$ is a unit vector normal to the surface of a moving body; ii) the fluid velocity $\vec{v}$ tends to zero at infinity, i.e., at large distances away from the body. Therefore, by choosing the coordinate system origin somewhere inside the body, an appropriate general solution of Laplace equation can be represented as a multipole series

$$\phi = \vec{A} \cdot \vec{\nabla}\left(\frac{1}{r}\right) + B_{ik}\frac{\partial^2}{\partial x_i \partial x_k}\left(\frac{1}{r}\right) + ..., \qquad (1.20)$$

i.e., as a dipole, a quadrupole, etc. contributions. The monopole one, which is $\phi \propto 1/r$, is excluded because it yields $v_r \propto 1/r^2$ with a non-zero total flux at large distances. The contributions of various multipoles in equation (1.20) are determined by the shape of the moving body, and in a general case, all of them are present. Nevertheless, it turns out that whatever the multipole series, the sought after total kinetic energy of the fluid is fully determined by the dipole term alone, which is specified by the vector $\vec{A}$ in equation (1.20). This can be demonstrated in the following way. Consider kinetic energy of the fluid inside a volume $V$, bounded by a remote sphere of a large radius, $R$, which contains the moving body inside it. Then

$$\int v^2 dV = \int u^2 dV + \int (\vec{v} + \vec{u})(\vec{v} - \vec{u})dV, \qquad (1.21)$$

with the first term in the right-hand side of (1.21) equal to $u^2(V - V_0)$, where $V_0$

is the volume of the body. In the second term the factor $(\vec{v}+\vec{u})$ can be written as $\vec{\nabla}(-\phi + \vec{u}\cdot\vec{r})$, which, by taking into account the fluid incompressibility equation $\vec{\nabla}\cdot\vec{v} = 0$, yields

$$\int dV(\vec{v}+\vec{u})(\vec{v}-\vec{u}) = \int dV(\vec{v}-\vec{u})\vec{\nabla}(-\phi+\vec{u}\cdot\vec{r})$$

$$= \int dV\vec{\nabla}\{(\vec{v}-\vec{u})(-\phi+\vec{u}\cdot\vec{r})\} = \int d\vec{S}(\vec{v}-\vec{u})(-\phi+\vec{u}\cdot\vec{r})$$

$$- \int d\vec{S}_0(\vec{v}-\vec{u})(-\phi+\vec{u}\cdot\vec{r})$$

Here $S$ and $S_0$ are, respectively, the surface of the remote sphere and the surface of the moving body. Since at the latter $d\vec{S}_0(\vec{v}-\vec{u}) = 0$ (the normal component of the fluid velocity is equal to that of the moving body), the second surface integral is equal to zero. In the first surface integral one can choose the radius $R$ of a remote sphere to be so large that it becomes sufficient to retain only the dipole contribution to $\phi$. Thus, by putting there $\phi \approx -(\vec{A}\cdot\vec{r})/r^3$, so that

$$\vec{v} = -\vec{\nabla}\phi = \frac{-3(\vec{A}\cdot\vec{r})\vec{r} + \vec{A}r^2}{r^5},$$

and taking into account that at the surface of a sphere $d\vec{S} = R^2 d\Omega\,\vec{n}$, where $d\Omega$ is an element of solid angle and $\vec{n} = \vec{R}/R$ is the unit normal vector, one gets

$$\int d\vec{S}(\vec{v}-\vec{u})(-\phi+\vec{u}\cdot\vec{r}) = \int d\Omega \left(\frac{\vec{A}\cdot\vec{n}}{R^2} + \vec{u}\cdot\vec{R}\right)\left(\frac{\vec{A}-3\vec{n}(\vec{A}\cdot\vec{n})}{R} - \vec{u}R^2\right)\cdot\vec{n}$$

By using tensor notations, it can be written as

$$\int d\Omega \left(\frac{A_i n_i}{R^2} + u_k n_k R\right)\left(-\frac{2A_l n_l}{R} - u_l n_l R^2\right)$$

Since, according to Problem 1.0.1, $\int d\Omega\, n_i n_k = 4\pi < n_i n_k > = \frac{4\pi}{3}\delta_{ik}$, it follows now that in the limit of a large $R$

$$\int dV(\vec{v}+\vec{u})(\vec{v}-\vec{u}) = -4\pi\vec{A}\cdot\vec{u} - \frac{4\pi}{3}R^3 u^2,$$

and, therefore, according to (1.21), the total kinetic energy of the fluid is equal to

$$E_{kin} = \frac{\rho}{2}\left[-4\pi(\vec{A}\cdot\vec{u}) - u^2 V_0\right] \qquad (1.22)$$

For example, if the moving body is a sphere of radius $a$, it follows from the symmetry consideration that the respective flow potential $\phi$ contains only a dipole term with vector $\vec{A}$ being directed along the sphere velocity vector $\vec{u}$: $\vec{A} = \alpha\vec{u}$. The constant $\alpha$ is determined by the boundary condition $v_n = u_n$

at $r = a$, which yields $\alpha = -a^3/2$. Therefore, equation (1.22) results in the following kinetic energy of the fluid:

$$E_{kin} = \frac{\pi \rho a^3 u^2}{3} = \frac{\rho V_0}{2} \frac{u^2}{2}.$$

## Problem 1.0.7

A plane slab of a stationary hot medium of temperature $T = T_0$ is instantly brought into contact with two cold plates of temperature $T_p \ll T_0$, which are located at $x = \pm L$. Estimate the medium cooling time, $\tau_c$, if its thermal conductivity coefficient varies with the temperature as $\kappa(T) = \kappa_0 (T/T_0)^\alpha$.

The cooling time can be defined as an interval of time during which, say, a half of the initial thermal energy is lost. To start with, consider the simplest case of the temperature-independent thermal conductivity coefficient, i.e., with the parameter $\alpha = 0$. In a stationary medium with a constant thermal conductivity the temperature evolution is governed by the standard heat conduction equation

$$\frac{\partial T}{\partial t} = \chi_0 \frac{\partial^2 T}{\partial^2 x}, \tag{1.23}$$

where the temperature diffusivity coefficient $\chi_0 = \kappa_0/C_0$. Here $C_0$ is the medium volumetric heat capacity. The contact with the cold plate results in immediate cooling of the adjacent material, hence, a transition layer with thickness $\Delta \ll L$ is formed, within which the temperature drops from that of the hot center, where $T \approx T_0$, to the plate temperature $T_p$. The temporal evolution of $\Delta(t)$ follows from the diffusion nature of equation (1.23), hence $\Delta(t) \approx \sqrt{\chi_0 t}$. Therefore, a significant fraction of the initial thermal energy becomes lost when $\Delta$ is of the order of $L$, which yields the cooling time

$$\tau_c^{(0)} \sim \tau_0 \equiv L^2/\chi_0 \tag{1.24}$$

Different stages of this cooling process are qualitatively depicted in Figure 1.2.

What a difference the heat conductivity, that varies with the temperature, can make? The answer can be deduced from the following consideration. The thermal energy transmitted to the plate is tapped from the temperature reduction inside the transition layer $\Delta$. Therefore, the heat flux $q_T = -\kappa(T)(\partial T/\partial x)$, being considered as a function of the coordinate $x$ at a fixed momemt of time, is increasing toward the plate, and reaches its maximum, $q_T^{(max)} = q_p(t)$, at $x = L$. Thus, close enough to the plate, where $(L - x) \ll \Delta$, the heat flux $q_T$ is approximately constant and equal to $q_p$:

$$-\kappa(T)\frac{\partial T}{\partial x} = q_p(t), \tag{1.25}$$

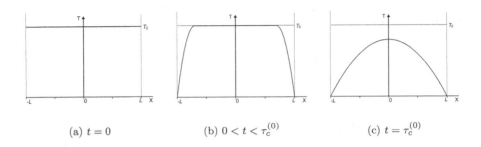

(a) $t = 0$            (b) $0 < t < \tau_c^{(0)}$            (c) $t = \tau_c^{(0)}$

**FIGURE 1.2**
Cooling stages of a stationary medium with a constant heat conductivity coefficient

which determines the temperature profile in this part of the transition layer. It follows then from equation (1.25) that the medium portion with the temperature $T$ within the interval $T_p < T < T_0$ is separated from the boundary plate by the distance

$$\delta x(T) = - \int_{T_p}^{T} dT \frac{\partial x}{\partial T} = \frac{1}{q_p(t)} \int_{T_p}^{T} dT \kappa(T)$$

Therefore, the total thickness of the transition layer, $\Delta(t)$, can be estimated as

$$\Delta(t) \approx \frac{1}{q_p(t)} \int_{T_p}^{T_0} dT \kappa(T) \tag{1.26}$$

As seen from this expression, in the case when the heat conductivity coefficient $\kappa(T)$ does not depend on the temperature, or is an increasing function of $T$, the main contribution to the integral in (1.26) comes from the upper limit (recall that $T_0 \gg T_p$). Then, the transition layer thickness is $\Delta \sim \kappa_0 T_0 / q_p$, where $\kappa_0 = \kappa(T_0)$ is heat conductivity of the hot medium, and, hence, the estimate (1.24) for the cooling time holds. The reason for that is quite apparent: under a fixed heat flux $q_p$, a decrease in the thermal conductivity of cooler medium is offset by an appropriate increase in the temperature gradient $\partial T / \partial x$, i.e., by steepening of the temperature profile. This, however, does not affect the total thickness of the transition layer and, hence, the total heat losses.

    Therefore, in order to make a difference, the heat conductivity has to increase with decreasing temperature in such a way that the transition layer thickness given by expression (1.26) becomes determined by the cold material with temperature $T \sim T_p$. For a power law dependence with $\kappa(T) \propto T^\alpha$, it requires $\alpha < -1$. It is interesting to note that in this case one can derive the exact analytical expression for the cooling time, which is not possible for the

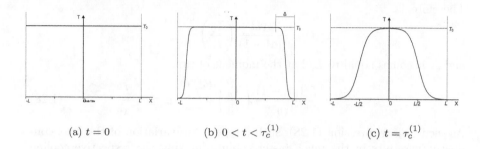

(a) $t = 0$          (b) $0 < t < \tau_c^{(1)}$          (c) $t = \tau_c^{(1)}$

**FIGURE 1.3**
Cooling of a stationary medium with the heat conductivity coefficient increasing with the decreasing temperature

seemingly much simpler equation (1.23). To do so, consider in more detail the respective cooling process, consecutive stages of which are illustrated in Figure 1.3. Since in this case the width of the transition layer is determined by the cold material with $T \sim T_p$, one can divide the total domain into two separate parts: the cold material with $T \sim T_p$ in the transition layer of thickness $\Delta$, and the hot one with $T = T_0$, which occupies the remaining part of the slab with the width equal to $(L - \Delta)$ (see Figure 1.3(b)). In the course of cooling, the width of the cold part, $\Delta(t)$, is growing, and a half of the initial thermal energy is lost when $\Delta$ becomes equal to $L/2$ (Figure 1.3(c)). Thus, the sought after cooling time can be derived by considering temporal evolution of $\Delta(t)$, which follows from the energy balance requirement. Indeed, the total thermal energy of the slab, which is equal to $Q = C_0 T_0 (L - \Delta)$, is decreasing due to the thermal flux transmitted to the plate, hence

$$\frac{dQ}{dt} = -C_0 T_0 \frac{d\Delta}{dt} = -q_p \tag{1.27}$$

On the other hand, the flux $q_p$ and width $\Delta$ are related to each other by equation (1.26) (which in the present case, when $\Delta$ is determined by the cold gas, becomes a rigorous equality). Thus, (1.26) yields

$$\Delta = \frac{1}{q_p} \int\limits_{T_p}^{T_0} \kappa(T) dT = \frac{\kappa_0}{q_p} \int\limits_{T_p}^{T_0} \left(\frac{T}{T_0}\right)^\alpha dT = \frac{\kappa_0 T_0}{q_p(|\alpha| - 1)} \left(\frac{T_0}{T_p}\right)^{(|\alpha|-1)}$$

By combining it with equation (1.27), one gets

$$\Delta \frac{d\Delta}{dt} = \frac{\chi_0}{(|\alpha| - 1)} \left(\frac{T_0}{T_p}\right)^{(|\alpha|-1)}$$

Therefore,

$$\Delta^2(t) = \frac{2\chi_0 t}{(|\alpha| - 1)} \left(\frac{T_0}{T_p}\right)^{(|\alpha|-1)},$$

and $\Delta$ becomes equal to $L/2$ at the moment of time

$$t = \tau_c^{(1)} = \frac{L^2(|\alpha| - 1)}{8\chi_0} \left(\frac{T_p}{T_0}\right)^{(|\alpha|-1)} \ll \tau_0 = \frac{L^2}{\chi_0} \qquad (1.28)$$

As seen from expression (1.28), such temperature variation of the heat conductivity results in the much faster cooling, making the respective cooling time $\tau_c$ much shorter (by a large factor of $(T_0/T_p)^{(|\alpha|-1)}$) than in the case of a constant $\kappa = \kappa_0$ (see equations (1.23) and (1.24)).

## Problem 1.0.8

Consider the setup of Problem 1.0.7, when the medium under cooling is an initially uniform hot gas of temperature $T_0$ and number density $n_0$, whose thermal conductivity coefficient varies with temperature as $\kappa(T) = \kappa_0(T/T_0)^{-\alpha}$ with $\alpha > 0$. Estimate the gas cooling time $\tau_c$, assuming that the cooling process is quasistatic, i.e., $\tau_c$ substantially exceeds the gas dynamic timescale $\tau_s = L/c_s$, where $c_s$ is the speed of sound in the gas.

Let us start with the above stated assumptions. Firstly, since the only source of gas cooling is thermal conduction, the respective cooling time scales as $\tau_c \propto L^2$. Therefore, the required quasistatic condition $\tau_c \gg \tau_s$ is always satisfied if the gas slab is thick enough, i.e., for large $L$. Secondly, although in a "normal" gas the thermal conductivity coefficient is increasing with the temperature, the reverse variation takes place in a fully ionized magnetized plasma. Its classical (under Coulomb collisions) thermal conductivity across the magnetic field scales as

$$\kappa_\perp \sim n v_T \lambda \frac{v_T^2}{\omega_B^2 \lambda^2} \propto \frac{n^2}{B^2 \sqrt{T}},$$

where $n$, $\lambda \propto T^2/n$, $v_T$ are, respectively, the number density, mean free path, and thermal velocity of electrons and ions, and $\omega_B$ is the gyro frequency of charged particles. Thus, one gets $\kappa \propto T^{-1/2}$, if the magnetic field and density remain constant, and $\kappa \propto T^{-5/2}$ in the case of the quasistatic cooling, when the plasma thermal pressure remains uniform and, hence, $n \propto 1/T$. Thus, the temperature drop close to the boundary plate has to be accompanied by the density rise there, which results in a persistent gas flow toward the plate, with the velocity $v_x = v(x, t)$. In this case the energy balance equation takes the following form

$$\frac{\partial}{\partial t}\left(\frac{3}{2}nT\right) = \frac{\partial}{\partial x}\left(\kappa \frac{\partial T}{\partial x} - \frac{5}{2}nTv\right) \qquad (1.29)$$

As seen, the gas flow results in the additional convective energy flux $q_v = 5nTv/2$, which acts on top of the heat conduction contribution $q_T = -\kappa(\partial T/\partial x)$. At first glance one may conclude that appearance of the convective flux of energy does not affect the overall gas cooling time, because $q_v = 0$ at the rigid boundary plate. It turns out, however, that the convective transport of energy makes the cooling time much shorter than it would be in a stationary gas. The reason is that the gas flow advects the temperature profile, making it more steep and, hence, increases the heat conduction energy flux $q_p$ transmitted to the boundary plate. Thus, in this case the energy balance equation (1.29) should be supplemented with the continuity equation

$$\frac{\partial n}{\partial t} + \frac{\partial}{\partial x}(nv) = 0, \tag{1.30}$$

and the quasistatic requirement

$$\frac{\partial}{\partial x}(nT) = 0 \rightarrow p = nT \equiv p(t) \tag{1.31}$$

In solving these equations, we first observe that the left-hand side of (1.29) does not depend on the coordinate $x$; hence, the total energy flux can be written as

$$q(x,\,t) = q_T + q_v = -\kappa\frac{\partial T}{\partial x} + \frac{5}{2}nTv = q_p(t)\frac{x}{L}, \tag{1.32}$$

where $q_p(t) \equiv q(L,\,t)$ is the heat flux transmitted to the boundary plate. It follows then from equations (1.29) and (1.32) that

$$\frac{dp}{dt} = -\frac{2}{3L}q_p(t), \tag{1.33}$$

and the cooling time $\tau_c$, which is defined as $p(t = \tau_c) = p_0/2 = n_0T_0/2$, can be derived from the relation

$$\int_0^{\tau_c} q_p(t)dt = \frac{3}{4}n_0T_0L \tag{1.34}$$

Since the temperature drop from that of the hot gas to $T = T_p$ occurs inside a narrow transition layer of width $\Delta \ll L$, the heat conduction is relevant only there, while in the main body of the gas its cooling is provided by the convective energy flux

$$q_v \approx q(x,\,t) = q_p(t)\frac{x}{L} = \frac{5}{2}p(t)v$$

Therefore, the gas velocity there is equal to

$$v(x,\,t) = \frac{2}{5}\frac{q_p(t)}{p(t)}\frac{x}{L}, \tag{1.35}$$

which, according to equations (1.30) and (1.33), indicates the adiabatic ($\gamma = 5/3$) cooling due to a uniform expansion of the gas:

$$\frac{1}{n}\frac{dn}{dt} = \frac{3}{5}\frac{1}{p}\frac{dp}{dt}$$

In what follows, the temperature and density of this uniform hot gas will be designated as $T_h(t)$ and $n_h(t)$, so that $p(t) = n_h T_h$, and $T_h(0) = T_0$, $n_h(0) = n_0$.

On the other hand, in the transition layer the gas velocity goes down rapidly (see below); hence, the required energy flux (1.32) is provided there by the heat conductivity:

$$q = q_T = -\kappa_0 \left(\frac{T}{T_0}\right)^{-\alpha} \frac{\partial T}{\partial x} = q_p(t) = \frac{5}{2}n_h T_h v_h, \qquad (1.36)$$

where $v_h$ is velocity of the hot gas inflow into the transition layer (note, that since $\Delta \ll L$, one can put $x \approx L$ in expressions (1.32) and (1.35)). This equation determines the gas temperature and density profiles in the transition layer. The important point here is that while the convective energy flux of the hot gas is ultimately transmitted to the boundary plate via heat conductivity, the associated flux of the gas material cannot leave the system since the velocity $v$ should be equal to zero at the boundary $x = L$. Therefore, this gas is accumulated in the transition layer, and if $\Delta N$ is the number of particles per unit area in the transition layer, the material balance equation yields

$$\frac{d(\Delta N)}{dt} = n_h v_h = \frac{2}{5}\frac{q_p}{T_h} \qquad (1.37)$$

Thus, in order to obtain a closed set of equations, one should relate $\Delta N$ and $q_p$ by using equation (1.36) and the quasistatic condition $n = p/T = n_h T_h/T$. Hence,

$$\Delta N = \int_{L-\Delta}^{L} dx\, n = \int_{T_h}^{T_p} dT \frac{n_h T_h}{T}\frac{dx}{dT} = \frac{\kappa_0}{q_p}\int_{T_p}^{T_h} \frac{n_h T_h}{T}\left(\frac{T}{T_0}\right)^{-\alpha} dT$$

If the heat conductivity coefficient is increasing with the temperature decreasing, i.e., for $\alpha > 0$, the main contribution to $\Delta N$ comes from the lower limit of the above integral, i.e., from a very dense gas with the temperature $T \sim T_p$, hence

$$\Delta N(t) = \frac{\kappa_0 p(t)}{q_p(t)\alpha}\left(\frac{T_0}{T_p}\right)^{\alpha} \qquad (1.38)$$

The fact that $\Delta N$ is determined by the cold gas with $T \sim T_p$ means that the flux of particles inside the transition layer remains unattenuated until reaching such a temperature. Therefore, for $T_p \ll T \ll T_h$ one gets $nv \approx const$, hence

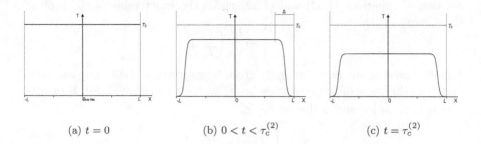

(a) $t = 0$        (b) $0 < t < \tau_c^{(2)}$        (c) $t = \tau_c^{(2)}$

**FIGURE 1.4**
Quasistatic cooling of a gas with the heat conductivity coefficient increasing with the decreasing temperature

$v$ slows down as $v \propto n^{-1} \propto T$ in this temperature interval, and then even faster, turning ultimately into zero at $T = T_p$.

Thus, by inserting expression (1.38) into equation (1.37) and taking into account the adiabatic law: $T_h = T_0(p/p_0)^{2/5}$, one gets

$$\frac{\kappa_0 T_0}{\alpha} \left(\frac{T_0}{T_p}\right)^\alpha \frac{d}{dt}\left(\frac{p}{q_p}\right) = \frac{2}{5} q_p \left(\frac{p_0}{p}\right)^{2/5} \qquad (1.39)$$

This equation, together with equation (1.33), constitutes a complete system of two ordinary differential equations for two unknown functions, $p(t)$ and $q_p(t)$. In order to estimate the sought after gas cooling time there is no need, however, to seek their exact solution: it is sufficient to ignore temporal evolution of the gas pressure, $p(t)$, which varies from $p_0$ to $p_0/2$ during this time interval. Thus, by putting $p \sim p_0$ in equation (1.39), one gets

$$\kappa_0 T_0 p_0 \left(\frac{T_0}{T_p}\right)^\alpha \frac{d}{dt}\left(\frac{1}{q_p}\right) \sim q_p,$$

which yields

$$q_p \sim n_0 T_0 \left(\frac{T_0}{T_p}\right)^{\alpha/2} \left(\frac{\chi_0}{t}\right)^{1/2}, \qquad (1.40)$$

where we put $\kappa_0 \sim n_0 \chi_0$. Then, the estimate of the cooling time, which follows from equations (1.40) and (1.34), reads

$$\tau_c^{(2)} \sim \frac{L^2}{\chi_0} \left(\frac{T_p}{T_0}\right)^\alpha \qquad (1.41)$$

As seen from expressions (1.28) and (1.41), in the quasistatic regime of cooling, the accompanied gas flow toward cold boundaries results in the further reduction of the gas cooling time by the factor of $(T_0/T_p) \gg 1$. Numerical solution of equations (1.33) and (1.39) yields the exact value of the cooling time

$$\tau_c^{(2)} = 0.56 \frac{L^2}{\chi_0} \left( \frac{T_p}{T_0} \right)^\alpha$$

Finally, knowing the heat flux $q_p(t)$ given by expression (1.40), one can estimate the thickness of the transition layer, $\Delta$, from equation (1.36). Straightforward derivation shows that at $t \sim \tau_c^{(2)}$

$$\frac{\Delta}{L} \sim \begin{cases} \left( \frac{T_p}{T_0} \right)^\alpha, & \alpha < 1 \\ \left( \frac{T_p}{T_0} \right), & \alpha > 1 \end{cases}$$

As seen, the transition layer still remains narrow when a significant fraction of the initial thermal energy of the gas is lost. Thus, all assumptions regarding this cooling process are now verified, and its consecutive stages are shown in Figure 1.4.

# 2

---

## Electrodynamics

---

### 2.1 The tensor of dielectric permeability. Electromagnetic waves in a medium.

In linear electrodynamics of a continuous medium the electric current density $\vec{j}(\vec{r}, t)$ is proportional to the electric field $\vec{E}(\vec{r}, t)$: $\vec{j} = \hat{\sigma}\vec{E}$, where $\hat{\sigma}$ is a linear operator. In the general case the current density takes the following form (see Problem 1.0.5):

$$j_\alpha(\vec{r}, t) = \int d\vec{r}' dt' \sigma_{\alpha\beta}(\vec{r}, \vec{r}', t, t') E_\beta(\vec{r}', t') \qquad (2.1)$$

By introducing the electric induction $\vec{D}$ as

$$\vec{D}(\vec{r}, t) = \vec{E}(\vec{r}, t) + 4\pi \int\limits_{-\infty}^{t} dt' \vec{j}(\vec{r}, t'),$$

Maxwell's equations can be written as follows:

$$\vec{\nabla} \times \vec{E} = -\frac{1}{c} \frac{\partial \vec{B}}{\partial t}, \qquad (2.2)$$

$$\vec{\nabla} \times \vec{B} = \frac{1}{c} \frac{\partial \vec{E}}{\partial t} + \frac{4\pi}{c}(\vec{j} + \vec{j}_{ext}) =$$

$$= \frac{1}{c} \frac{\partial \vec{D}}{\partial t} + \frac{4\pi}{c} \vec{j}_{ext} \equiv \frac{1}{c} \frac{\partial(\hat{\epsilon}\vec{E})}{\partial t} + \frac{4\pi}{c} \vec{j}_{ext}, \qquad (2.3)$$

where $\vec{j}_{ext}$ is the electric current density of external source. The dielectric permeability operator, $\hat{\epsilon}$ in equation (2.3), has a form analogous to that in (2.1):

$$D_\alpha(\vec{r}, t) = \int d\vec{r}' dt' \epsilon_{\alpha\beta}(\vec{r}, \vec{r}', t, t') E_\beta(\vec{r}', t') \qquad (2.4)$$

In the case of a homogeneous and steady medium the tensor functions, $\sigma_{\alpha\beta}$ and $\epsilon_{\alpha\beta}$ in equations (2.1) and (2.4), must have the following form:

$$\sigma_{\alpha\beta}(\vec{r}, \vec{r}', t, t') = \sigma_{\alpha\beta}(\vec{\rho}, \tau),$$

$$\epsilon_{\alpha\beta}(\vec{r}, \vec{r}', t, t') = \epsilon_{\alpha\beta}(\vec{\rho}, \tau),$$

$$\vec{\rho} = \vec{r} - \vec{r}', \ \tau = t - t'$$

This makes it convenient to explore Fourier transforms, because in the $(\vec{k}, \omega)$ representation the integral operators $\hat{\sigma}$ and $\hat{\epsilon}$ in equations (2.1) and (2.4) are replaced by simple multipliers: the **conductivity tensor** $\sigma_{\alpha\beta}(\vec{k}, \omega)$, and the **dielectric permeability tensor** $\epsilon_{\alpha\beta}(\vec{k}, \omega)$:

$$j_\alpha(\vec{k}, \omega) = \sigma_{\alpha\beta}(\vec{k}, \omega) E_\beta(\vec{k}, \omega), \tag{2.5}$$

$$D_\alpha(\vec{k}, \omega) = \epsilon_{\alpha\beta}(\vec{k}, \omega) E_\beta(\vec{k}, \omega) \tag{2.6}$$

According to equations (2.1) and (2.4),

$$\sigma_{\alpha\beta}(\vec{k}, \omega) = \int \sigma_{\alpha\beta}(\vec{\rho}, \tau) \exp[-i(\vec{k} \cdot \vec{\rho} - \omega\tau)] d\vec{\rho} d\tau$$

$$\epsilon_{\alpha\beta}(\vec{k}, \omega) = \int \epsilon_{\alpha\beta}(\vec{\rho}, \tau) \exp[-i(\vec{k} \cdot \vec{\rho} - \omega\tau)] d\vec{\rho} d\tau,$$

with

$$\epsilon_{\alpha\beta}(\vec{k}, \omega) = \delta_{\alpha\beta} + \frac{4\pi i}{\omega} \sigma_{\alpha\beta}(\vec{k}, \omega) \tag{2.7}$$

In the $(\vec{k}, \omega)$ representation Maxwell's equations (2.2) and (2.3) in the absence of external current are equivalent to the following system of equations for the electric field $\vec{E}$:

$$L_{\alpha\beta} E_\beta = \left( k_\alpha k_\beta - k^2 \delta_{\alpha\beta} + \frac{\omega^2}{c^2} \epsilon_{\alpha\beta} \right) E_\beta = 0; \tag{2.8}$$

therefore, the **dispersion equation**, which determines a link between the frequency, $\omega$, of the electromagnetic wave in a medium and its wave vector, $\vec{k}$, takes the form

$$det||L_{\alpha\beta}(\vec{k}, \omega)|| = 0 \tag{2.9}$$

In the case of an isotropic and inversion-invariant medium (the latter means that the medium is identical with its stereoisomeric counterpart) the most general form of the dielectric permeability tensor is as follows:

$$\epsilon_{\alpha\beta}(\vec{k}, \omega) = \epsilon_{||}(k, \omega) \frac{k_\alpha k_\beta}{k^2} + \epsilon_\perp(k, \omega) \left( \delta_{\alpha\beta} - \frac{k_\alpha k_\beta}{k^2} \right) \tag{2.10}$$

The electromagnetic waves in such a medium can be separated into the **longitudinal waves** (l), with $\vec{E} \parallel \vec{k}$, and the **transverse waves** (t), for which $\vec{E} \perp \vec{k}$. The respective dispersion equations, which follow from equations (2.9) and (2.10), read:

$$\epsilon_{||}(k, \omega_l) = 0, \quad k^2 = \frac{\omega_t^2}{c^2} \epsilon_\perp(k, \omega_t) \tag{2.11}$$

The damping of the electromagnetic waves is determined by the antihermitian part of the dielectric permeability tensor. Thus, the energy dissipation power per unit volume is equal to

$$Q = -\frac{i\omega}{8\pi} \epsilon_{\alpha\beta}^{(A)} E_{0\alpha}^* E_{0\beta}, \tag{2.12}$$

where $\vec{E}_0$ is the electric field amplitude of the wave. In the so-called **transparency range**, where the wave damping is weak, i.e., $\epsilon_{\alpha\beta}^{(A)} \ll \epsilon_{\alpha\beta}^{(H)}$, the volumetric energy , $W$, and the volumetric linear momentum , $\vec{P}$, of the wave can be defined as

$$W = \frac{1}{16\pi\omega} \frac{\partial}{\partial\omega} [\omega^2 \epsilon_{\alpha\beta}^{(H)}(\vec{k}, \omega)] E_{0\alpha}^* E_{0\beta}, \quad \vec{P} = \frac{\vec{k}}{\omega} W, \qquad (2.13)$$

where $\epsilon_{\alpha\beta}^{(H)}$ is the hermitian part of the dielectric permeability tensor.

## Problem 2.1.1

The electromagnetic properties of a homogeneous isotropic medium without spatial dispersion can be described by the "traditional" electric permittivity, $\epsilon(\omega)$, and the magnetic permeability, $\mu(\omega)$. Express $\epsilon(\omega)$ and $\mu(\omega)$ in terms of the limiting as $k \to 0$ values of $\epsilon_{||}(k, \omega)$ and $\epsilon_{\perp}(k, \omega)$, introduced in equation (2.10).

In terms of $\epsilon$ and $\mu$, the electric current in a medium is equal to

$$\vec{j} = \frac{\partial \vec{P}}{\partial t} + c\vec{\nabla} \times \vec{M},$$

where $\vec{P}$ and $\vec{M}$ are, respectively, the polarization and the magnetization vectors given by the following relations:

$$\vec{P} = \frac{(\epsilon - 1)}{4\pi} \vec{E}, \quad \vec{M} = \frac{(\mu - 1)}{4\pi\mu} \vec{B}$$

Then, in the $(\vec{k}, \omega)$ representation, using $\vec{B} = c(\vec{k} \times \vec{E})/\omega$ from (2.2), one gets

$$\vec{j} = -i\omega\vec{P} + ic(\vec{k} \times \vec{M}) =$$
$$-\frac{i\omega(\epsilon - 1)}{4\pi} \vec{E} + \frac{ic^2(\mu - 1)}{4\pi\mu\omega} [\vec{k}(\vec{k} \cdot \vec{E}) - k^2\vec{E}]$$

This yields the conductivity tensor

$$\sigma_{\alpha\beta} = -\frac{i\omega(\epsilon - 1)}{4\pi} \delta_{\alpha\beta} - \frac{ik^2 c^2(\mu - 1)}{4\pi\mu\omega} \left( \delta_{\alpha\beta} - \frac{k_\alpha k_\beta}{k^2} \right),$$

and, according to relation (2.7), the dielectric permeability tensor

$$\epsilon_{\alpha\beta} = \epsilon\delta_{\alpha\beta} + \frac{k^2 c^2(\mu - 1)}{\mu\omega^2} \left( \delta_{\alpha\beta} - \frac{k_\alpha k_\beta}{k^2} \right) \qquad (2.14)$$

It follows now from equations (2.10) and (2.14), that in the limit of $k \to 0$, when spatial dispersion is absent,

$$\lim_{k \to 0} \epsilon_\parallel(k, \omega) = \lim_{k \to 0} \epsilon_\perp(k, \omega),$$

so that

$$\epsilon(\omega) = \lim_{k \to 0} \epsilon_\parallel(k, \omega), \quad \mu^{-1}(\omega) = 1 - \frac{\omega^2}{c^2} \lim_{k \to 0} \frac{(\epsilon_\perp - \epsilon_\parallel)}{k^2}$$

## Problem 2.1.2

Derive the dielectric permeability tensor for a "cold" plasma (a gas of immobile heavy ions and initially resting electrons with the number density $n$), which is immersed into an external magnetic field $\vec{B}$.

Since the ions are immobile, they do not contribute to the electric current (their only role is to compensate the space charge of electrons). Thus, in the linear approximation the electric current $\vec{j} = -ne\vec{v}$, where $\vec{v}$, the velocity of electrons, is determined from the linearized equation of motion:

$$m\frac{\partial \vec{v}}{\partial t} = -e\left[\vec{E} + \frac{1}{c}(\vec{v} \times \vec{B})\right] \tag{2.15}$$

By introducing the unit vector along the external magnetic field, $\vec{h} = \vec{B}/B$, and the electron gyrofrequency $\omega_B = eB/mc$, one gets from equation (2.15) that

$$\vec{v} = -\frac{ie}{m\omega}\vec{E} - \frac{i\omega_B}{\omega}(\vec{v} \times \vec{h}) \tag{2.16}$$

By taking the dot and cross products of (2.16) with $\vec{h}$, one eliminates the factor $(\vec{v} \times \vec{h})$ in (2.16):

$$(\vec{v} \times \vec{h}) = \frac{ie}{m\omega}(\vec{h} \times \vec{E}) + \frac{i\omega_B}{\omega}[\vec{v} - \vec{h}(\vec{h} \cdot \vec{v})],$$

$$(\vec{v} \cdot \vec{h}) = -\frac{ie}{m\omega}(\vec{h} \cdot \vec{E}),$$

so that

$$\vec{v} = \left(1 - \frac{\omega_B^2}{\omega^2}\right)^{-1}\left[-\frac{ie}{m\omega}\vec{E} + \frac{e\omega_B}{m\omega^2}(\vec{h} \times \vec{E}) + \frac{ie\omega_B^2}{m\omega^3}(\vec{h} \cdot \vec{E})\vec{h}\right]$$

Using tensor notation yields the conductivity tensor

$$\sigma_{\alpha\beta} = \left(1 - \frac{\omega_B^2}{\omega^2}\right)^{-1}\left(\frac{ine^2}{m\omega}\delta_{\alpha\beta} - \frac{ine^2\omega_B^2}{m\omega^3}h_\alpha h_\beta - \frac{ne^2\omega_B}{m\omega^2}e_{\alpha\beta\gamma}h_\gamma\right)$$

Then, according to relation (2.7), one gets the dielectric permeability tensor

$$\epsilon_{\alpha\beta} = \left[1 - \frac{\omega_{pe}^2}{(\omega^2 - \omega_B^2)}\right]\delta_{\alpha\beta} + \frac{\omega_{pe}^2\omega_B^2}{\omega^2(\omega^2 - \omega_B^2)}h_\alpha h_\beta + i\frac{\omega_{pe}^2\omega_B}{\omega(\omega^2 - \omega_B^2)}e_{\alpha\beta\gamma}h_\gamma,$$

(2.17)

where $\omega_{pe} = (4\pi n e^2/m)^{1/2}$ is the so-called **electron plasma frequency**. Note, that this dielectric tensor is a hermitian one, as it should be for a medium without any dissipation process.

## Problem 2.1.3

Find the dispersion law for electromagnetic waves in a "cold" plasma without external magnetic field.

According to expression (2.17), the dielectric permeability tensor of such a plasma takes a simple form

$$\epsilon_{\alpha\beta} = \epsilon(\omega)\delta_{\alpha\beta}, \quad \epsilon(\omega) = 1 - \frac{\omega_{pe}^2}{\omega^2}$$

(2.18)

Therefore, a general rule (2.11) for the transverse waves yields

$$k^2 = \frac{\omega_t^2}{c^2}\epsilon(\omega_t) = \frac{\omega_t^2}{c^2}\left(1 - \frac{\omega_{pe}^2}{\omega_t^2}\right),$$

so that

$$\omega_t(k) = (\omega_{pe}^2 + k^2c^2)^{1/2}$$

(2.19)

For the longitudinal (electrostatic) plasma oscillations the necessary condition reads $\epsilon(\omega_l) = 1 - \omega_{pe}^2/\omega_l^2 = 0$, hence, $\omega_l = \omega_{pe}$. This is the so-called **electron Langmuir wave**.

## Problem 2.1.4

Each molecule of a rarefied gas comprises two oppositely directed dipoles $\pm\vec{d}$, separated by a distance $\vec{a}$ which is proportional to the electric force exerted on the dipoles: $\vec{a} = \gamma\vec{F} = \gamma(\vec{d}\cdot\vec{\nabla})\vec{E}$. Derive the dielectric permeability tensor for such a gas, if the number density of molecules is equal to $n$, and the orientation of their dipoles is random and isotropic.

It is helpful to start with some general procedure concerning the averaged (macroscopic) electric charges and currents in a medium. From the microscopic viewpoint the charge and current volumetric densities are as follows:

$$\rho(\vec{r}) = \Sigma_{i,a} q_{ia} \delta(\vec{r} - \vec{r}_a - \vec{\xi}_{ia}), \ \vec{j}(\vec{r}) = \Sigma_{i,a} q_{ia} \dot{\vec{\xi}}_{ia} \delta(\vec{r} - \vec{r}_a - \vec{\xi}_{ia}) \qquad (2.20)$$

Here the medium is represented as a collection of atoms, where $\vec{r}_a$ is the location of the nucleus of the a-th atom, and $\vec{r}_a + \vec{\xi}_{ia}$ are the coordinates of the i-th electron in this atom (so that $i = 0$ corresponds to the nucleus with $q_{0a} = Ze$). The macroscopic electric charge, $< \rho >$, and the electric current, $< \vec{j} >$, are the result of averaging the respective expressions in (2.20) over the so-called "physically infinitesimal" volume, which contains a large number of atoms but is small compared to the variation length scale $\lambda$ of the macroscopic electromagnetic field. Since the size of an atom, which is of the order of $\xi_i$, is small compared to $\lambda$, by the very meaning of the macroscopic description, the typical values of $|\vec{r} - \vec{r}_a|$ will be much larger than $\xi_i$ while executing the above mentioned averaging. Therefore, the following expansions can be used in calculating $< \rho >$ and $< \vec{j} >$ from (2.20):

$$< \rho > \approx < \Sigma_{i,a} q_{ia} \delta(\vec{r} - \vec{r}_a) > - < \Sigma_{i,a} q_{ia} \xi_{ia\alpha} \frac{\partial}{\partial x_\alpha} \delta(\vec{r} - \vec{r}_a) > +$$

$$\frac{1}{2} < \Sigma_{i,a} q_{ia} \xi_{ia\alpha} \xi_{ia\beta} \frac{\partial^2}{\partial x_\alpha \partial x_\beta} \delta(\vec{r} - \vec{r}_a) > + ...$$

$$< j_\alpha > = < \Sigma_{i,a} q_{ia} \dot{\xi}_{ia\alpha} \delta(\vec{r} - \vec{r}_a) > - < \Sigma_{i,a} q_{ia} \dot{\xi}_{ia\alpha} \xi_{ia\beta} \frac{\partial}{\partial x_\beta} \delta(\vec{r} - \vec{r}_a) > + ...$$

The first term in the expression for $< \rho >$ is equal to zero because the atoms are neutral, while the contribution of the second term can be written as

$$< \rho > = -\vec{\nabla} \cdot \vec{P}, \ \vec{P} = < \Sigma_{i,a} q_{ia} \vec{\xi}_{ia} \delta(\vec{r} - \vec{r}_a) >,$$

where $\vec{P}$ is the dielectric polarization of the medium. However, $\vec{P} = 0$ in this particular case of the medium and, hence, one needs to consider the next term in the expansion series, which yields:

$$< \rho > = \frac{1}{2} \frac{\partial^2 D_{\alpha\beta}}{\partial x_\alpha \partial x_\beta}, \ D_{\alpha\beta} = < \Sigma_{i,a} q_{ia} \xi_{ia\alpha} \xi_{ia\beta} \delta(\vec{r} - \vec{r}_a) >$$

In this case the averaged electric charge is determined by the quadrupolar electric moment induced in the medium. It can be straightforwardly verified now that the expression for the macroscopic electric current $< \vec{j} >$ takes the form

$$< \vec{j} > = \frac{\partial \vec{P}}{\partial t} + c(\vec{\nabla} \times \vec{M}) - \frac{1}{2} \frac{\partial}{\partial t} \frac{\partial D_{\alpha\beta}}{\partial x_\beta},$$

where

$$\vec{M} = < \frac{1}{2c} \Sigma_{i,a} q_{ia} (\vec{\xi}_{ia} \times \dot{\vec{\xi}}_{ia} \delta(\vec{r} - \vec{r}_a) >$$

is the magnetization of the medium. Therefore, in order to obtain the conductivity tensor of this medium and, hence, its dielectric permeability, one needs to derive first the parameters $\vec{M}$ and $D_{\alpha\beta}$. In doing so it is convenient to consider the dipole $\vec{d}$ as a limit of two opposite charges $\pm q$ separated by a distance $\vec{l}$ from each other, when $\vec{l} \to 0$, $q \to \infty$, and $q\vec{l} \to \vec{d}$. Since the magnetic moment of a system of point charges is equal to

$$\vec{m} = \frac{1}{2c}\Sigma_i q_i(\vec{r}_i \times \dot{\vec{r}}_i),$$

it is easy to show that the magnetic moment of the two oppositely directed electric dipoles, $\pm\vec{d}$, separated by a distance $\vec{a}$, is equal to

$$\vec{m} = \frac{1}{2c}(\vec{d} \times \dot{\vec{a}})$$

Since in this case $\vec{a} = \gamma(\vec{d} \cdot \vec{\nabla})\vec{E} = i\gamma(\vec{k} \cdot \vec{d})\vec{E}$ (it is assumed that $\vec{E} \propto \exp[i(\vec{k} \cdot \vec{r} - \omega t)]$), one gets

$$m_\alpha = \frac{\gamma\omega}{2c}e_{\alpha\beta\mu}d_\beta d_\nu k_\nu E_\mu$$

By averaging this expression over all possible orientations of the dipole moment $\vec{d}$ (so that $< d_\beta d_\nu >= \delta_{\beta\nu}d^2/3$), the magnetization vector takes the form

$$\vec{M} = \frac{\gamma\omega nd^2}{6c}(\vec{k} \times \vec{E}) \qquad (2.21)$$

A similar derivation for $D_{\alpha\beta}$ goes as follows. By starting from the quadrupolar moment for a system of point charges, that is

$$D_{\alpha\beta} = \Sigma_i q_i x_{i\alpha} x_{i\beta},$$

one gets

$$D_{\alpha\beta} = d_\alpha a_\beta + d_\beta a_\alpha = i\gamma(d_\alpha d_\mu k_\mu E_\beta + d_\beta d_\mu k_\mu E_\alpha),$$

which yields

$$D_{\alpha\beta} = \frac{i\gamma nd^2}{3}(k_\alpha E_\beta + k_\beta E_\alpha) \qquad (2.22)$$

Thus, according to expressions (2.21) and (2.22), the macroscopic electric current is equal to

$$<\vec{j}> = c(\vec{\nabla} \times \vec{M}) - \frac{1}{2}\frac{\partial}{\partial t}\frac{\partial D_{\alpha\beta}}{\partial x_\beta} = ic(\vec{k} \times \vec{M}) - \frac{\omega k}{2}D_{\alpha\beta} = -\frac{i\gamma\omega nd^2 k^2}{3}E_\alpha,$$

which yields the conductivity tensor

$$\sigma_{\alpha\beta} = -\frac{i\gamma\omega nd^2 k^2}{3}\delta_{\alpha\beta},$$

and, hence,

$$\epsilon_{\alpha\beta} = \delta_{\alpha\beta} + \frac{4\pi i}{\omega}\sigma_{\alpha\beta} = \left(1 + \frac{4\pi}{3}\gamma nd^2 k^2\right)\delta_{\alpha\beta}$$

As seen from this expression, the dielectric permittivity of such a medium differs from that of free space only due to the spatial dispersion ($\vec{k} \neq 0$).

## Problem 2.1.5

Derive the electric field potential $\phi(\vec{r})$ created by a resting point charge $q$ in a uniaxial crystal.

A uniaxial crystal is an anisotropic dielectric, whose permittivity, $\epsilon_\parallel$, along a distinguished direction (called the optical axis ) differs from its permittivity, $\epsilon_\perp$, in the plane perpendicular to the optical axis. If $\vec{n}$ is a unit vector along the optical axis, the dielectric tensor for such a crystal takes the form

$$\epsilon_{\alpha\beta} = \epsilon_\perp(\delta_{\alpha\beta} - n_\alpha n_\beta) + \epsilon_\parallel n_\alpha n_\beta \tag{2.23}$$

Since $\vec{E} = -\vec{\nabla}\phi$, and $D_\alpha = \epsilon_{\alpha\beta}E_\beta$, the electrostatic Maxwell's equation $\vec{\nabla} \cdot \vec{D} = 4\pi\rho$ yields:

$$\frac{\partial}{\partial x_\alpha}\left(\epsilon_{\alpha\beta}\frac{\partial\phi}{\partial x_\beta}\right) = -4\pi q\delta(\vec{r}) \tag{2.24}$$

(it is assumed that the charge is located at $\vec{r} = 0$). By taking the coordinate $z$-axis to be directed along the optical axis, one gets from equations (2.23) and (2.24) that

$$\epsilon_\parallel\frac{\partial^2\phi}{\partial^2 z} + \epsilon_\perp\left(\frac{\partial^2\phi}{\partial^2 x} + \frac{\partial^2\phi}{\partial^2 y}\right) = -4\pi q\delta(x)\delta(y)\delta(z)$$

By introducing the new variable, $z_* = z(\epsilon_\perp/\epsilon_\parallel)^{1/2}$, this equation transforms into

$$\frac{\partial^2\phi}{\partial^2 x} + \frac{\partial^2\phi}{\partial^2 y} + \frac{\partial^2\phi}{\partial^2 z_*} = -\frac{4\pi q}{\epsilon_\perp}\delta(x)\delta(y)\delta(z) = -\frac{4\pi q}{(\epsilon_\perp\epsilon_\parallel)^{1/2}}\delta(x)\delta(y)\delta(z_*)$$

It now looks like a "standard" Poisson equation for a point charge; hence, its solution is

$$\phi(\vec{r}) = \frac{q}{r_*(\epsilon_\perp\epsilon_\parallel)^{1/2}} = \frac{q}{(\epsilon_\perp\epsilon_\parallel)^{1/2}[x^2 + y^2 + z^2(\epsilon_\perp/\epsilon_\parallel)]^{1/2}}$$

## Problem 2.1.6

Find the electrostatic potential of a point charge $q$ in a medium with the dielectric tensor equal to

$$\epsilon_{\alpha\beta} = \epsilon(k)\delta_{\alpha\beta}, \quad \epsilon(k) = 1 + \frac{1}{(ka)^2}$$

The Fourier transform of the Poisson equation (2.24) in this case yields:

$$k^2\epsilon(k)\phi(\vec{k}) = 4\pi\rho(\vec{k}) = \frac{4\pi q}{(2\pi)^{3/2}}\int dV \delta(\vec{r})\exp(-i\vec{k}\cdot\vec{r}) = \frac{4\pi q}{(2\pi)^{3/2}}$$

Thus, $\phi(\vec{k}) = 2q/(2\pi)^{1/2}(k^2 + a^{-2})$, and

$$\phi(\vec{r}) = \frac{1}{(2\pi)^{3/2}}\int \phi(\vec{k})\exp(i\vec{k}\cdot\vec{r})d\vec{k} =$$

$$\frac{q}{\pi}\int_0^\pi \sin\theta d\theta \int_0^\infty k^2 dk \frac{\exp(ikr\cos\theta)}{(k^2 + a^{-2})}$$

(spherical coordinates with the polar axis along $\vec{r}$ are used in the above integrand). A simple integration over the angle $\theta$ yields

$$\phi(\vec{r}) = \frac{2q}{\pi r}\int_0^\infty kdk \frac{\sin(kr)}{k^2 + a^{-2}} = \frac{2q}{\pi r}\int_0^\infty \kappa d\kappa \frac{\sin(\kappa r/a)}{1 + \kappa^2} \qquad (2.25)$$

In order to calculate the above integral one can use the odd and even properties of the functions $\sin(\kappa z)$ and $\cos(\kappa z)$, so that

$$\int_0^\infty \frac{\sin(\kappa z)}{1 + \kappa^2}\kappa d\kappa = \frac{1}{2i}\int_{-\infty}^\infty \kappa d\kappa \frac{\exp(i\kappa z)}{1 + \kappa^2}$$

The latter integral can be calculated with the help of the theory of residues in the complex plane for the variable $\kappa$, as shown in Figure 2.1. Indeed, since in the upper half-plane of $\kappa$ the integrand tends to zero exponentially, one can supplement the required integration along the real axis with the integral along a remote semicircle in the upper half-plane, which yields

$$\int_{-\infty}^\infty \kappa d\kappa \frac{\exp(i\kappa z)}{1 + \kappa^2} = 2\pi i Res\left(\frac{\kappa\exp(i\kappa z)}{1 + \kappa^2}\right)_{\kappa=i} = i\pi\exp(-z)$$

Thus, according to equation (2.25),

$$\phi(\vec{r}) = \frac{q}{r}\exp(-r/a)$$

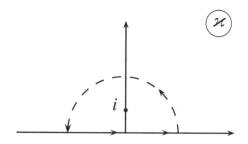

**FIGURE 2.1**
Integration in the complex plane of variable $\kappa$

The exponential decrease of the electrostatic potential at large distances, $r \gg a$, has a simple physical explanation. Indeed, since the permittivity grows sharply for small values of $k$, namely $k \ll a^{-1}$, it results in a strong shielding of the Coulomb field (a small $k$ corresponds to a large $r$ in accordance to the relation $kr \sim 1$). Such an effect, known as the Debye shielding, takes place in the ionized gas (plasma).

## Problem 2.1.7

A point charge $q$ is immersed into a dielectric fluid with permittivity equal to $\epsilon$, which is flowing as a whole with velocity $\vec{v}$. Derive the resulting electrostatic potential, assuming that $v < c/\epsilon^{1/2}$.

It is convenient to find first the scalar and vector potentials in the reference frame of the fluid, where the latter is at rest, while the charge is moving with velocity equal to $-\vec{v}$. Then, Maxwell's equations read

$$\vec{\nabla} \cdot (\epsilon \vec{E}) = 4\pi \rho_{ext}, \ \vec{\nabla} \times \vec{E} = -\frac{1}{c}\frac{\partial \vec{B}}{\partial t},$$

$$\vec{\nabla} \cdot \vec{B} = 0, \ \vec{\nabla} \times \vec{B} = \frac{4\pi}{c}\vec{j}_{ext} + \frac{1}{c}\frac{\partial(\epsilon \vec{E})}{\partial t},$$

where

$$\rho_{ext} = q\delta(\vec{r} + \vec{v}t), \ \vec{j}_{ext} = -q\vec{v}\delta(\vec{r} + \vec{v}t)$$

Thus,

$$\vec{B} = \vec{\nabla} \times \vec{A}, \ \vec{E} = -\vec{\nabla}\phi - \frac{1}{c}\frac{\partial \vec{A}}{\partial t},$$

and in the Lorentz gauge

$$\vec{\nabla} \cdot \vec{A} + \frac{\epsilon}{c}\frac{\partial \phi}{\partial t} = 0,$$

the potentials $\vec{A}$ and $\phi$ satisfy the following wave equations:

$$\vec{\nabla}^2\vec{A} - \frac{\epsilon}{c^2}\frac{\partial^2\vec{A}}{\partial^2 t} = -\frac{4\pi}{c}\vec{j}_{ext}, \quad \vec{\nabla}^2\phi - \frac{\epsilon}{c^2}\frac{\partial^2\phi}{\partial^2 t} = -\frac{4\pi}{\epsilon}\rho_{ext}$$

After Fourier transformation the solutions read:

$$\vec{A}'(\vec{k}', \omega') = -\frac{2q\vec{v}}{c}\frac{\delta(\omega' + \vec{k}'\cdot\vec{v})}{[(k')^2 - \epsilon(\omega')^2/c^2]}, \tag{2.26}$$

$$\phi'(\vec{k}', \omega') = \frac{2q}{\epsilon}\frac{\delta(\omega' + \vec{k}'\cdot\vec{v})}{[(k')^2 - \epsilon(\omega')^2/c^2]}, \tag{2.27}$$

where the symbol prime indicates that the respective quantities are related to the reference frame of the fluid. To return to the laboratory frame, where the charge is at rest, one needs to use the Lorentz transformation for the four-vectors $A_\alpha = (\vec{A}, i\phi)$ and $k_\alpha = (\vec{k}, i\omega/c)$:

$$A_\alpha = L_{\alpha\beta}A'_\beta, \quad k_\alpha = L_{\alpha\beta}k'_\beta$$

If vector $\vec{v}$ is directed along the $x$-axis, the matrix $L_{\alpha\beta}$ is as follows:

$$L_{\alpha\beta} = \begin{pmatrix} \gamma & 0 & 0 & -i\gamma v/c \\ 0 & 1 & 0 & 0 \\ 0 & 0 & 1 & 0 \\ i\gamma v/c & 0 & 0 & \gamma \end{pmatrix},$$

which yields

$$A_x = \gamma\left(A'_x + \frac{v}{c}\phi'\right), \quad A_y = A_z = 0, \quad \phi = \gamma\left(\phi' + \frac{v}{c}A'_x\right),$$

with

$$k'_x = \gamma(k_x - v\omega/c^2), \quad k'_y = k_y, \quad k'_z = k_z, \quad \omega' = \gamma(\omega - k_x v)$$

It follows then from expressions (2.26) and (2.27) that

$$\phi(\vec{k}, \omega) = \frac{2q\gamma^2}{\epsilon\gamma_\epsilon^2}\frac{\delta(\omega)}{[k_x^2(\gamma^2/\gamma_\epsilon^2) + k_\perp^2]},$$

$$A_x(\vec{k}, \omega) = -\frac{v(\epsilon - 1)\gamma_\epsilon^2}{c}\phi(\vec{k}, \omega), \tag{2.28}$$

where

$$\gamma^2 = (1 - v^2/c^2)^{-1}, \quad \gamma_\epsilon^2 = (1 - \epsilon v^2/c^2)^{-1}$$

Since both $\phi$ and $A_x$ are proportional to $\delta(\omega)$, $\omega = 0$ is put in these expressions. Then, in the $(\vec{r}, t)$ representation, one gets:

$$\phi(\vec{r}, t) = \frac{1}{(2\pi)^2}\int\phi(\vec{k}, \omega)\exp[i(\vec{k}\cdot\vec{r} - \omega t)]d\vec{k}d\omega =$$

$$\frac{1}{(2\pi)^2}\frac{2q\gamma^2}{\epsilon\gamma_\epsilon^2}\int\frac{d\vec{k}\exp(i\vec{k}\cdot\vec{r})}{[k_x^2(\gamma^2/\gamma_\epsilon^2) + k_\perp^2]}$$

This integral can be calculated in the same way as the one in the Problem 2.1.3. Indeed, by introducing new variables $\vec{k}_* = (k_x\gamma/\gamma_\epsilon, \vec{k}_\perp)$ and $\vec{r}_* = (x\gamma/\gamma_\epsilon, \vec{r}_\perp)$, it reduces to the standard Coulomb integral, so that

$$\phi(\vec{r}) = \frac{q\gamma}{\epsilon\gamma_\epsilon} \frac{1}{[x^2(\gamma_\epsilon/\gamma)^2 + y^2 + z^2]^{1/2}} \tag{2.29}$$

The important point here is that the presented solution holds only when $\gamma_\epsilon^2 > 0$, i.e., $v < c/\sqrt{\epsilon}$. If the fluid velocity exceeds this critical value, which is equal to the propagation speed of the electromagnetic waves in this medium, the electromagnetic field becomes confined inside the so-called Cherenkov cone with the angle $\theta = \cos^{-1}(c/v\epsilon^{1/2})$ (see, e.g., V. L. Ginzburg, *Applications of Electrodynamics in Theoretical Physics and Astrophysics*, Ch.7, Gordon and Breach, 1989).

By comparing expression (2.29) with the potential derived in Problem 2.1.5, it is seen that from the electrostatic point of view such a moving dielectric is analogous to a uniaxial crystal with the optical axis along $\vec{v}$ and the permittivities $\epsilon_\parallel = \epsilon$, $\epsilon_\perp = \epsilon(\gamma_\epsilon/\gamma)^2$. However, there is a difference, since in the case of a moving dielectric fluid the magnetic field is also induced. Indeed, according to equation (2.28),

$$A_x(\vec{r}) = -\frac{v(\epsilon - 1)\gamma_\epsilon^2}{c}\phi(\vec{r}),$$

so that

$$\vec{B}(\vec{r}) = \vec{\nabla} \times \vec{A} = (\epsilon - 1)\gamma_\epsilon^2\left(\vec{E} \times \frac{\vec{v}}{c}\right)$$

Finally, it is worth noting that this problem can be solved without the transition to the reference frame of the fluid by using Minkowski's equations for a moving dielectric (see, e.g., L. D. Landau, E. M. Lifshitz, and L. P. Pitaevskii, *Electrodynamics of Continuous Media*, §76, Pergamon Press, 1984).

## Problem 2.1.8

Determine the dielectric permeability tensor for a cold electron-positron plasma immersed into an external magnetic field $\vec{B}$.

The sought after result follows straightforwardly from the electron contribution to the tensor $\epsilon_{\alpha\beta}$ obtained in Problem 2.1.2, because electrons and positrons have the same mass and equal but opposite electric charges. Therefore, the contribution of positrons cancels the last term on the right-hand side of equation (2.17), while doubling the two other terms. Hence, in this case

$$\epsilon_{\alpha\beta} = \left[1 - 2\frac{\omega_{pe}^2}{(\omega^2 - \omega_B^2)}\right]\delta_{\alpha\beta} + \frac{2\omega_{pe}^2\omega_B^2}{\omega^2(\omega^2 - \omega_B^2)}h_\alpha h_\beta,$$

where all notations are the same as in Problem 2.1.2. As seen, this tensor is structurally analogous to the dielectric permeability tensor of a uniaxial crystal given by expression (2.23). The optical axis of such a plasma is directed along the external magnetic field, with the parallel, $\epsilon_\|$, and the perpendicular, $\epsilon_\perp$, permittivities equal to

$$\epsilon_\| = 1 - \frac{2\omega_{pe}^2}{\omega^2}, \quad \epsilon_\perp = 1 - \frac{2\omega_{pe}^2}{(\omega^2 - \omega_B^2)}$$

## Problem 2.1.9

Determine the dispersion equations for the electromagnetic waves in a uniaxial crystal.

One can always choose the coordinate in such a way that the $z$-axis is directed along the optical axis of the crystal, while the wave vector $\vec{k}$ is lying in the $(x, z)$ plane: $\vec{k} = (k_x, 0, k_z)$. Then Maxwell's system of equations (2.8) with the dielectric tensor of the form (2.23) reads:

$$\left(-k_z^2 + \frac{\omega^2}{c^2}\epsilon_\perp\right) E_x + k_x k_z E_z = 0$$

$$\left(-k^2 + \frac{\omega^2}{c^2}\epsilon_\perp\right) E_y = 0 \qquad (2.30)$$

$$\left(-k_x^2 + \frac{\omega^2}{c^2}\epsilon_\perp\right) E_z + k_z k_x E_x = 0$$

The respective dispersion equation, $Det\|L_{ik}\| = 0$, takes the form

$$\left(-k^2 + \frac{\omega^2}{c^2}\epsilon_\perp\right)\left[\left(-k_z^2 + \frac{\omega^2}{c^2}\epsilon_\perp\right)\left(-k_x^2 + \frac{\omega^2}{c^2}\epsilon_\|\right) - k_x^2 k_z^2\right] = 0,$$

which yields two possible solutions. The first one is $-k^2 + (\omega^2/c^2)\epsilon_\perp = 0$. It corresponds to the **ordinary wave**, whose refractive index is equal to $n = \epsilon_\perp^{1/2}$ and does not depend on the direction of the wave vector $\vec{k}$. As seen from equations (2.30), such a wave is linearly polarized: the electric field has only one non-zero component, $E_y$, so that the magnetic field $\vec{B}$ is lying in the $(x, z)$ plane and is directed perpendicular to $\vec{k}$.

The second solution can be written as

$$\frac{k_x^2}{\epsilon_\|} + \frac{k_z^2}{\epsilon_\perp} = \frac{\omega^2}{c^2},$$

which corresponds to the **extraordinary wave**. Its refractive index does depend on the angle $\theta$ between the wave vector $\vec{k}$ and the optical axis:

$$n = \left(\frac{\sin^2\theta}{\epsilon_\parallel} + \frac{\cos^2\theta}{\epsilon_\perp}\right)^{-1/2}$$

This wave is also linearly polarized, but its magnetic field vector $\vec{B}$ is directed along the $y$-axis, while the electric field $\vec{E}$ lies in the $(x, z)$ plane.

## Problem 2.1.10

The extraordinary electromagnetic wave propagates in a uniaxial crystal with given permittivities $\epsilon_\parallel$ and $\epsilon_\perp$. Its wave vector $\vec{k}$ makes angle $\theta$ with the optical axis of the crystal. Find the direction of the ray vector of this wave.

As in the previous problem, it is convenient to direct the $z$-axis of the coordinate system along the optical axis, and make the wave vector $\vec{k}$ lying in the $(x, z)$ plane. Then the vectors $\vec{E}$ and $\vec{D}$ are also lying in this plane, while the magnetic field $\vec{B}$ is directed along the $y$-axis. Since the electric induction $\vec{D}$ is perpendicular to $\vec{k}$, its components can be written as follows (see Figure 2.2):

$$D_x = D\cos\theta, \quad D_z = -D\sin\theta$$

Therefore, the electric field components are

$$E_x = \frac{D_x}{\epsilon_\perp} = \frac{1}{\epsilon_\perp}D\cos\theta, \quad E_z = \frac{D_z}{\epsilon_\parallel} = -\frac{1}{\epsilon_\parallel}D\sin\theta \qquad (2.31)$$

The direction of the ray vector is given by the Poynting vector $\vec{S} = \frac{c}{4\pi}(\vec{E}\times\vec{B})$ and, in general cases, its direction is different from that of the wave vector $\vec{k}$. Indeed, as it follows from expressions (2.31), the angle $\theta_1$ between the vector $\vec{S}$ and the optical axis is given by

$$\tan\theta_1 = -\frac{E_z}{E_x} = \frac{\epsilon_\perp}{\epsilon_\parallel}\tan\theta$$

Therefore, if $\epsilon_\perp \neq \epsilon_\parallel$, the two angles are equal to each other only when $\theta = 0, \pi/2$.

## Problem 2.1.11

Derive the reflection coefficient for a linearly polarized electromagnetic wave at a plane boundary between free space and a dielectric medium with dielectric permittivity $\epsilon$ and magnetic permeability $\mu$.

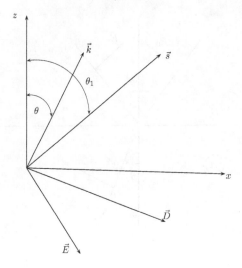

**FIGURE 2.2**

Geometry of the extraordinary electromagnetic wave

Let the boundary be the plane $z = 0$, and the plane of incidence being the $(x, z)$ plane with the angle of incidence equal to $\theta_0$ as shown in Figure 2.3. The geometry of reflection and refraction is completely determined by the spatial invariance of the system in the $(x, y)$ plane and its temporal invariance (no variation with time). The former implies that all three waves involved, the incident one (0), the reflected one (1), and the refracted wave (2), have the same $(x, y)$ components of their wave vectors:

$$k_{1y} = k_{2y} = k_{0y} = 0, \; k_{1x} = k_{2x} = k_{0x} = k_0 \sin\theta_0,$$

while the latter requires that they share the same frequency $\omega$. Together with their dispersion equations: $k_1 = k_0 = \omega/c$, $k_2 = n\omega/c$, where $n = (\epsilon\mu)^{1/2}$ is the refraction index of the dielectric, these yield that $\theta_1 = \theta_0$ and the Snell's law

$$\sin\theta_2 = \frac{1}{n}\sin\theta_0$$

The reflection (and refraction) coefficient follows from the boundary conditions at the vacuum-dielectric interface and depends on the polarization of the incident wave. Consider, for example, a particular case when the magnetic field vector $\vec{B}$ is perpendicular to the plane of incidence (the directions of $\vec{E}$ for all three waves are shown in Figure 2.3). In this case Maxwell's equations require continuity at $z = 0$ of the tangential components for vectors $\vec{E}$ and $\vec{H} = \vec{B}/\mu$:

$$B_0 + B_1 = \frac{B_2}{\mu}, \; E_0\cos\theta_0 - E_1\cos\theta_0 = E_2\cos\theta_2$$

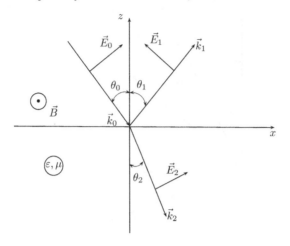

**FIGURE 2.3**
Reflection and refraction of the electromagnetic wave

By taking into account that $B_0 = E_0$, $B_1 = E_1$, $B_2 = E_2/n$, one gets:

$$E_1 = E_0 \frac{n\cos\theta_0 - (1 - \sin^2\theta_0/n^2)^{1/2}}{n\cos\theta_0 + (1 - \sin^2\theta_0/n^2)^{1/2}}$$

$$E_2 = E_0 \frac{2\cos\theta_0}{n\cos\theta_0 + (1 - \sin^2\theta_0/n^2)^{1/2}}$$

The reflection coefficient is equal to $R = (E_1/E_0)^2$.

## Problem 2.1.12

Determine the condition under which the surface electromagnetic wave can propagate along the plane boundary between free space and a dielectric with the permittivity $\epsilon(\omega)$. Derive the respective dispersion relation for a particular case of the cold electron plasma with $\epsilon(\omega) = 1 - \omega_{pe}^2/\omega^2$ (see Problem 2.1.2).

Without loss of generality one can assume that the dielectric occupies the domain $z < 0$, and the surface wave propagates along the $x$-axis at the interface. Thus, the spatial and temporal dependence of the electromagnetic field components has the form $f(z)\exp[i(kx - \omega t)]$, where $\omega$ is the frequency of the wave and $k$ is its wave vector. Then, Maxwell's equations

$$\vec{\nabla} \times \vec{B} = \frac{1}{c}\frac{\partial \vec{D}}{\partial t} = -\frac{i\omega\epsilon}{c}\vec{E}, \ \vec{\nabla} \times \vec{E} = -\frac{1}{c}\frac{\partial \vec{B}}{\partial t} = \frac{i\omega}{c}\vec{B}, \qquad (2.32)$$

yield the wave equation

$$\vec{\nabla}^2 \vec{E} + \frac{\omega^2}{c^2} \epsilon \vec{E} = 0 \tag{2.33}$$

It has to be solved separately in free space, $z > 0$, and in the dielectric, $z < 0$. Since the interest is in the surface wave, which is localized at the interface $z = 0$, the appropriate solutions of equation (2.33) take the form:

$$z > 0, \ \vec{E} = \vec{E}_1 \exp(-\kappa_1 z) \exp[i(kx - \omega t)], \tag{2.34}$$
$$\kappa_1 = (k^2 - \omega^2/c^2)^{1/2} > 0$$

$$z < 0, \ \vec{E} = \vec{E}_2 \exp(\kappa_2 z) \exp[i(kx - \omega t)], \tag{2.35}$$
$$\kappa_2 = (k^2 - \epsilon\omega^2/c^2)^{1/2} > 0$$

In order to find the polarization of the surface wave and its dispersion relation, consider the separate components of equations (2.32):

$$-\frac{\partial E_y}{\partial z} = \frac{i\omega}{c} B_x, \ \frac{\partial B_y}{\partial z} = \frac{i\omega}{c} \epsilon E_x,$$

$$\frac{\partial E_x}{\partial z} - ikE_z = \frac{i\omega}{c} B_y, \ \frac{\partial B_x}{\partial z} - ikB_z = \frac{i\omega}{c} \epsilon E_y, \tag{2.36}$$

$$ikE_y = \frac{i\omega}{c} B_z, \ ikB_y = -\frac{i\omega}{c} \epsilon E_z$$

Since both $E_y$ and $B_x$ should be continuous across the boundary surface $z = 0$, it follows from the first equation in the upper line of equations (2.36) that $E_y = B_x = 0$. Indeed, if otherwise, the left-hand side of this equation would be discontinuous, according to the solutions (2.34) and (2.35), while the right-hand side would not. Then, the first equation in the bottom line of (2.36) yields $B_z = 0$. Thus, only the field components $B_y, E_x, E_z$ are present. Consider now the second equation in the upper line of equations (2.36), which in free space and in the dielectric reads

$$-\kappa_1 B_y = \frac{i\omega}{c} E_x, \ \kappa_2 B_y = \frac{i\omega}{c} \epsilon(\omega) E_x$$

Since the field components $B_y$ and $E_x$ are both continuous at the boundary $z = 0$, the solvability condition of the above relation is $\kappa_2 = -\epsilon(\omega)\kappa_1$. Therefore, such a surface wave can exist only in the frequency range where the permittivity of the dielectric, $\epsilon(\omega)$, is negative.

By substituting here expressions (2.34) and (2.35) for $\kappa_1$ and $\kappa_2$, one gets the following dispersion equation for the surface electromagnetic wave:

$$[k^2 - \epsilon(\omega)\omega^2/c^2]^{1/2} = -\epsilon(\omega)(k^2 - \omega^2/c^2)^{1/2} \tag{2.37}$$

In the case of a cold electron plasma with $\epsilon(\omega) = 1 - \omega_{pe}^2/\omega^2$ it yields

$$\omega(k) = \frac{\omega_{pe}}{\sqrt{2}} \{1 + 2(kc/\omega_{pe})^2 - [1 + 4(kc/\omega_{pe})^4]^{1/2}\}^{1/2}$$

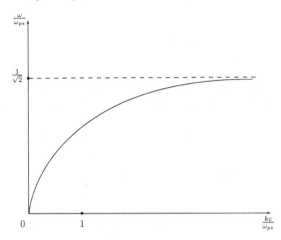

**FIGURE 2.4**
Dispersion law for the surface electromagnetic wave in a plasma

A sketch of this dispersion curve is drawn in Figure 2.4. In the limiting case of short wavelength, when $kc/\omega_{pe} \gg 1$, such a wave becomes almost electrostatic, so that the magnetic field $\vec{B}$ is much weaker than the electric one, $\vec{E}$. The dispersion equation in this limit takes a simple form $\epsilon(\omega) = -1$, which becomes evident from the general dispersion law (2.37) as $c \to \infty$.

### Problem 2.1.13

Derive the volumetric density of energy and linear momentum associated with the electromagnetic wave propagating in a transparent medium.

This part of the energy and the linear momentum can be defined as the difference of these quantities in a medium with and without the electromagnetic wave when other macroscopic parameters such as its density and temperature are fixed. The difference can be obtained in the following way. Assume that very weak dissipation is present in the medium, which leads to the damping of the wave, so that its amplitude decreases with time and tends to zero at $t \to \infty$. In the course of the damping the volumetric densities of the energy, $W$, and of the linear momentum, $\vec{P}$, of the wave decrease due to their absorption in the medium:

$$\frac{dW}{dt} = -Q, \quad \frac{d\vec{P}}{dt} = -\vec{\Pi}, \tag{2.38}$$

where $Q$ and $\vec{\Pi}$ are the rates of the energy and the linear momentum absorption per unit volume of the medium. Thus, by integrating these relations over time from the initial moment up to the infinity, when the wave completely disappeared, one gets

$$W = \int_0^\infty Q\,dt, \quad \vec{P} = \int_0^\infty \vec{\Pi}\,dt \qquad (2.39)$$

The energy absorption power is equal to $Q = <\vec{j} \cdot \vec{E}>$, where the symbol $<>$ means averaging over the period of the wave. Since $Q$ is proportional to the square of the field amplitude, its calculation requires real forms of $\vec{j}$ and $\vec{E}$ as

$$\vec{E} = \frac{1}{2}[\vec{E}_0 \exp(-i\omega t) + \vec{E}_0^* \exp(i\omega t)], \quad \vec{j} = \frac{1}{2}[\vec{j}_0 \exp(-i\omega t) + \vec{j}_0^* \exp(i\omega t)],$$

so that

$$Q = <\vec{j} \cdot \vec{E}> = \frac{1}{4}(\vec{E}_0 \cdot \vec{j}_0^* + \vec{j}_0 \cdot \vec{E}_0^*) = \frac{1}{4}(E_{0\alpha}\sigma_{\alpha\beta}^* E_{0\beta}^* + E_{0\beta}^* \sigma_{\alpha\beta} E_{0\alpha}) =$$

$$\frac{1}{4}E_{0\alpha}E_{0\beta}^*(\sigma_{\beta\alpha} + \sigma_{\alpha\beta}^*) = -\frac{i\omega}{16\pi}(\epsilon_{\alpha\beta} - \epsilon_{\beta\alpha}^*)E_{0\alpha}^* E_{0\beta}$$

(the equation (2.7), that relates $\epsilon_{\alpha\beta}$ and $\sigma_{\alpha\beta}$, has been used here). Thus, the energy dissipation in a medium is determined by the antihermitian part of its dielectric permeability tensor and is equal to

$$Q = -\frac{i\omega}{16\pi}\epsilon_{\alpha\beta}^{(A)} E_{0\alpha}^* E_{0\beta} \qquad (2.40)$$

This dissipation results in wave damping with some decrement $\gamma$, so that $\vec{E}(t) = \vec{E}_0 \exp(-\gamma t)$. The above derivation assumed that $\gamma \ll \omega$, or, equivalently, that $\epsilon_{\alpha\beta}^{(A)} \ll \epsilon_{\alpha\beta}^{(H)}$. Under a given value of the decrement $\gamma$ the dissipation power $Q \propto \exp(-2\gamma t)$, therefore it follows from equations (2.38-2.39) after integration there that

$$W = \frac{-i\omega}{16\pi\gamma}\epsilon_{\alpha\beta}^{(A)} E_{0\alpha}^* E_{0\beta} \qquad (2.41)$$

It is clear from the very definition of the energy $W$ that this quantity must not depend on the damping rate, which, nevertheless, does appear in equation (2.41). However, it can be eliminated by taking into account the relation between $\gamma$ and $\epsilon_{\alpha\beta}^{(A)}$ that follows from the dispersion equations (2.8-2.9). In the absence of damping the tensor $\epsilon_{\alpha\beta}$ is a hermitian one, the frequency of the wave, $\omega$, is real ($\gamma = 0$), and the electric field amplitude $\vec{E}_0$ satisfies the equations (2.8):

$$L_{\alpha\beta}E_{0\beta} = \left[k_\alpha k_\beta - k^2\delta_{\alpha\beta} + \frac{\omega^2}{c^2}\epsilon_{\alpha\beta}^{(H)}\right]E_{0\beta} = 0$$

If a small but finite $\epsilon_{\alpha\beta}^{(A)}$ is present, the frequency acquires a small imaginary part $\delta\omega = -i\gamma$, and the operator $L_{\alpha\beta}$ takes the form

$$L_{\alpha\beta}(\omega - i\gamma) \approx L_{\alpha\beta}(\omega) - i\gamma\frac{\partial L_{\alpha\beta}}{\partial\omega} = L_{\alpha\beta}^{(H)} + L_{\alpha\beta}^{(A)},$$

$$L_{\alpha\beta}^{(H)} = k_\alpha k_\beta - k^2\delta_{\alpha\beta} + \frac{\omega^2}{c^2}\epsilon_{\alpha\beta}^{(H)},$$

$$L_{\alpha\beta}^{(A)} = \frac{\omega^2}{c^2}\epsilon_{\alpha\beta}^{(A)} - \frac{i\gamma}{c^2}\frac{\partial(\omega^2\epsilon_{\alpha\beta}^{(H)})}{\partial\omega}$$

A small damping also results in a slightly changed eigenvector, so that now $E_\beta = E_{0\beta} + \delta E_\beta$, with $\delta E_\beta$ satisfying the following equation:

$$L_{\alpha\beta}E_\beta = (L_{\alpha\beta}^{(H)} + L_{\alpha\beta}^{(A)})(E_{0\beta} + \delta E_\beta) \approx L_{\alpha\beta}^{(H)}\delta E_\beta + L_{\alpha\beta}^{(A)}E_{0\beta} = 0$$

This correction describes a small change in the polarization of the wave, which is not of interest here. Therefore, in order to find the sought after decrement $\gamma$, it is sufficient to take the scalar product of the above equation with the vector $E_{0\alpha}^*$:

$$L_{\alpha\beta}^{(A)}E_{0\alpha}^*E_{0\beta} = -L_{\alpha\beta}^{(H)}\delta E_\beta E_{0\alpha}^* = -L_{\beta\alpha}^{(H)*}E_{0\alpha}^*\delta E_\beta = 0$$

(here it is taken into account that $L_{\beta\alpha}^{(H)*} = L_{\alpha\beta}^{(H)}$, and also that $E_{0\alpha}$ is the eigenvector of this operator: $L_{\alpha\beta}^{(H)}E_{0\beta} = 0$). This yields

$$\omega^2\epsilon_{\alpha\beta}^{(A)}E_{0\alpha}^*E_{0\beta} = i\gamma\frac{\partial(\omega^2\epsilon_{\alpha\beta}^{(H)})}{\partial\omega}E_{0\alpha}^*E_{0\beta}$$

By substituting this expression into equation (2.41), one finally gets that

$$W = \frac{1}{16\pi\omega}\frac{\partial(\omega^2\epsilon_{\alpha\beta}^{(H)})}{\partial\omega}E_{0\alpha}^*E_{0\beta} \tag{2.42}$$

Although this expression features only the electric field amplitude, it describes the entire energy of the electromagnetic wave, which includes both the "electric" and the "magnetic" contributions. It is worth noting, however, that in a general case one cannot define these two energy contributions separately.

The energy (2.42) can be written in another form as

$$W = \frac{|\vec{B}_0|^2}{16\pi} + \frac{1}{16\pi}\frac{\partial(\omega\epsilon_{\alpha\beta}^{(H)})}{\partial\omega}E_{0\alpha}^*E_{0\beta} \tag{2.43}$$

Indeed, it is easy to verify that the difference between the two expressions for the energy, which is equal to

$$\frac{1}{16\pi}(\epsilon_{\alpha\beta}^{(H)}E_{0\alpha}^*E_{0\beta} - |\vec{B}_0|^2),$$

vanishes for the electromagnetic fields $\vec{E}_0$ and $\vec{B}_0$ that satisfy Maxwell's equations in a transparent medium:

$$(\vec{k} \times \vec{E}_0) = \frac{\omega}{c}\vec{B}_0, \quad (\vec{k} \times \vec{B}_0)_\alpha = -\frac{\omega}{c}\epsilon_{\alpha\beta}^{(H)}E_{0\beta}$$

By taking the scalar product of the second of these equations with $\vec{E}_0^*$, and by using also the first one, one gets:

$$\epsilon_{\alpha\beta}^{(H)}E_{0\alpha}^*E_{0\beta} = -\frac{c}{\omega}(\vec{k} \times \vec{B}_0) \cdot \vec{E}_0^* = \frac{c}{\omega}(\vec{k} \times \vec{E}_0^*) \cdot \vec{B}_0 = \vec{B}_0^* \cdot \vec{B}_0 = |\vec{B}_0|^2$$

In order to determine the linear momentum of the wave, $\vec{P}$, one notes that the absorbed momentum per unit time, $\vec{\Pi}$, is equal to the averaged electromagnetic force exerted on the medium. Thus,

$$\vec{\Pi} = <\rho\vec{E} + \frac{1}{c}(\vec{j} \times \vec{B})>$$

Since $\vec{B} = c(\vec{k} \times \vec{E})/\omega$, and in the $(\vec{k}, \omega)$ representation the space charge $\rho = (\vec{k} \cdot \vec{j})/\omega$ (it follows from the charge conservation requirement), one gets

$$\vec{\Pi} = \frac{\vec{k}}{\omega} <\vec{j} \cdot \vec{E}> = \frac{\vec{k}}{\omega}Q$$

Therefore, it follows at once that

$$\vec{P} = \frac{\vec{k}}{\omega}W$$

## Problem 2.1.14

Verify the general expressions (2.13) for the energy and linear momentum of the electromagnetic wave in a medium by a straightforward calculation of these quantities for the electron Langmuir oscillations in a cold plasma (see Problem 2.1.3).

In the absence of the Langmuir wave there is no macroscopic electric field, and the electrons are at rest; hence, the energy and the momentum of the plasma are equal to zero. Consider now the Langmuir wave propagating along the $z$-axis with the electric field:

$$\vec{E} = E_0 \cos(kz - \omega_{pe}t)\vec{e}_z$$

The energy of the medium comprises now the energy of the electric field and the kinetic energy of electrons (heavy immobile ions remain at rest):

$$W = \frac{<E^2>}{8\pi} + \frac{nm}{2}<v^2>$$

The $z$-component of the equation of motion for electrons,

$$m\frac{\partial v}{\partial t} = -eE_z,$$

yields $v = eE_0 \sin(kz - \omega_{pe}t)/m\omega_{pe}$. Thus, the energy of the wave is equal to $W = E_0^2/8\pi$.

Consider now its linear momentum. Since the wave is electrostatic, there is no magnetic field present; hence, the electric field alone does not contribute to the momentum of the wave, and the latter is entirely due to the motion of electrons:

$$\vec{P} = <mv(n + \delta n)> \vec{e_z} = m <v\delta n> \vec{e_z}$$

Here $\delta n$ is the perturbation of the electron density, which has to be derived from the linearized continuity equation

$$\frac{\partial \delta n}{\partial t} + n\frac{\partial v}{\partial z} = 0$$

It yields

$$\delta n = \frac{knE_0}{m\omega_{pe}^2}\sin(kz - \omega_{pe}t),$$

so that

$$\vec{P} = m <v\delta n> \vec{e_z} = \frac{\vec{k}E_0^2}{8\pi\omega_{pe}} = \frac{\vec{k}}{\omega_{pe}}W$$

(the relation $\omega_{pe}^2 = 4\pi ne^2/m$ has been used here). On the other hand, the dielectric tensor for such a plasma (see equation (2.18)) is equal to $\epsilon_{\alpha\beta} = \delta_{\alpha\beta}(1 - \omega_{pe}^2/\omega^2)$; therefore, according to equation (2.13),

$$W = \frac{E_0^2}{16\pi\omega}\frac{\partial[\omega^2\epsilon(\omega)]}{\partial\omega}\Big|_{\omega=\omega_{pe}} = \frac{E_0^2}{8\pi},$$

which confirms the result of the direct calculation of the energy.

## Problem 2.1.15

Show, that in the reference frame moving relative to the resting plasma with the velocity exceeding the phase velocity of the Langmuir oscillation, the energy of the wave becomes negative.

Consider the linearized Poisson equation, the continuity equation, and the equation of motion for the electrons, which in the absence of the wave are moving with the velocity $u$ along the $z$-axis:

$$\frac{\partial E}{\partial z} = -4\pi e \delta n,$$

$$\frac{\partial \delta n}{\partial t} + \frac{\partial}{\partial z}(n\delta v + u\delta n) = 0,$$

$$\frac{\partial \delta v}{\partial t} + u\frac{\partial \delta v}{\partial z} = -\frac{e}{m}E,$$

where $\delta n$ and $\delta v$, are perturbations of, respectively, the density and the velocity of electrons. By seeking a solution in the form of a wave propagating along the $z$-axis, when $E = E_0 \cos(kz - \omega t)$, one gets from the above equations that

$$\delta n = \frac{eE_0 kn}{m\omega_{pe}^2} \sin(kz - \omega t), \ \delta v = \frac{eE_0}{m(\omega - ku)} \sin(kz - \omega t),$$

so the the dispersion equation takes the form $(\omega - ku)^2 = \omega_{pe}^2$, i.e., $\omega = ku \pm \omega_{pe}$. It shows the frequency Doppler shift equal to $ku$, while the two signs, $\pm\omega_{pe}$, correspond to the waves propagating to the right and to the left along the $z$-axis in the reference frame where the plasma is at rest. Consider now the wave that propagates backwards in this reference frame, i.e., for which $(\omega - ku) = -\omega_{pe}$, and calculate its energy. The contribution of the electric field is $W_e =< E^2/8\pi >= E_0^2/16\pi$, while the change of the kinetic energy of electrons due to the wave is equal to

$$W_k =< \frac{1}{2}m(n + \delta n)(v + \delta v)^2 > -\frac{1}{2}mnu^2 =$$

$$\frac{1}{2}mn < (\delta v)^2 > +mu < (\delta n)(\delta v) >= \frac{E_0^2}{16\pi} - \frac{E_0^2}{8\pi}\frac{ku}{\omega_{pe}}$$

Thus, the total energy of the wave, which is equal to

$$W = W_e + W_K = \frac{E_0^2}{8\pi}\left(1 - \frac{ku}{\omega_{pe}}\right),$$

becomes negative if $u > \omega_{pe}/k$.

## Problem 2.1.16

Determine the energy of the surface electromagnetic wave considered in Problem 2.1.12.

By following the procedure explored in Problem 2.1.13, one assumes that a small antihermitian part is present in the dielectric tensor, and proceeds to calculate the corresponding decrement of the wave damping. In the case of an isotropic dielectric without the spatial dispersion, when $\delta_{\alpha\beta} = \epsilon(\omega)\delta_{\alpha\beta}$, the antihermitian part of this tensor is determined by the imaginary part of the permittivity $\epsilon(\omega)$ : $\epsilon_{\alpha\beta}^{(A)} = i\epsilon''(\omega)\delta_{\alpha\beta}$. Therefore, according to equation (2.40), the dissipated power per unit volume is equal to $Q = \omega\epsilon''(\omega)|E|^2/8\pi$. However, for the surface wave under consideration an adequate characteristic is the dissipation power per unit area of the interface $z = 0$ rather than per unit volume. Thus, one should integrate the above-given $Q$ over the volume of the dielectric per unit boundary area, i.e., over the coordinate $z$ from $-\infty$ to zero. Since the amplitude of the electric field decreases there as $\exp(\kappa_2 z)$, the power $Q \propto \exp(2\kappa_2 z)$, so that the sought after

$$\tilde{Q} = \int_{-\infty}^{0} Q dz = \omega\epsilon''(\omega)\frac{|E_0|^2}{16\pi\kappa_2},$$

where $|E_0|$ is the electric field amplitude in the dielectric at $z = 0$. This dissipation leads to the wave damping with the decrement $\gamma$, in analogy with equation (2.39),

$$\tilde{W} = \int_{0}^{\infty} \tilde{Q}dt = \frac{\tilde{Q}_0}{2\gamma} = \frac{\omega\epsilon''}{2\gamma}\frac{|E_0|^2}{16\pi\kappa_2}$$

The relation between $\epsilon''$ and $\gamma$ follows from the dispersion equation (2.37). Its expansion up to the first order of these quantities yields

$$\gamma\left[\frac{\omega\epsilon}{c^2}\left(\frac{1}{\kappa_1} + \frac{1}{\kappa_2}\right) + \frac{\partial\epsilon}{\partial\omega}\left(\frac{\omega^2}{2c^2\kappa_2} - \kappa_1\right)\right] = \epsilon''\left(\frac{\omega^2}{2c^2\kappa_2} - \kappa_1\right)$$

Thus, the energy of the surface electromagnetic wave takes the form

$$\tilde{W} = \frac{|E_0|^2}{8\pi}\frac{\omega}{4\kappa_2}\left[\frac{\omega\epsilon}{c^2}\left(\frac{1}{\kappa_1} + \frac{1}{\kappa_2}\right) + \frac{\partial\epsilon}{\partial\omega}\left(\frac{\omega^2}{2c^2\kappa_2} - \kappa_1\right)\right]\left(\frac{\omega^2}{2c^2\kappa_2} - \kappa_1\right)^{-1}$$

This, quite a cumbersome expression, becomes rather simple in the limit of a short wavelength quasielectrostatic oscillation, when

$$\tilde{W} \approx \frac{|E_0|^2}{8\pi}\frac{\omega_0}{4k}\left(\frac{\partial\epsilon}{\partial\omega}\right)_{\omega_0},$$

where $\omega_0$ is the oscillation frequency, which is determined by the condition $\epsilon(\omega_0) = -1$. Thus, in the case of the cold electron plasma, when $\epsilon(\omega) = 1 - \omega_{pe}^2/\omega^2$, the frequency $\omega_0 = \omega_{pe}/\sqrt{2}$, and $\tilde{W} = |E_0|^2/16\pi k$.

### Problem 2.1.17

Derive the rate of collisonless damping of Langmuir waves in a plasma with a finite temperature of electrons (the **Landau damping**).

Following Problem 2.1.14, consider the Langmuir wave propagating along the $z$-axis, whose electric field is equal to

$$E_z(z,\,t) = E_0 \sin(kz - \omega t) \tag{2.44}$$

As shown in Problem 2.1.14, such a wave possesses the energy per unit volume $W_w = E_0^2/8\pi$, and the linear momentum $P_z = kW/\omega_{pe} = kE_0^2/8\pi\omega_{pe}$. In order to derive the wave damping rate, one can consider the linear momentum exchange between the wave and the electrons, assuming that the field (2.44) has been switched on at some instance $t = 0$, with the initial velocity distribution function of electrons, $f(z,\,v_z, t=0) = f_0(v)$. Then, the rate of the linear momentum transfer from the wave to electrons, $dP_{ez}/dt \equiv dP_e/dt$, is equal to

$$\frac{dP_e}{dt} = -e\langle n(z,\,t)E(z,\,t)\rangle = -e\int\limits_{-\infty}^{+\infty} dv\langle f(z,\,v,\,t)E(z,\,t)\rangle \tag{2.45}$$

where the symbol $\langle\rangle$ means averaging over the variations along $z$. It is apparent from expression (2.45) that a non-zero result there requires a variation of the distribution function $f$ that is in phase with the electric field. The appearance of such a variation, $\delta f(z,\,v,\,t)$, follows from Liouville's theorem (see, e.g., H. Goldstein, *Classical Mechanics*, p.436, Addison-Wesley, 1981), which implies that in a collisionless plasma the distribution function, $f(\vec{r},\,\vec{v},\,t)$, is constant along the particle trajectory in the phase space $(\vec{r},\,\vec{v})$. Thus, it is useful to follow the trajectory of an individual electron, say, the one with initial (i.e., at $t = 0$) position $z = z_0$ and the velocity $v_z = v_0$. Due to the electric field (2.44) its velocity varies with time, $v(t) = v_0 + \delta v(z_0,\,v_0,\,t)$, with the perturbation $\delta v$ being determined by the equation of motion

$$\frac{d\delta v}{dt} \approx -\frac{e}{m}E_0 \sin[kz_0 - (\omega - kv_0)t]$$

Its solution with the initial condition $\delta v(0) = 0$ reads

$$\delta v = -\frac{eE_0}{m(\omega - kv_0)}\{\cos[kz_0 - (\omega - kv_0)t] - \cos(kz_0)\} \tag{2.46}$$

Then, the above-mentioned conservation of the distribution function along the electron trajectory yields

$$f(z,\,v,\,t) = f_0(v_0) = f_0[v - \delta v(z,\,t)]$$
$$\approx f_0(v) + \frac{df_0}{dv}\frac{eE_0}{m(\omega - kv)}\cos[kz_0 - (\omega - kv_0)t] - \cos(kz_0)\}$$

By inserting this expression for $f$, as well as (2.44) for $E$, into equation (2.45), and averaging the result over $z_0$, one finds that the first term in expression (2.46) makes no contribution, while the second one results in

$$\langle Ef\rangle = \frac{eE_0^2}{2m}\frac{df_0}{dv}\frac{\sin(\omega - kv)t}{(\omega - kv)},$$

yielding the following momentum exchange rate:

$$\frac{dP_e}{dt} = -\frac{e^2 E_0^2}{2m} \int\limits_{-\infty}^{+\infty} dv \frac{df_0}{dv} \frac{\sin(\omega - kv)t}{(\omega - kv)} \tag{2.47}$$

Since

$$\lim_{t \to \infty} \frac{\sin(\omega - kv)t}{(\omega - kv)} = \pi\delta(\omega - kv),$$

after several wave periods expression (2.47) becomes equal to

$$\frac{dP_e}{dt} = -\frac{\pi e^2 E_0^2}{2mk} \left(\frac{df_0}{dv}\right)_{v=\omega/k} \tag{2.48}$$

Then, by using the linear momentum conservation law,

$$\frac{dP_w}{dt} = -\frac{dP_e}{dt} = 2\gamma P_w,$$

and the expression $P_w = kE_0^2/8\pi\omega_{pe}$, one gets from relation (2.48) the sought after decrement

$$\gamma = \frac{2\pi^2 e^2 E_0^2 \omega_{pe}}{mk^2} \left(\frac{df_0}{dv}\right)_{v=\omega/k} \tag{2.49}$$

This derivation demonstrates that such a damping, ($\gamma < 0$), or amplification, ($\gamma > 0$), of the Langmuir wave is determined by its interaction with a small fraction of the so-called resonant electrons, whose velocity is close to the phase velocity of the wave $v_{ph} = \omega_{pe}/k$. In the particular case of the Maxwellian distribution function, when

$$f_0(v) = \frac{n}{\sqrt{\pi}v_{te}} \exp(-v^2/v_{te}^2),$$

where $v_{te}$ is the thermal velocity of electrons, it follows from expression (2.49) that

$$\gamma = -\sqrt{\pi}\omega_{pe} \frac{v_{ph}^3}{v_{te}^3} \exp(-v_{ph}^2/v_{te}^2) \tag{2.50}$$

As seen from (2.50), the Langmuir waves can actually propagate in a plasma only if their phase velocity substantially exceeds the thermal velocity of electrons, i.e., the number of the resonant electrons is small; otherwise the wave becomes strongly damped. However, the situation can be reversed by injecting into a thermal plasma a beam of fast electrons. Then, in the respective interval of phase velocities the derivative in equation (2.49) changes its sign (becomes positive), resulting in the excitation of Langmuir waves, the effect known in plasma physics as the "bump-on-tail" instability.

## 2.2 Natural optical activity. The Faraday and Kerr effects.

If a transparent isotropic medium is not invariant with respect to the inversion transformation (i.e., a medium is not identical to its stereoisomeric counterpart), the first two terms of the expansion of its dielectric tensor, that account for a weak spatial dispersion, take the form

$$\epsilon_{\alpha\beta} = \epsilon(\omega)\delta_{\alpha\beta} + if(\omega)e_{\alpha\beta\gamma}k_{\gamma}, \tag{2.51}$$

where $\vec{k}$ is the wave vector, and the functions $\epsilon(\omega)$ and $f(\omega)$ are real (so that the tensor $\epsilon_{\alpha\beta}$ is a hermitian one). Such a medium possesses the property of natural optical activity, which means rotation of the polarization plane of the electromagnetic wave propagating in the medium. The angle of rotation per unit path length of the ray is equal to

$$\frac{d\phi}{dl} = f\frac{\omega^2}{2c^2}$$

The dielectric permeability tensor of an isotropic transparent medium without spatial dispersion immersed into a weak external magnetic field $\vec{B}$ can be written as

$$\epsilon_{\alpha\beta} = \epsilon(\omega)\delta_{\alpha\beta} + ib(\omega)e_{\alpha\beta\gamma}B_{\gamma}, \tag{2.52}$$

where $\epsilon(\omega)$ and $b(\omega)$ are real. The polarization plane of the electromagnetic wave propagating in such a medium rotates with the following rate (the Faraday effect):

$$\frac{d\phi}{dl} = \frac{b\omega}{2c\sqrt{\epsilon}}B\cos\theta,$$

where $\theta$ is the angle between the wave vector $\vec{k}$ and $\vec{B}$.

The dielectric permeability tensor of an isotropic transparent medium without spatial dispersion immersed into a weak external electric field $\vec{E}$ takes the form

$$\epsilon_{\alpha\beta} = \epsilon(\omega)\delta_{\alpha\beta} + a_1E^2\delta_{\alpha\beta} + a_2E_{\alpha}E_{\beta}, \tag{2.53}$$

which makes it analogous to a uniaxial crystal (the Kerr effect) with the optical axis directed along $\vec{E}$, and with the permittivities

$$\epsilon_{\parallel} = \epsilon + (a_1 + a_2)E^2, \quad \epsilon_{\perp} = \epsilon + a_1E^2$$

## Problem 2.2.1

The natural optical activity of a medium is determined by the quantity $f(\omega)$ in equation (2.51). Express $f(\omega)$ in terms of the general dielectric tensor $\epsilon_{\alpha\beta}(\vec{r}, \vec{r}', t, t')$ of equation (2.4).

If a homogeneous, steady, and isotropic medium is not invariant with respect to the inversion transformation, the only "buiding blocks" available for constructing its dielectric tensor are vector $(\vec{r} - \vec{r}')$, and the invariant tensors $\delta_{\alpha\beta}$ and $e_{\alpha\beta\gamma}$. Therefore, the most general form of its dielectric tensor is as follows:

$$\epsilon_{\alpha\beta}(\vec{r}, \vec{r}', t, t') = \epsilon_{\alpha\beta}(\vec{\rho}, \tau) =$$
$$a_1(\rho, \tau)\delta_{\alpha\beta} + a_2(\rho, \tau)\rho_\alpha\rho_\beta + a_3(\rho, \tau)e_{\alpha\beta\gamma}\rho_\gamma, \qquad (2.54)$$
$$\vec{\rho} = \vec{r} - \vec{r}', \ \tau = t - t'$$

Note, that the constant $a_3$ is a pseudoscalar and, therefore, can be non-zero only in a medium that is not invariant under inversion. Thus,

$$\epsilon_{\alpha\beta}(\vec{k}, \omega) = \delta_{\alpha\beta} \int d\vec{\rho} d\tau a_1 \exp[-i(\vec{k} \cdot \vec{\rho} - \omega\tau)] +$$
$$\int d\vec{\rho} d\tau a_2 \rho_\alpha\rho_\beta \exp[-i(\vec{k} \cdot \vec{\rho} - \omega\tau) + \qquad (2.55)$$
$$e_{\alpha\beta\gamma} \int d\vec{\rho} d\tau a_3 \rho_\gamma \exp[-i(\vec{k} \cdot \vec{\rho} - \omega\tau)$$

In an "ordinary" medium the functions $a_{1,2,3}(\rho, \tau)$ in equation (2.55) differ appreciably from zero only at distances $\rho \leq r_0$, where the order of $r_0$ is the same as the molecular size. On the other hand, the very applicability of the macroscopic description requires that $r_0 \ll \lambda$—the characteristic spatial scale of the electromagnetic field variation. Since $k = 2\pi/\lambda$, it means that the quantity $\vec{k} \cdot \vec{\rho}$ remains small, $(k\rho \ll 1)$, throughout the spatial intervals that provide the main contribution to the integrals in (2.55). In other words, it means that in macroscopic electrodynamics spatial dispersion is normally weak. Thus, the following approximation holds:

$$\exp(-i\vec{k} \cdot \vec{\rho}) \approx 1 - i\vec{k} \cdot \vec{\rho} + ...,$$

By substituting it into equation (2.55) and subsequently integrating over the entire solid angle $d\vec{\rho} = \rho^2 d\rho d\vec{\Omega}$, one gets

$$\epsilon_{\alpha\beta}(\vec{k}, \omega) = \delta_{\alpha\beta}\left(1 + 4\pi \int\limits_0^\infty \exp(i\omega\tau)d\tau \int\limits_0^\infty d\rho\rho^2 a_1(\rho, \tau)+\right.$$

$$\frac{4\pi}{3} \int\limits_0^\infty \exp(i\omega\tau)d\tau \int\limits_0^\infty d\rho\rho^4 a_2(\rho, \tau)\left.\right) - \qquad (2.56)$$

$$ie_{\alpha\beta\gamma}k_\gamma\left(\frac{4\pi}{3} \int\limits_0^\infty \exp(i\omega\tau)d\tau \int\limits_0^\infty d\rho\rho^4 a_3(\rho, \tau)\right),$$

where the following has been used:

$$\int \rho_\alpha d\vec{\Omega} = 0, \quad \int \rho_\alpha\rho_\beta\rho_\gamma d\vec{\Omega} = o, \quad \int \rho_\alpha\rho_\beta d\vec{\Omega} = \frac{4\pi}{3}\rho^2\delta_{\alpha\beta}$$

Thus, the dielectric tensor $\epsilon_{\alpha\beta}(\vec{k}, \omega)$ has the structure postulated in expression (2.51), with the functions $\epsilon(\omega)$ and $f(\omega)$ given by the expressions inside the brackets in equation (2.56). This represents the first two terms in the expansion of the dielectric tensor in a series of the small parameter $r_0/\lambda$.

## Problem 2.2.2

Find the change of the polarization plane for a linearly polarized electromagnetic wave in the course of its propagation in a medium with the dielectric tensor given in equation (2.51).

One should start from Maxwell's equations (2.8),

$$L_{\alpha\beta}E_\beta = \left(k_\alpha k_\beta - k^2\delta_{\alpha\beta} + \frac{\omega^2}{c^2}\epsilon_{\alpha\beta}\right)E_\beta = 0,$$

which in the case of a wave propagating along the $z$-axis, when $\vec{k} = (0, 0, k)$, take the form:

$$\left(-k^2 + \frac{\omega^2}{c^2}\epsilon\right)E_x + i\frac{\omega^2}{c^2}fkE_y = 0$$

$$\left(-k^2 + \frac{\omega^2}{c^2}\epsilon\right)E_y - i\frac{\omega^2}{c^2}fkE_x = 0$$

$$\frac{\omega^2}{c^2}\epsilon E_z = 0$$

For the transverse wave with $E_z = 0$, one then gets the following dispersion equation:

$$\left(-k^2 + \frac{\omega^2}{c^2}\epsilon\right)^2 = \frac{\omega^4}{c^4}f^2k^2,$$

$$\left(-k^2 + \frac{\omega^2}{c^2}\epsilon\right) = \pm\frac{\omega^2}{c^2}fk \tag{2.57}$$

The two solutions, that differ by sign in equation (2.57), correspond to the polarizations $E_x = \pm iE_y$; therefore, these are, respectively, the clockwise and the counterclockwise circularly polarized electromagnetic waves. Thus, according to (2.57), under a given frequency $\omega$ these two waves have slightly different wave vectors:

$$k_\pm \approx \frac{\omega}{c}\sqrt{\epsilon} \pm \Delta k, \quad \Delta k = \frac{\omega^2}{2c^2}f \ll \frac{\omega}{c}\sqrt{\epsilon}$$

This difference leads to rotation of the polarization plane for a linearly polarized electromagnetic wave propagating in this medium. Assume, for example, that its electric field is directed along the $x$-axis at $z = 0$. In order to determine the subsequent evolution of the polarization, one can represent this wave as a superposition of the two circularly polarized eigenwaves derived above. Thus,

$$\vec{E} = \frac{E_0}{\sqrt{2}}\exp(-i\omega t)[\vec{e}_+ \exp(ik_+z) + \vec{e}_- \exp(ik_-z)],$$

$$\vec{e}_\pm = (\vec{e}_x \pm i\vec{e}_y)/\sqrt{2},$$

which, in components, reads

$$E_x = \frac{E_0}{2}\exp\left[i\left(\frac{\omega}{c}\sqrt{\epsilon}z - \omega t\right)\right][\exp(i\Delta kz) + \exp(-i\Delta kz)] =$$

$$E_0\cos(\Delta kz)\exp\left[i\left(\frac{\omega}{c}\sqrt{\epsilon}z - \omega t\right)\right], \tag{2.58}$$

$$E_y = \frac{iE_0}{2}\exp\left[i\left(\frac{\omega}{c}\sqrt{\epsilon}z - \omega t\right)\right][\exp(i\Delta kz) - \exp(-i\Delta kz)] =$$

$$-E_0\sin(\Delta kz)\exp\left[i\left(\frac{\omega}{c}\sqrt{\epsilon}z - \omega t\right)\right] \tag{2.59}$$

Thus, the ratio $E_y/E_x = -\tan(\Delta kz)$, which indicates that the wave remains linearly polarized at any location $z$. However, the polarization plane makes an angle $\Delta\phi = \Delta kz$ with the $x$-axis. It means that such a weak spatial dispersion results in the rotation of the polarization plane with the rate equal to

$$\frac{d\phi}{dz} = \Delta k = \frac{\omega^2}{2c^2}f \tag{2.60}$$

## Problem 2.2.3

An electromagnetic wave propagates in an optically active medium with polarization rotation constant $d\phi/dl$ (see Problem 2.2.2) equal to $0.02\pi/cm$. Due to a different absorption of the clockwise and the counter-clockwise circularly polarized eigenwaves, the initially linearly polarized wave becomes polarized elliptically, so that after progating on $1m$ the polarization ellipse has a semiaxes ratio equal to 2. What will be the polarization of the wave after propagating another $1m$ in this medium?

Let $l_\pm$ be the absoption lengths for the clockwise and the counterclockwise circularly polarized waves. Then, by taking into account the waves' absorption, one gets, instead of equations (2.58) and (2.59), the following expressions for the electric field components:

$$E_x = \frac{E_0}{2}\exp[i(\frac{\omega}{c}\sqrt{\epsilon}z - \omega t)][\exp(i\Delta kz - z/l_+) + \exp(-i\Delta kz - z/l_-]$$

$$E_y = \frac{iE_0}{2}\exp[i(\frac{\omega}{c}\sqrt{\epsilon}z - \omega t)][\exp(i\Delta kz - z/l_+) - \exp(-i\Delta kz - z/l_-)]$$

Then, the ratio $E_y/E_x$, which determines the polarization of the wave, is equal to

$$\frac{E_y}{E_x} = i\frac{[\exp(i\Delta kz - z/l_+) - \exp(-i\Delta kz - z/l_-)]}{[\exp(i\Delta kz - z/l_+) + \exp(-i\Delta kz - z/l_-)]}$$

According to relation (2.60), and for the case of $\Delta k = d\phi/dl = 0.02\pi/cm$, at $z = L = 1m$ the quantity $\Delta kL = 2\pi$. Hence, at this location

$$\frac{E_y}{E_x} = i\frac{[\exp(-L/l_+) - \exp(-L/l_-)]}{[\exp(-L/l_+) + \exp(-L/l_-)]} = i\frac{(1 - \alpha)}{(1 + \alpha)},$$

with $\alpha = \exp[L(l_+^{-1} - l_-^{-1})]$. It indicates elliptic polarization with the main axes along $x$ and $y$. The semiaxes ratio is equal to $(1 + \alpha)/(1 - \alpha)$, and since at $L = 1m$ it is equal to 2, one gets $\alpha = 1/3$. Therefore, at the location $z = 2L$ the wave is elliptically polarized with the same main axes and the semiaxes ratio equal to $(1 + \alpha^2)/(1 - \alpha^2) = 5/4$.

## Problem 2.2.4

The molecules of a rarefied gas are single-spiral right-handed "double helices": two rigid uniformly charged ($\pm\rho$ per unit length each) helical threads (Figure 2.5), which are randomly oriented in space. Under the action of the electric field $\vec{E}$ the threads become displaced relative to each other along the joint helical line by the distance $\Delta$, which is proportional

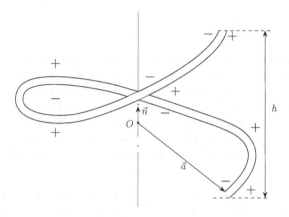

**FIGURE 2.5**
The "double helix" molecule

to the net projection of the electric field on the helical line: $\Delta = \beta \int_0^L \vec{E} \cdot d\vec{l}$.

Show that such a gas possesses natural optical activity, and derive the respective rotation rate of the electromagnetic wave polarization plane. Assume that the spatial dispersion is weak, i.e., the size of molecules is small compared to the wavelength: $(a, h \ll \lambda)$.

Following from a general theory (see equation (2.51) and Problems 2.2.1 and 2.2.2), one now needs to determine the dielectric permeability tensor of this gas up to the first order in the small parameters $a/\lambda$ and $h/\lambda$. Thus, while calculating the response of a molecule to the electric field of the electromagnetic wave with the wave vector $\vec{k}$ and frequency $\omega$, it is sufficient to use the following approximation:

$$\vec{E}(\vec{r}, t) = \vec{E}_0 \exp[i(\vec{k} \cdot \vec{r} - \omega t)] \approx \vec{E}_0 (1 + i\vec{k} \cdot \vec{r}) \exp(-i\omega t), \qquad (2.61)$$

where $\vec{E}_0$ is the electric field at the origin (point $O$ in Figure 2.5). The orientation of a molecule is completely determined by two vectors: the unit vector $\vec{n}$ directed along the axis of the molecule's helix, and the vector $\vec{a}$, which specifies locations of the molecule's ends in the plane perpendicular to $\vec{n}$ (see Figure 2.5). Then, each point of the helix can be labeled by the azimuthal angle $\phi$ as

$$\vec{R}(\phi) = \frac{h\phi}{2\pi}\vec{n} + \vec{a}\cos\phi + (\vec{n} \times \vec{a})\sin\phi, \qquad (2.62)$$

with $0 \leq \phi \leq 2\pi$. By using equations (2.61-2.62), one gets

$$\int_0^L \vec{E} \cdot d\vec{l} \approx \vec{E}_0 \exp(-i\omega t) \int_0^{2\pi} d\phi \left[ 1 + i \left( \frac{h\phi}{2\pi} \vec{k} \cdot \vec{n} + \vec{k} \cdot \vec{a} \cos\phi + \right. \right.$$

$$\left. \left. \vec{k} \cdot (\vec{n} \times \vec{a}) \sin\phi \right) \right] \cdot \left( \frac{h}{2\pi} \vec{n} - \vec{a} \sin\phi + (\vec{n} \times \vec{a}) \cos\phi \right)$$

Thus, the displacement of the helical threads is equal to

$$\Delta(\vec{n}, \vec{a}, \vec{E}_0) = \beta \exp(-i\omega t)\{h(\vec{E}_0 \cdot \vec{n}) + i(\vec{E}_0 \cdot \vec{n})(\vec{k} \cdot \vec{n})h^2/2 +$$
$$i(\vec{E}_0 \cdot \vec{a})(\vec{k} \cdot \vec{n}h) - i\pi(\vec{E}_0 \cdot \vec{a}[\vec{k} \cdot (\vec{n} \times \vec{a})] + i\pi(\vec{k} \cdot \vec{a})[\vec{E}_0 \cdot (\vec{n} \times \vec{a})]\} \quad (2.63)$$

This displacement of the oppositely charged threads relative to each other results in the electric charges $\pm q = \pm \rho \Delta$ appearing at the ends of the molecule, and the electric current $I = \rho \dot{\Delta} = -i\omega\rho\Delta$ along the helix line. Thus, the molecule acquires electric, $\vec{d}$, and magnetic, $\vec{m}$, dipole moments equal to

$$\vec{d} = hq\vec{n} = h\rho\Delta\vec{n}, \ \vec{m} = \frac{I}{c}\pi a^2 \vec{n} = \frac{-i\omega\rho\Delta}{c}\pi a^2 \vec{n} \quad (2.64)$$

Therefore, the dielectric polarization of the gas, $\vec{P}$, and its magnetization, $\vec{M}$, can be now written as

$$\vec{P} = N\langle\vec{d}\rangle, \ \vec{M} = N\langle\vec{m}\rangle,$$

where $N$ is the number density of molecules, and the symbol $\langle\rangle$ means the average over all their possible orientations. Then, the electric current density in the gas, which is equal to

$$\vec{j} = \frac{\partial\vec{P}}{\partial t} + c(\vec{\nabla} \times \vec{M}) = -i\omega\vec{P} + ic(\vec{k} \times \vec{M}), \quad (2.65)$$

can be used in order to find the conductivity tensor, $\sigma_{\alpha\beta}$, of such a gas, and, hence, its dielectric permeability, $\epsilon_{\alpha\beta}$.

It is convenient to proceed with the required averaging in two steps. First, consider the vector $\vec{n}$ as fixed, and average over all possible orientations of vector $\vec{a}$ in the plane orthogonal to $\vec{n}$. For example, by substituting expression (2.63) into equation (2.64) for $\vec{d}$, and by using tensor notation, the last term in expression (2.63) yields:

$$i\pi\langle\vec{n}(\vec{k} \cdot \vec{a})[\vec{E}_0 \cdot (\vec{n} \times \vec{a})]\rangle_a =$$

$$i\pi\langle n_i k_l a_l E_{0m} e_{mpq} n_p a_q\rangle = i\pi\langle a_l a_q\rangle n_i k_l E_{0m} e_{mpq} n_p =$$

$$i\pi n_i k_l E_{0m} e_{mpq} n_p \frac{a^2}{2}(\delta_{lq} - n_l n_q) = \frac{i\pi a^2}{2} k_l E_{0m} e_{mpl} n_i n_p$$

The fourth term in the right-hand side of expression (2.63) makes the same contribution, while the third one vanishes after this averaging. By proceeding now to the second step, the averaging over the vector $\vec{n}$, whose direction in space is distributed randomly, so that

$$\langle n_i n_k \rangle = \frac{1}{3}\delta_{ik}, \quad \langle n_i n_k n_l \rangle = 0,$$

one gets the polarization of the gas $\vec{P}$:

$$\vec{P} = Nh\rho\beta \exp(-i\omega t)\left[\frac{h}{3}\vec{E}_0 + \frac{i\pi a^2}{3}(\vec{k} \times \vec{E}_0)\right] \qquad (2.66)$$

Since, according to equation (2.65), the magnetization contribution to the electric current is proportional to the wave vector $\vec{k}$ and, therefore, already contains the small parameter $ka, kh \ll 1$, it is sufficient to derive the magnetization $\vec{M}$ in the zeroth-order approximation. It means, that while inserting (2.63) into the expression (2.64) for $\vec{m}$, only the first term in (2.63) should be retained. Thus, after the averaging over the orientations of $\vec{n}$, one gets

$$\vec{M} = -\frac{i\pi a^2 h}{3c}\omega N\rho\beta\vec{E}_0 \exp(-i\omega t) \qquad (2.67)$$

It follows then from equations (2.65-2.67) that the macroscopic electric current in the gas is equal to

$$\vec{j} = -\frac{i\omega N\rho\beta}{3}[h^2\vec{E} + 2i\pi ha^2(\vec{k} \times \vec{E})],$$

which yields the conductivity tensor

$$\sigma_{\alpha\beta} = -\frac{iNh^2\omega\rho\beta}{3}\delta_{\alpha\beta} - \frac{2\pi Nha^2\omega\rho\beta}{3}e_{\alpha\beta\gamma}k_\gamma$$

Finally, one gets the dielectric permeability tensor

$$\epsilon_{\alpha\beta} = \delta_{\alpha\beta} + \frac{4\pi i}{\omega}\sigma_{\alpha\beta} =$$

$$\epsilon\delta_{\alpha\beta} + ife_{\alpha\beta\gamma}k_\gamma, \quad \epsilon = 1 + \frac{4\pi}{3}Nh^2\rho\beta, \quad f = \frac{8\pi^2}{3}Nha^2\rho\beta \qquad (2.68)$$

Therefore, such a gas of "double helices" indeed possesses natural optical activity, the strength of which is determined by the constant $f$ in equation (2.68). One may notice that the polarization and the magnetization of the gas provide equal contributions to this constant. If $a = 0$ or $h = 0$, the constant $f = 0$ and, hence, there is no natural optical activity. This is not a surprise, because in these cases the helix reduces, respectively, to a rod or to a ring, and the molecules become identical to their stereoisomeric counterpart.

It is worth demonstrating that the same result can be obtained directly by averaging the microscopic electric currents without referring to the polarization and the magnetization of the gas. In order to avoid a singularity in such a calculation, it is convenient to introduce an infinitely small cross-section area of the molecule, $S$. Then, the electric current density due to the above derived displacement can be written as

$$\vec{j}_{micro} = \rho \dot{\Delta} \vec{t}/S = -i\omega \Delta(\vec{n},\, \vec{a},\, \vec{E}_0)\vec{t}/S,$$

where $\vec{t}$ is a unit vector along the helix line, which, according to equation (2.62), is equal to

$$\vec{t} = (a^2 + h^2/4\pi^2)^{-1/2}\left(\frac{h}{2\pi}\vec{n} - \vec{a}\sin\phi + (\vec{n}\times\vec{a})\cos\phi\right)$$

In the above expression the electric current at the location $\vec{R}(\phi)$ is proportional to $\vec{E}_0$ — the electric field at point $O$ of Figure 2.5. However, in order to find the conductivity of the medium, one needs to compare the electric current and the electric field at the same location. Therefore, it is necessary to express $\vec{E}_0$ in terms of $\vec{E}(\vec{R})$, which yields $\vec{E}_0 = \vec{E}\exp(-i\vec{k}\cdot\vec{R}) \approx \vec{E}(1 - i\vec{k}\cdot\vec{R})$. Furthermore, since all terms in the expression (2.63), except the first one, are proportional to $k$, i.e., are already first order in the small parameter $ka$, $kh \ll 1$, the difference between the vectors $\vec{E}$ and $\vec{E}_0$ is significant only in the first term. Elsewhere, one can put $\vec{E}_0 = \vec{E}$. Therefore, with the help of relations (2.62-2.63), one gets

$$\vec{j}_{micro} = \frac{-i\omega\rho\beta}{S(a^2 + h^2/4\pi^2)^{1/2}}\left(\frac{h}{2\pi}\vec{n} - \vec{a}\sin\phi + (\vec{n}\times\vec{a})\cos\phi\right)\cdot$$
$$\left\{h(\vec{E}\cdot\vec{n}) - ih(\vec{E}\cdot\vec{n}\left[\frac{h\phi}{2\pi}(\vec{k}\cdot\vec{n}) + (\vec{k}\cdot\vec{a})\cos\phi + \vec{k}\cdot(\vec{n}\times\vec{a})\sin\phi\right] + \right.$$
$$ih^2/2(\vec{E}\cdot\vec{n})(\vec{k}\cdot\vec{n}) + ih(\vec{E}\cdot\vec{a})(\vec{k}\cdot\vec{n}) - i\pi(\vec{E}\cdot\vec{a})[\vec{k}\cdot(\vec{n}\times\vec{a})] + $$
$$\left. i\pi(\vec{k}\cdot\vec{a})[\vec{E}\cdot(\vec{n}\times\vec{a})]\right\}$$

In order to find the macroscopic conductivity of this gas as a continuous medium, one needs to average the above-written microscopic current over a "physically infinitesimal" volume, that is over a domain that is small compared to the wavelength $\lambda = 2\pi/k$, but which contains a large number of molecules. This procedure can be carried out step by step. First, the tensor coefficient that relates $\vec{j}_{micro}$ with $\vec{E}$ has be to averaged over the parameter $\phi$, i.e., over the different points of a molecule with the fixed vectors $\vec{n}$ and $\vec{a}$. Then, the obtained tensor should be averaged over the orientation vectors $\vec{n}$ and $\vec{a}$, similarly to what has been performed above. Since the microscopic current is present only inside the molecules, the result should be multiplied by the fraction of volume occupied by molecules, that is by $NSL$, where $L = 2\pi(a^2 + h^2/4\pi^2)^{1/2}$ is the length of the molecule. The final result for the tensor $\sigma_{\alpha\beta}$ and $\epsilon_{\alpha\beta}$ confirms the findings of (2.68).

## Problem 2.2.5

Show that if a homogeneous medium with the dielectric permeability tensor (2.51) becomes slightly non-homogeneous (i.e., the characteristic length $l$ of the spatial variation of its parameters is large compared to the molecular size $r_0$), the relation between the vectors $\vec{D}$ and $\vec{E}$ takes the following form:

$$\vec{D}(\vec{r}) = \epsilon(\vec{r})\vec{E}(\vec{r}) + f(\vec{r})\vec{\nabla} \times \vec{E}(\vec{r}) + \frac{1}{2}[\vec{\nabla}f(\vec{r}) \times \vec{E}(\vec{r})] \qquad (2.69)$$

The most general relation between the vectors $\vec{D}$ and $\vec{E}$ in an isotropic medium can be written as

$$D_\alpha(\vec{r}) = \int d\vec{r}' \epsilon_{\alpha\beta}(\vec{r}, \vec{r}') E_\beta(\vec{r}'),$$

$$\epsilon_{\alpha\beta}(\vec{r}, \vec{r}') = a_1(\vec{r}, \vec{r}')\delta_{\alpha\beta} + a_2(\vec{r}, \vec{r}')(r_\alpha - r'_\alpha)(r_\beta - r'_\beta) +$$
$$a_3(\vec{r}, \vec{r}')e_{\alpha\beta\gamma}(r_\gamma - r'_\gamma) \qquad (2.70)$$

(compare it with equation (2.54)).

It is important that the functions $a_{1,2,3}$ in (2.70) cannot be arbitrary. As follows from a rather general thermodynamical consideration, the tensor $\epsilon_{\alpha\beta}(\vec{r}, \vec{r}')$ must satisfy the following constraint: $\epsilon_{\alpha\beta}(\vec{r}, \vec{r}') = \epsilon_{\beta\alpha}(\vec{r}', \vec{r})$, which is a consequence of the Onsager symmetry relations for the kinetic coefficients (see, e.g., E. M. Lifshitz and L. P. Pitaevskii, *Statistical Physics*, Part 2, Ch.8, Oxford, 1995). Thus, by introducing the new variables, $\vec{R} = (\vec{r} + \vec{r}')/2$ and $\vec{\rho} = \vec{r} - \vec{r}'$, these constraints become satisfied identically if $a_{1,2,3}(\vec{r}, \vec{r}') \equiv a_{1,2,3}(\vec{R}, \rho)$. In a homogeneous medium they do not depend on $\vec{R}$, and expression (2.70) reduces to that of (2.54). If the non-homogeneity is weak, one can explore the following expansion procedure: $\vec{R} = \vec{r} - \vec{\rho}/2$, thus

$$a_{1,2,3}(\vec{R}, \rho) \approx a_{1,2,3}(\vec{r}, \rho) - \frac{1}{2}\rho\vec{\nabla}a_{1,2,3}$$

By also taking into account that

$$E_\beta(\vec{r}') \approx E_\beta(\vec{r}) - \rho_\gamma \frac{\partial E_\beta}{\partial x_\gamma},$$

one gets from expression (2.70) that

$$D_\alpha(\vec{r}) = E_\alpha(\vec{r}) \left( \int d\vec{\rho}a_1(\vec{r}, \rho) - \frac{1}{2} \int d\vec{\rho}(\vec{\nabla}a_1 \cdot \vec{\rho}) \right) +$$

$$E_\beta(\vec{r}) \left( \int d\vec{\rho}a_2(\vec{r}, \rho)\rho_\alpha\rho_\beta - \frac{1}{2} \int d\vec{\rho}\rho_\alpha\rho_\beta(\vec{\nabla}a_2 \cdot \vec{\rho}) \right) +$$

$$E_\beta(\vec{r}) \left( \int d\vec{\rho}a_3(\vec{r}, \rho)e_{\alpha\beta\gamma}\rho_\gamma - \frac{1}{2} \int d\vec{\rho}(\vec{\nabla}a_3 \cdot \vec{\rho})e_{\alpha\beta\gamma}\rho_\gamma \right) -$$

$$\int d\vec{\rho}\frac{\partial E_\beta}{\partial x_\delta}\rho_\delta[a_1(\vec{r}, \rho)\delta_{\alpha\beta} + a_2(\vec{r}, \rho)\rho_\alpha\rho_\beta + a_3(\vec{r}, \rho)e_{\alpha\beta\gamma}\rho_\gamma]$$

After integration over the entire solid angle $d\vec{\Omega}$ as $d\vec{\rho} = \rho^2 d\rho d\vec{\Omega}$, one gets that, similarly to equation (2.56),

$$D_\alpha(\vec{r}) = \epsilon(\vec{r})E_\alpha(\vec{r}) - f(\vec{r})e_{\alpha\beta\gamma}\frac{\partial E_\beta}{\partial x_\gamma} -$$

$$\frac{1}{2}e_{\alpha\beta\gamma}E_\beta\frac{\partial f}{\partial x_\gamma}$$

Thus, the relation (2.69) is confirmed, with

$$\epsilon(\vec{r}) = 4\pi \left( \int d\rho a_1(\vec{r}, \rho)\rho^2 + \frac{1}{3} \int d\rho a_2(\vec{r}, \rho)\rho^4 \right),$$

$$f(\vec{r}) = \frac{4\pi}{3} \int d\rho a_3(\vec{r}, \rho)\rho^4$$

## Problem 2.2.6

Determine the variation of polarization for an electromagnetic wave propagating in an isotropic medium without spatial dispersion in the presence of an external magnetic field $\vec{B}$ (the Faraday effect).

In an external magnetic field the dielectric permeability tensor of such a medium has the form (2.52):

$$\epsilon_{\alpha\beta} = \epsilon\delta_{\alpha\beta} + ibe_{\alpha\beta\gamma}B_\gamma$$

Therefore, in a coordinate system with the $z$-axis directed along the wave vector $\vec{k}$, and with the external magnetic field lying in the $(x, z)$ plane, Maxwell's equations (2.8) read:

$$\left(-k^2 + \frac{\omega^2}{c^2}\epsilon\right)E_x + i\frac{\omega^2}{c^2}bB_zE_y = 0,$$

$$\left(-k^2 + \frac{\omega^2}{c^2}\epsilon\right)E_y - i\frac{\omega^2}{c^2}bB_zE_x + i\frac{\omega^2}{c^2}bB_xE_z = 0, \qquad (2.71)$$

$$\frac{\omega^2}{c^2}\epsilon E_z - i\frac{\omega^2}{c^2}bB_xE_y = 0$$

It is worth noting here that the dielectric tensor (2.52) is actually the first two terms in the expansion of $\epsilon_{\alpha\beta}(\vec{B})$ as a power series of an external magnetic field. Therefore, the same level of accuracy should be adopted in dealing with equations (2.71). As seen from the third equation in (2.71), the electric field component $E_z$ is of the first order in $B$; hence, the last term in the second of these equations is of the second order in $B$ and, therefore, can be neglected. Then, the solvability condition for the first two equations in (2.71) for the electric field components $E_x$, $E_y$ reads

$$k^2 - \frac{\omega^2}{c^2}\epsilon = \pm b\frac{\omega^2}{c^2}B_z = \pm b\frac{\omega^2}{c^2}B\cos\theta,$$

where $\theta$ is the angle between the vectors $\vec{k}$ and $\vec{B}$. The two signs correspond to the circularly polarized electromagnetic waves ($E_x = \pm iE_y$) with slightly different wave vectors

$$k_\pm = \frac{\omega}{c}\sqrt{\epsilon} \pm \Delta k, \quad \Delta k = \frac{b}{2\sqrt{\epsilon}}\frac{\omega}{c}B\cos\theta$$

Thus, after calculations similar to those in Problem 2.2.2, one finds that presence of the external magnetic field results in a rotation of the polarization plane with the rate equal to

$$\frac{d\phi}{dl} = \Delta k = \frac{b}{2\sqrt{\epsilon}}\frac{\omega}{c}B\cos\theta \qquad (2.72)$$

The important difference with the natural optical activity is that in this case the sense and the rate of rotation depend on the direction in which the wave is propagating.

## Problem 2.2.7

Derive the rate of the Faraday rotation in a cold electron plasma (see Problem 2.1.2).

In the case of a weak external magnetic field the general expression (2.17) for the plasma dielectric permeability tensor reduces to

$$\epsilon_{\alpha\beta} \approx \left(1 - \frac{\omega_{pe}^2}{\omega^2}\right)\delta_{\alpha\beta} + i\frac{\omega_{pe}^2\omega_B}{\omega^3}e_{\alpha\beta\gamma}h_\gamma, \ \vec{h} = \frac{\vec{B}}{B}$$

Its comparison with that of (2.52) reveals the Faraday constant equal to $b = e\omega_{pe}^2/mc\omega^3$, yielding, according to equation (2.72), the Faraday rotation rate

$$\frac{d\phi}{dl} = \frac{b}{2\sqrt{\epsilon}}\frac{\omega}{c}B\cos\theta = \frac{\omega_B}{2c}\frac{\omega_{pe}^2\cos\theta}{\omega^2(1 - \omega_{pe}^2/\omega^2)^{1/2}}$$

Consider now the meaning of a "weak" external magnetic field in this particular case. According to the dispersion relation (2.19), for the electromagnetic wave in a plasma the frequency $\omega \geq \omega_{pe}$. Thus, it becomes evident from expression (2.17) that the weak field approximation holds if $\omega_B \ll \omega_{pe}$ (the electron gyrofrequency is small compared to the electron plasma frequency).

## Problem 2.2.8

A linearly polarized electromagnetic wave propagating in free space is incident normally on the plane boundary of an isotropic dielectric occupying the half-space $z < 0$. The dielectric is immersed into a uniform external electric field $\vec{E}_0$, which is lying in the $(x, y)$ plane. Determine the polarization of the reflected wave.

In the coordinate frame with the $x$-axis directed along $\vec{E}_0$, the dielectric tensor (2.53) takes the form:

$$\epsilon_{xx} = \epsilon_{\parallel} = \epsilon + (a_1 + a_2)E_0^2, \ \epsilon_{yy} = \epsilon_{zz} = \epsilon_{\perp} = \epsilon + a_1 E_0^2$$

If the electric field of the incident wave, $\vec{E}^{(i)}$, which also lies in the $(x, y)$ plane, makes the angle $\theta_i$ with $\vec{E}_0$, its components can be written as

$$E_x^{(i)} = E^{(i)}\cos\theta_i, \ E_y^{(i)} = E^{(i)}\sin\theta_i$$

A change in the polarization of the reflected wave is due to a small difference in $\epsilon_{xx}$ and $\epsilon_{yy}$, which results in the different reflection coefficients for the $x$ and $y$ linear polarizations. Thus, by reducing the general expressions obtained in Problem 2.1.11 for the case of the normal incidence, and assuming the external electric field being rather weak ($a_{1,2}E_0^2 \ll \epsilon$), one gets for the electric field $\vec{E}^{(r)}$ of the reflected wave:

$$E_x^{(r)} = E_x^{(i)} \frac{\sqrt{\epsilon_\parallel} - 1}{\sqrt{\epsilon_\parallel} + 1} = E^{(i)} \cos\theta_i \frac{\sqrt{\epsilon_\parallel} - 1}{\sqrt{\epsilon_\parallel} + 1} \approx$$

$$E^{(i)} \cos\theta_i \left(1 + \frac{(a_1 + a_2)E_0^2}{\epsilon(\epsilon - 1)}\right) \frac{\sqrt{\epsilon} - 1}{\sqrt{\epsilon} + 1},$$

$$E_y^{(r)} = E_y^{(i)} \frac{\sqrt{\epsilon_\perp} - 1}{\sqrt{\epsilon_\perp} + 1} \approx E^{(i)} \sin\theta_i \left(1 + \frac{(a_1)E_0^2}{\epsilon(\epsilon - 1)}\right) \frac{\sqrt{\epsilon} - 1}{\sqrt{\epsilon} + 1}$$

For a transparent dielectric, when $\epsilon$, $a_{1,2}$ are real, the reflected wave remains linearly polarized, but with a slightly different polarization plane, i.e., $\theta_r = \theta_i + \Delta\theta$. Indeed, one gets from the above equations that

$$\tan\theta_r = \frac{E_y^{(r)}}{E_x^{(r)}} \approx \tan\theta_i \left(1 - \frac{a_2 E_0^2}{\epsilon(\epsilon - 1)}\right),$$

which yields

$$\Delta\theta \approx -\frac{a_2 E_0^2}{\epsilon(\epsilon - 1)} \sin\theta_i \cos\theta_i$$

## 2.3   The frequency dispersion of the electric permittivity. The propagation of electromagnetic waves.

In an isotropic medium without spatial dispersion the relation between the electric induction, $\vec{D}(t)$, and the electric field, $\vec{E}(t)$, takes the form:

$$\vec{D}(t) = \vec{E}(t) + \int_0^\infty f(\tau)\vec{E}(t - \tau)d\tau,$$

which yields the dielectric permeability tensor $\epsilon_{\alpha\beta} = \epsilon(\omega)\delta_{\alpha\beta}$, with the electric permittivity

$$\epsilon(\omega) = 1 + \int_0^\infty f(\tau)\exp(i\omega\tau)d\tau \tag{2.73}$$

The function $\epsilon(\omega)$, defined by equation (2.73), is an analytical function in the upper half-plane of the complex variable $\omega$, and $\epsilon(\omega) \to 1$ at $|\omega| \to \infty$.

The real, $\epsilon'$, and the imaginary, $\epsilon''$, parts of this function satisfy the **Kramers-Kronig relations**

$$\epsilon'(\omega) = 1 + \frac{1}{\pi}P\int_{-\infty}^{+\infty} \frac{\epsilon''(\omega')}{(\omega' - \omega)}d\omega', \quad \epsilon''(\omega) = -\frac{1}{\pi}P\int_{-\infty}^{+\infty} \frac{\epsilon'(\omega') - 1}{(\omega' - \omega)}d\omega' \tag{2.74}$$

This relation should be modified for a conducting medium, for which $\epsilon(\omega)$ has a pole at $\omega = 0$ (see Problem 2.3.2).

## Problem 2.3.1

Find the "memory" function, $f(\tau)$, introduced in equation (2.73), for a medium with the electric permittivity

$$\epsilon(\omega) = 1 - \frac{\omega_p^2}{\omega(\omega + i\gamma)}, \ \gamma > 0$$

According to (2.73),

$$\int_0^\infty f(\tau) \exp(i\omega\tau) d\tau = -\frac{\omega_p^2}{\omega(\omega + i\gamma)}$$

The "memory" function $f(\tau)$ makes physical sense only for $\tau \geq 0$ (the principle of causality). Therefore, one can extend it formally to $\tau < 0$ by putting there $f(\tau) \equiv 0$, and, hence, re-write the above relation as

$$\int_{-\infty}^\infty f(\tau) \exp(i\omega\tau) d\tau = -\frac{\omega_p^2}{\omega(\omega + i\gamma)}$$

Then, by considering it as the Fourier transform for the function $f(\tau)$, the inverse transformation yields

$$f(\tau) = \frac{\omega_p^2}{2\pi} \int_{-\infty}^\infty d\omega \frac{\exp(-i\omega\tau)}{\omega(\omega + i\gamma)} \tag{2.75}$$

The integrand in (2.75) has a singularity at $\omega = 0$, so one needs to define a recipe for dealing with it. In order to do so, consider the plane of the complex variable $\omega$. Then, for $\tau < 0$ the integrand in (2.75) tends to zero exponentially in the upper half-plane (see Figure 2.6). Therefore, the integral along the real axis in (2.75) can be supplemented with the integral along a remote semicircle there. After that, it becomes evident that the integration path must pass above the singular point $\omega = 0$ as shown in Figure 2.6. In this case there are not any singularities inside the integration path, which ensures that $f(\tau) = 0$ for $\tau < 0$, as it has been postulated in deriving the expression (2.75).

In order to find $f(\tau)$ for $\tau > 0$, one can supplement the integral in equation (2.75) with the integral along a remote semicircle in the lower half-plane of

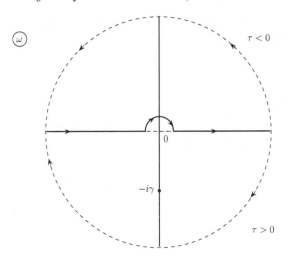

**FIGURE 2.6**
The "passing rule" in derivation of the memory function $f(\tau)$

the complex variable $\omega$. Then, the entire integral is reduced to the sum of the residues at the points $\omega = 0$ and $\omega = -i\gamma$:

$$f(\tau) = \frac{\omega_p^2}{2\pi} 2\pi i \left[ \frac{1}{i\gamma} + \frac{\exp(-\gamma\tau)}{(-i\gamma)} \right] = \frac{\omega_p^2}{\gamma}[1 - \exp(-\gamma\tau)]$$

Note, that this "memory" function does not tend to zero at $\tau \to +\infty$. The reason is that the medium under consideration is a conductor, for which $\epsilon(\omega)$ has a pole at $\omega = 0$.

## Problem 2.3.2

Retrieve the entire dielectric permittivity $\epsilon(\omega)$ of a medium from its imaginary part, which is equal to

$$\epsilon''(\omega) = \frac{\gamma\alpha^2}{\omega(\omega^2 + \gamma^2)}, \quad \gamma > 0$$

In order to find the real part of the permittivity, $\epsilon'(\omega)$, one can use the Kramers-Kronig relations (2.74). However, since the medium under consideration is a conductor ($\epsilon''(\omega)$ has a pole at $\omega = 0$), the relations (2.74) should be applied to the "regularized" permittivity

$$\hat{\epsilon}(\omega) = \epsilon(\omega) - \frac{4\pi i}{\omega}\sigma_0,$$

where $\sigma_0$ is the static electric conductivity of the medium:

$$\sigma_0 = \lim_{\omega \to 0} [\omega \epsilon''(\omega)/4\pi].$$

In this case $\sigma_0 = \alpha^2/4\pi\gamma$, hence

$$\hat{\epsilon}''(\omega) = \frac{\gamma\alpha^2}{\omega(\omega^2 + \gamma^2)} - \frac{\alpha^2}{\omega\gamma} = \frac{\alpha^2\omega}{\gamma(\omega^2 + \gamma^2)}$$

Then, the first of the relations (2.74) yields:

$$\hat{\epsilon}'(\omega) = \epsilon'(\omega) = 1 + \frac{1}{\pi}P\int\limits_{-\infty}^{+\infty} d\omega' \frac{\hat{\epsilon}''(\omega')}{\omega' - \omega} =$$

$$1 - \frac{\alpha^2}{\pi\gamma}P\int\limits_{-\infty}^{+\infty} \frac{\omega' d\omega'}{(\omega' - \omega)[(\omega')^2 + \gamma^2]}$$

This integral can be calculated by using the plane of the complex variable $\omega$ (see Figure 2.7), where the above integral is supplemented by the integral along a remote semicircle (which tends to zero), and along an infinitesimal semicircle around the point $\omega' = \omega$ (which is equal to the half-residue at this point). Since the only singularity inside the so formed integral path is a pole at $\omega' = i\gamma$, one gets:

$$P\int\limits_{-\infty}^{+\infty} \frac{\omega' d\omega'}{(\omega' - \omega)[(\omega')^2 + \gamma^2]} = 2\pi i Res|_{\omega'=i\gamma} + i\pi Res|_{\omega'=\omega} = \pi\frac{\gamma}{\omega^2 + \gamma^2}$$

Thus, $\epsilon'(\omega) = 1 - \alpha^2/(\omega^2 + \gamma^2)$, and the permittivity of the medium is equal to

$$\epsilon(\omega) = 1 - \frac{\alpha^2}{\omega(\omega + i\gamma)}$$

## Problem 2.3.3

Determine the dielectric permittivity $\epsilon(\omega)$ of a medium from the reflection coefficient $R(\omega)$ of the normal incident electromagnetic waves in the entire frequency range $0 \leq \omega < +\infty$.

According to Problem 2.1.11, for a normally incident electromagnetic wave the reflection coefficient is equal to

$$R(\omega) = \left| \frac{\sqrt{\epsilon(\omega)} - 1}{\sqrt{\epsilon(\omega)} + 1} \right|^2$$

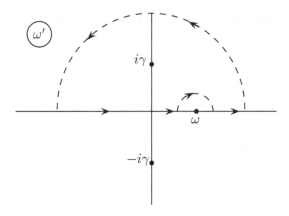

**FIGURE 2.7**

Integration contour in the complex plane of variable $\omega'$

In order to retrieve the permittivity $\epsilon(\omega)$ from this expression, one can apply the method used in the derivation of the Kramers-Kronig relations (2.74). Thus, consider the complex function

$$\Phi(\omega) = \frac{\sqrt{\epsilon(\omega)} - 1}{\sqrt{\epsilon(\omega)} + 1} \equiv \sqrt{R(\omega)} \exp[i\phi(\omega)] \qquad (2.76)$$

Then, the function

$$\Psi(\omega) \equiv \ln \Phi(\omega) = \frac{1}{2} \ln R(\omega) + i\phi(\omega)$$

So the problem is now reduced to the determination of the imaginary part of $\Psi(\omega)$, when its real part is known.

Since the function $\epsilon(\omega)$ has no singularities and null points in the upper half-plane of the complex variable $\omega$, and it tends to unity only at $|\omega| \to \infty$, the above-defined functions $\Phi(\omega)$ and $\Psi(\omega)$ are also analytical functions in this domain. Thus, consider the integral

$$\int_\Gamma \frac{\Psi(\omega)}{\omega^2 - \omega_0^2} d\omega,$$

along the path $\Gamma$ shown in Figure 2.8. Although the function $\Psi(\omega)$ diverges logarithmically at $|\omega| \to \infty$ (since at this limit $R(\omega) \to 0$), the contribution from a remote semicircle remains negligible due to the denominator of the integrand. It follows then from the Cauchy's theorem that

$$P \int_{-\infty}^{+\infty} \frac{\Psi(\omega)}{\omega^2 - \omega_0^2} d\omega + \frac{i\pi}{2\omega_0} \Psi(-\omega_0) - \frac{i\pi}{2\omega_0} \Psi(\omega_0) = 0$$

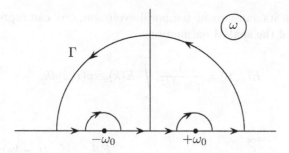

**FIGURE 2.8**
Integration in the complex plane of variable $\omega$

By taking into account that the real and the imaginary parts of $\epsilon(\omega)$ are, respectively, an even and an odd function of $\omega$ on the real axis (see expression (2.73)), one concludes that the same is the case for the function $\Psi(\omega)$. Thus, the real part of the above-written equation (its imaginary part is identically equal to zero) yields

$$\Im\Psi(\omega_0) = \frac{\omega_0}{\pi} P \int_{-\infty}^{+\infty} \frac{\Psi(\omega)}{\omega^2 - \omega_0^2} d\omega$$

therefore

$$\phi(\omega) = \frac{\omega}{\pi} P \int_0^\infty \frac{\ln R(\omega')}{\omega^2 - (\omega')^2} d\omega'$$

This, together with equation (2.76), enables one, in principle, to derive the permittivity $\epsilon(\omega)$.

## Problem 2.3.4

Determine the propagation velocity and the shape evolution for a quasi-monochromatic train of electromagnetic waves in a transparent medium with the dielectric permittivity $\epsilon(\omega)$.

Consider a one-dimensional electromagnetic wave train, which at the initial moment, $t = 0$, has the form

$$E(z, 0) = E_0 \exp(-z^2/l^2) \exp(ik_0 z), \tag{2.77}$$

where $E$ is one of the transverse components of the electric field, say, $E_x$. In order to obtain its subsequent temporal evolution, one can represent it as a superposition of the spatial harmonics:

$$E(z, 0) = \frac{1}{(2\pi)^{1/2}} \int\limits_{-\infty}^{+\infty} E(k) \exp(ikz) dk,$$

where

$$E(k) = \frac{1}{(2\pi)^{1/2}} \int\limits_{-\infty}^{+\infty} E(z, 0) \exp(-ikz) dz = \frac{E_0 l}{\sqrt{2}} \exp\left[-\frac{(k - k_0)^2 l^2}{4}\right] \quad (2.78)$$

As seen from equation (2.78), the spectrum of this wave train has the Gaussian shape with a characteristic width $\Delta k \sim l^{-1}$. Thus, it can be considered as a quasimonochromatic one if $\Delta k \ll k_0$, i.e., $k_0 l \gg 1$. Each spatial harmonic corresponds to a frequency $\omega(k)$, which is specifield by the dispersion equation $k = \omega \sqrt{\epsilon(\omega)}/c$. Thus,

$$E(z, t) = \frac{1}{(2\pi)^{1/2}} \int\limits_{-\infty}^{+\infty} E(k) \exp[i(kz - \omega(k)t)] dk \quad (2.79)$$

For a quasimonochromatic wave train with a narrow spectrum, $\Delta k \ll k_0$, the following expansion for $\omega(k)$ can be used:

$$\omega(k) \approx \omega(k_0) + \left(\frac{\partial \omega}{\partial k}\right)_{k_0} (k - k_0) + \frac{1}{2} \left(\frac{\partial^2 \omega}{\partial^2 k}\right)_{k_0} (k - k_0)^2 =$$

$$\omega_0 + v_g(k - k_0) + \alpha(k - k_0)^2$$

By substituting this expression and the spectrum (2.78) into equation (2.79), one gets

$$E(z, t) = \Phi(z, t) \exp[i(k_o z - \omega_0 t)],$$

$$\Phi(z, t) = \frac{E_0 l}{2\sqrt{\pi}} \int\limits_{-\infty}^{+\infty} dk \exp\left(-\frac{k^2 l^2}{4} + ik(z - v_g t) - ik^2 \alpha t\right) =$$

$$\frac{E_0}{(1 + 4i\alpha t/l^2)^{1/2}} \exp\left[-\frac{(z - v_g t)^2}{(l^2 + 4i\alpha t)}\right]$$

The function $\Phi(z, t)$ is the so-called **complex envelope** of the wave train, and its modulus determines the intensity of the signal:

$$|\Phi(z, t)| = \frac{E_0}{\sqrt{L(t)/l}} \exp\left[-\frac{(z - v_g t)^2}{L^2(t)}\right], \quad (2.80)$$

$$L(t) = \left(l^2 + \frac{16\alpha^2 t^2}{l^2}\right)^{1/2} \quad (2.81)$$

As seen from expressions (2.80-2.81), the wave train propagates with the **group velocity** $v_g = \partial\omega/\partial k$, while its width, $L(t)$, is increasing with time due to the frequency dispersion. In a transparent dielectric

$$v_g = c\frac{\sqrt{\epsilon}}{(\epsilon + \frac{1}{2}\omega\frac{d\epsilon}{d\omega})},$$

and it can be proved (see, e.g., L. D. Landau and E. M. Lifshitz, *Electrodynamics of Continuous Media*, §84, Pergamon Press,1984) that $v_g$ cannot exceed $c$, the speed of light in free space.

In the example under consideration, when the initial signal has the form (2.77), the width of the signal, $L(t)$, given by expression (2.81), is monotonically increasing with time. However, this is not a general rule, so it is possible to construct the initial signal in such a way that its width will be decreasing with time for some interval. This issue is about the phase synchronism between the harmonics contributing to the signal. In the case of the wave train (2.77) the width has a minimum at $t = 0$, because at this instant all the harmonics in equation (2.78) have the same phase. Therefore, one can always choose the initial phases in such a way that they become equal at any given moment of time, $t_0 > 0$, so that the width $L(t)$ would have a minimum at $t = t_0$. Indeed, consider, instead of (2.78), the following spectrum of the wave train:

$$E(k) = \frac{E_0 l}{\sqrt{2}} \exp\left[-\frac{(k - k_0)^2 l^2}{4} + i\alpha(k - k_0)2t_0\right]$$

Then, by repeating the derivation performed above, one gets the similar expression for the signal envelope but with

$$L(t) = \left(l^2 + \frac{16\alpha^2(t - t_0)^2}{l^2}\right)^{1/2}$$

Thus, the width of the wave train is decreasing during the time interval $0 < t < t_0$, gets its minimum, $L_{min} = l$ at $t = t_0$, and is increasing afterwards. Clearly, whatever the $t_0$, the minimum of $L(t)$ is determined by the "uncertainty" relation $\Delta k \cdot \Delta z \sim 1$.

## Problem 2.3.5

Prove, that in a transparent medium described by the dielectric permeability tensor $\epsilon_{\alpha\beta}(\omega)$, the energy of the electromagnetic wave propagates with the group velocity $\vec{v}_g = \partial\omega(\vec{k})/\partial\vec{k}$.

The flux of energy in the electromagnetic wave is determined by the Poynting vector $\vec{S} = c(\vec{E} \times \vec{B})/4\pi$, so the propagation velocity of energy can be defined as $\vec{v}_E = \langle \vec{S} \rangle / W$, where $\langle \vec{S} \rangle$ is the flux of energy averaged over the period of the wave, and $W$ is the energy of the wave (per unit volume) given by expression (2.43). Thus, one has to demonstrate that $\vec{v}_E = \vec{v}_g$.

To start with, one represents the fields $\vec{E}$ and $\vec{B}$ in the real form, which is

$$\vec{E} = \frac{1}{2}[\vec{E}_0 \exp(-i\omega t) + \vec{E}_0^* \exp(i\omega t)], \ \vec{B} = \frac{1}{2}[\vec{B}_0 \exp(-i\omega t) + \vec{B}_0^* \exp(i\omega t)],$$

where $\vec{E}_0$ and $\vec{B}_0$ are the electric and magnetic field amplitudes. Then, the energy flux $\langle \vec{S} \rangle$ takes the form

$$\langle \vec{S} \rangle = \frac{c}{16\pi}[(\vec{E}_0 \times \vec{B}_0^*) + (\vec{E}_0^* \times \vec{B}_0)] \tag{2.82}$$

Consider now the $(\vec{k}, \omega)$ representation of Maxwell's equations (2.2-2.3) for the field amplitudes under a small variation of the wave vector $\delta \vec{k}$ and, respectively, the wave frequency $\delta\omega(\vec{k})$:

$$(\delta \vec{k} \times E_0) + (\vec{k} \times \delta \vec{E}_0) = \frac{\delta\omega}{c}\vec{B}_0 + \frac{\omega}{c}\delta \vec{B}_0, \tag{2.83}$$

$$(\delta \vec{k} \times B_0) + (\vec{k} \times \delta \vec{B}_0) = -\frac{\delta\omega}{c}\frac{\partial(\omega\epsilon_{\alpha\beta}^{(H)})}{\partial\omega}E_{0\beta} - \frac{\omega}{c}\epsilon_{\alpha\beta}^{(H)}\delta E_{0\beta} \tag{2.84}$$

By taking the scalar product of (2.83) with $\vec{B}_0^*$, and of (2.84) with $-\vec{E}_0^*$, one gets:

$$(\delta \vec{k} \times E_0) \cdot \vec{B}_0^* + (\vec{k} \times \delta \vec{E}_0) \cdot \vec{B}_0^* = \frac{\delta\omega}{c}\vec{B}_0 \cdot \vec{B}_0^* + \frac{\omega}{c}\delta \vec{B}_0 \cdot \vec{B}_0^*,$$

$$-(\delta \vec{k} \times B_0) \cdot \vec{E}_0^* - (\vec{k} \times \delta \vec{B}_0) \cdot \vec{E}_0^* =$$

$$\frac{\delta\omega}{c}\frac{\partial(\omega\epsilon_{\alpha\beta}^{(H)})}{\partial\omega}E_{0\beta}E_{0\alpha}^* + \frac{\omega}{c}\epsilon_{\alpha\beta}^{(H)}\delta E_{0\beta}E_{0\alpha}^* \tag{2.85}$$

Furthermore, the scalar product of the complex-conjugate of equations (2.2-2.3) with, respectively, $-\delta \vec{B}_0$ and $\delta \vec{E}_0$ yields

$$-(\vec{k} \times \vec{E}_0^*) \cdot \delta \vec{B}_0 = -\frac{\omega}{c}\vec{B}_0^* \cdot \delta \vec{B}_0, \tag{2.86}$$

$$(\vec{k} \times \vec{B}_0^*) \cdot \delta \vec{E}_0 = -\frac{\omega}{c}\epsilon_{\alpha\beta}^{(H)*}E_{0\beta}^*\delta E_{0\alpha} =$$

$$-\frac{\omega}{c}\epsilon_{\beta\alpha}^{(H)}E_{0\beta}^*\delta E_{0\alpha} = -\frac{\omega}{c}\epsilon_{\alpha\beta}^{(H)}E_{0\alpha}^*\delta E_{0\beta} \tag{2.87}$$

(it has been used in the last equation that $\epsilon_{\beta\alpha}^{(H)} = \epsilon_{\alpha\beta}^{(H)*}$). Then, by summing up the left-hand and the right-hand parts of equations (2.85-2.87) and using the standard vector identities, one gets

$$\delta \vec{k} \cdot [(\vec{E}_0 \times \vec{B}_0^*) + (\vec{E}_0^* \times \vec{B}_0)] =$$

$$\frac{\delta\omega}{c}[\vec{B}_0 \cdot \vec{B}_0^* + \frac{\partial(\omega\epsilon_{\alpha\beta}^{(H)})}{\partial\omega}E_{0\alpha}^*E_{0\beta}]$$

By comparing this relation with the given above expressions (2.82) for $\langle \vec{S} \rangle$ and (2.43) for $W$, it can be written as

$$\delta \vec{k} \cdot \langle \vec{S} \rangle = \delta \omega W,$$

which yields that

$$\vec{v}_g = \frac{\partial \omega}{\partial \vec{k}} = \frac{\langle \vec{S} \rangle}{W} = \vec{v}_E$$

## Problem 2.3.6

A plane electromagnetic wave train with a sharp front is normally incident on the surface of a transparent dielectric, which occupies the half-space $z > 0$, so that the electric field, $E_i(t)$, of the incident signal at $z = 0$ is equal to zero at $t < 0$, and $E_0 \sin(\omega_0 t)$, $t > 0$. Determine the electric field, $E(z, t)$ in the dielectric, if its permittivity is $\epsilon(\omega)$.

The incident signal can be represented as a superposition of the Fourier harmonics:

$$E_i(t) = \frac{1}{\sqrt{2\pi}} \int\limits_{-\infty}^{+\infty} d\omega \, E_i(\omega) \exp(-i\omega t),$$

$$E_i(\omega) = \frac{1}{\sqrt{2\pi}} \int\limits_{-\infty}^{+\infty} dt \, E_i(t) \exp(i\omega t) = \frac{E_0}{\sqrt{2\pi}} \int\limits_{0}^{+\infty} dt \sin(\omega_0 t) \exp(i\omega t - \delta t) =$$

$$\frac{E_0}{2\sqrt{2\pi}} \left( \frac{1}{\omega + \omega_0 + i\delta} - \frac{1}{\omega - \omega_0 + i\delta} \right) \tag{2.88}$$

(it is convenient to introduce here an infinitesimal damping of the signal, $\delta > 0$, which will be eliminated in the final results). For each of the harmonics in (2.88) the solution is already known (see Problem 2.1.11): the amplitude of the transmitted harmonic is equal to $2E_i(\omega)/[1 + \sqrt{\epsilon(\omega)}]$. Thus, the electric field in the dielectric takes the form

$$E(z, t) = \frac{1}{\sqrt{2\pi}} \int\limits_{-\infty}^{+\infty} d\omega \, \frac{2E_i(\omega)}{1 + \sqrt{\epsilon(\omega)}} \exp\left( \frac{i\omega}{c} \sqrt{\epsilon(\omega)} z - i\omega t \right) = \frac{E_0}{2\pi} \times$$

$$\int\limits_{-\infty}^{+\infty} d\omega \frac{\exp\left( \frac{i\omega}{c} \sqrt{\epsilon(\omega)} z - i\omega t \right)}{1 + \sqrt{\epsilon(\omega)}} \left( \frac{1}{\omega + \omega_0 + i\delta} - \frac{1}{\omega - \omega_0 + i\delta} \right) \tag{2.89}$$

Starting from this general expression, one can first prove that, as expected,

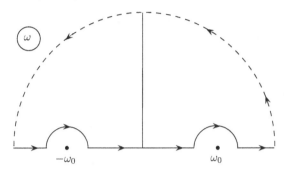

**FIGURE 2.9**
The "passing rule" in derivation of the electric field

$E(z, t) = 0$ for $z > ct$. To do so, consider the plane of the complex variable $\omega$, where the field $E(z, t)$ in equation (2.89) is determined by the integral along the real axis. Thus, it is seen now that the introduced above small damping provides a "rule of passing" for the singularities at $\omega = \pm\omega_0$: $\delta > 0$ moves them down from the real axis; therefore, they must be passed above as shown in Figure 2.9. Furthermore, since the permittivity $\epsilon(\omega) \to 1$ at $|\omega| \to \infty$, the integrand in (2.89) tends exponentially to zero at $|\omega| \to \infty$ in the upper half-plane of $\omega$ when $z > ct$. Therefore, one can supplement the integral along the real axis in (2.89) with the integral along a remote semicircle (see Figure 2.9). Then, the analytical properties of $\epsilon(\omega)$ in the upper half-plane ensure absence of any singularity inside this integration contour, which makes in this case the integral in (2.89) equal to zero: thus, $E(z, t) = 0$ at $z > ct$.

What is the electric field equal to at $z < ct$? The general expression (2.89) does not provide a visually explicit answer; therefore, one needs to consider separately various intervals of $z$, where a contribution of some particular frequencies to the integral in (2.89) dominates. As seen from (2.89), the spectrum has two sharp peaks at $\omega = \pm\omega_0$, which relates to propagation of the "main" signal. In order to find it out, consider, for simplicity, only one of them, say, at $\omega = \omega_0$, which corresponds to the incident signal $E_i(t) = E_0 \exp(-i\omega_0 t)$ at $t > 0$. Then

$$E_i(\omega) = \frac{E_0}{\sqrt{2\pi}} \int\limits_0^\infty d\tau \exp[i(\omega - \omega_0)\tau],$$

and, instead of expression (2.89), one gets

$$E(z, t) = \frac{E_0}{2\pi} \int\limits_0^\infty d\tau \int\limits_{-\infty}^{+\infty} d\omega \exp[i(\omega - \omega_0)\tau] \frac{2\exp[i(k(\omega)z - \omega t)]}{1 + \sqrt{\epsilon(\omega)}}$$

Since the interest is in the contribution of frequencies close to $\omega_0$, the following expansion can be inserted into the exponential factors there:

$$\omega = \omega_0 + \xi,$$

$$k(\omega) \approx k(\omega_0) + \left(\frac{dk}{d\omega}\right)_{\omega_0} \xi + \frac{1}{2}\left(\frac{d^2k}{d^2\omega}\right)_{\omega_0} \xi^2 = k_0 + \frac{\xi}{v_g} - \frac{\xi^2}{2}\frac{v_g'}{v_g^2}$$

This yields

$$E(z, t) = E_0 \exp[i(k_0 z - \omega_0 t)]\frac{2}{1 + \sqrt{\epsilon(\omega_0)}} F(z, t),$$

where the envelope is

$$F(z, t) = \frac{1}{2\pi} \int\limits_0^\infty d\tau \int\limits_{-\infty}^{+\infty} d\xi \exp\left[i\xi\tau + i\left(\frac{\xi}{v_g}(z - v_g t) - \frac{\xi^2}{2}\frac{v_g'}{v_g^2}\right)\right]$$

After a simple integration over $\xi$, one gets

$$F(z, t) = \frac{1}{\sqrt{i\pi}} \int\limits_{y_0}^\infty \exp(iy^2)dy, \quad y_0 = \frac{(z - v_g t)}{\sqrt{2z|v_g'|}}$$

As expected, the front of the main signal propagates into the dielectric with the group velocity $v_g$, while the front width is increasing with time due to dispersion:

$$\Delta z \sim \sqrt{z|v_g'|} \sim \sqrt{v_g|v_g'|t}$$

Far enough behind the front, i.e., for $y_0 < 0$, $|y_0| \gg 1$, the function $F(z, t) \to 1$, which means that the Fresnel solution for a monochromatic electromagnetic wave with the frequency $\omega = \omega_0$ is established in this region.

Although the frequency spectrum of the signal has a maximum at $\omega = \pm\omega_0$, $E_i(\omega)$ is non-zero at any given frequency. In particular, very high frequencies are also present in the spectrum. Since the permittivity of the dielectric, $\epsilon(\omega)$, tends to unity at these frequencies, their propagation velocity is very close to the speed of light in free space, $c$. Therefore, their contribution to the transmitted signal brings about the so-called "precursor," which propagates with the velocity equal to $c$ well ahead of the main signal discussed above. In order to derive the respective electric field, consider the asymptotic form of $\epsilon(\omega)$ at high frequencies: $\epsilon(\omega) \approx 1 - \omega_{pe}^2/\omega^2$, where $\omega_{pe}$ is the electron plasma frequency of a medium. It follows then from equation (2.89) that in the high-frequency limit

$$E(z, t) \approx -\frac{E_0\omega_0}{2\pi} \int_\Gamma \frac{d\omega}{\omega^2} \exp\left[i\left(\frac{\omega}{c}(z - ct) - \frac{\omega_{pe}^2}{2\omega}t\right)\right], \qquad (2.90)$$

with the integration along the path $\Gamma$ shown in Figure 2.10. Since there are

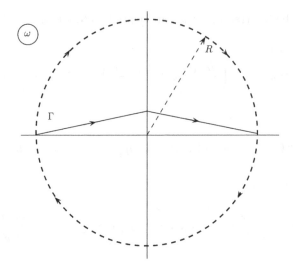

**FIGURE 2.10**
Integration contour in derivation of the "precursor"

no singularities in the upper half-plane of the complex variable $\omega$, the path $\Gamma$ can be replaced there by a semicircle of a sufficiently large radius. On the other hand, if $z - ct < 0$, the integrand in (2.90) tends exponentially to zero at $|\omega| \to \infty$ in the lower half-plane. Therefore, by supplementing expression (2.90) with the integral along a remote semicircle in the lower half-plane, the integration in (2.90) can be carried out along a circle of a sufficiently large radius. Thus, it is convenient to choose this radius to be equal to $R = [\omega_{pe}^2 ct/2(ct-z)]^{1/2}$, which is large indeed, since $R \gg \omega_{pe}$ for the precursor, where $\xi \equiv (ct - z)/ct \ll 1$. Hence, by putting $\omega = R\exp(i\phi)$, one gets from (2.90) that

$$E(z, t) \approx -\frac{iE_0\omega_0}{2\pi} \int\limits_0^{2\pi} \frac{d\phi}{R\exp(i\phi)} \left[ -i\left( \omega_{pe}t\sqrt{\frac{\xi}{2}}\exp(i\phi) + \right.\right.$$

$$\left.\left. \omega_{pe}t\sqrt{\frac{\xi}{2}}\exp(-i\phi) \right) \right] = \frac{E_0\omega_0}{\omega_{pe}}\sqrt{2\xi}J_1(\omega_{pe}t\sqrt{2\xi}) =$$

$$\frac{E_0\omega_0}{\omega_{pe}}\sqrt{2\left(1 - \frac{z}{ct}\right)}J_1\left(\omega_{pe}t\sqrt{2(1 - \frac{z}{ct})}\right), \qquad (2.91)$$

where $J_1$ is the Bessel function of the first kind (see, e.g., M. Abramowitz and I. Stegun, *A Handbook of Mathematical Functions*, §9, Dover). By taking into account the well-known properties of the Bessel functions, it follows from equation (2.91) that the precursor comprises a sequence of pulses (see Figure 2.11), whose amplitude and width are decreasing in the course of their

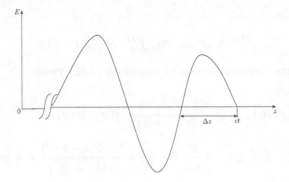

**FIGURE 2.11**
The pulse shape of the "precursor"

propagation in the dielectric as

$$E \sim E_0 \frac{\omega_0}{\omega_{pe}^2 t}, \quad \Delta z \sim \frac{c}{\omega_{pe}^2 t}$$

## 2.4 Cherenkov radiation. Transition radiation.

### Problem 2.4.1

A Dirac monopole with a magnetic charge $q$ is moving with a constant velocity $\vec{v}$ in a transparent dielectric of permittivity $\epsilon(\omega)$. By considering the flux of electromagnetic energy at large distances from the monopole, determine the condition under which Cherenkov radiation will be emitted, and derive the total power of this radiation.

In the presence of magnetic charges Maxwell's equations in such a medium should be modified to the following form:

$$\vec{\nabla} \cdot (\hat{\epsilon}\vec{E}) = 4\pi \rho_{ex}^{(E)}, \quad \vec{\nabla} \cdot \vec{B} = 4\pi \rho_{ex}^{(M)},$$

$$\vec{\nabla} \times \vec{B} = \frac{1}{c}\frac{\partial \hat{\epsilon}\vec{E}}{\partial t} + \frac{4\pi}{c}\vec{j}_{ex}^{(E)}, \quad \vec{\nabla} \times \vec{E} = -\frac{1}{c}\frac{\partial \vec{B}}{\partial t} - \frac{4\pi}{c}\vec{j}_{ex}^{(M)}, \qquad (2.92)$$

where $\rho_{ex}^{(E)}$, $\vec{j}_{ex}^{(E)}$ and $\rho_{ex}^{(M)}$, $\vec{j}_{ex}^{(M)}$ are the external charge densities and currents of, respectively, the electric and the magnetic charges. In our case $\rho_{ex}^{(E)}$, $\vec{j}_{ex}^{(E)}$

are absent, while

$$\rho_{ex}^{(M)} = q\delta(\vec{r} - \vec{v}t), \quad \vec{j}_{ex}^{(M)} = q\vec{v}\delta(\vec{r} - \vec{v}t)$$

Thus, the Fourier transformation of equations (2.92) yields

$$\vec{B}(\vec{k}, \omega) = \frac{2iq\left[\left(\frac{\epsilon\omega\vec{v}}{c^2}\right) - \vec{k}\right]}{\left(k^2 - \frac{\epsilon\omega^2}{c^2}\right)}\delta(\omega - \vec{k} \cdot \vec{v}), \qquad (2.93)$$

$$\vec{E}(\vec{k}, \omega) = -\frac{c}{\epsilon\omega}(\vec{k} \times \vec{B}) = \frac{2iq\delta(\omega - \vec{k} \cdot \vec{v})}{c\left(k^2 - \frac{\epsilon\omega^2}{c^2}\right)}(\vec{v} \times \vec{k})$$

Then, in the $(\vec{r}, t)$ representation, one gets the electric field

$$\vec{E}(\vec{r}, t) = \frac{iq}{2\pi^2 c}\int d\vec{k}d\omega\frac{\exp[i(\vec{k} \cdot \vec{r} - \omega t)]}{\left(k^2 - \frac{\epsilon\omega^2}{c^2}\right)}(\vec{v} \times \vec{k})\delta(\omega - \vec{k} \cdot \vec{v})$$

By directing the $z$-axis along the velocity vector $\vec{v}$, introducing the components of $\vec{r}$, $\vec{k}$ that are perpendicular to $\vec{v}$ as $\vec{r} = (\vec{r}_\perp, z)$, $\vec{k} = (\vec{k}_\perp, k_\parallel)$, and by taking into account that $k_\parallel = \omega/v$ due to the delta function, it can be written as

$$\vec{E}(\vec{r}, t) = \frac{iq}{2\pi^2 cv}\int d\vec{k}_\perp d\omega\frac{\exp(i\vec{k}_\perp \cdot \vec{r}_\perp)\exp[i\omega(z/v - t)]}{\left(k_\perp^2 + \frac{\omega^2}{v^2} - \frac{\epsilon\omega^2}{c^2}\right)}(\vec{v} \times \vec{k}_\perp) \qquad (2.94)$$

It is now convenient to use polar coordinates $(k_\perp, \theta)$ in the plane of $\vec{k}_\perp$, where $\theta$ is the angle between the vectors $\vec{k}_\perp$ and $\vec{r}_\perp$, and the cylindrical coordinates $(\rho, \phi, z)$ in space $\vec{r}$. It follows then from equation (2.94) that $E_z = 0$ from the outset,

$$E_\rho = \frac{iq}{2\pi^2 c}\int\limits_{-\infty}^{+\infty} d\omega \int\limits_{0}^{\infty} k_\perp^2 dk_\perp\frac{\exp[i\omega(z/v - t)]}{\left(k_\perp^2 + \frac{\omega^2}{v^2} - \frac{\epsilon\omega^2}{c^2}\right)}\int\limits_{0}^{2\pi} d\theta\sin\theta\exp(ik_\perp\rho\cos\theta) = 0$$

due to the last integral over $\theta$, which vanishes, and

$$E_\phi = \frac{iq}{2\pi^2 c}\int\limits_{-\infty}^{+\infty} d\omega \int\limits_{0}^{\infty} k_\perp^2 dk_\perp\frac{\exp[i\omega(z/v - t)]}{\left(k_\perp^2 + \frac{\omega^2}{v^2} - \frac{\epsilon\omega^2}{c^2}\right)}\int\limits_{0}^{2\pi} d\theta\cos\theta\exp(ik_\perp\rho\cos\theta)$$

Here the last integral over $\theta$ is equal to $2\pi iJ_1(k_\perp\rho)$, where $J_1$ is the Bessel function of the first kind (see Problem 2.3.6). Therefore,

$$E_\phi = -\frac{q}{\pi c}\int\limits_{-\infty}^{+\infty} d\omega \int\limits_{0}^{\infty} k_\perp^2 dk_\perp\frac{\exp[i\omega(z/v - t)]}{(k_\perp^2 + p^2)}J_1(k_\perp\rho),$$

where

$$p^2 = \left[\frac{\omega^2}{v^2} - \frac{\epsilon(\omega)\omega^2}{c^2}\right]$$

According to the well-known properties of the Bessel functions, the following relation holds:

$$\int\limits_0^\infty dk_\perp \frac{k_\perp^2 J_1(k_\perp \rho)}{(k_\perp^2 + p^2)} = pK_1(p\rho),$$

where $K_1$ is the modified Bessel function of the first kind. Thus,

$$E_\phi(\rho, z, t) = -\frac{q}{\pi c} \int\limits_{-\infty}^{+\infty} d\omega \exp[i\omega(z/v - t)]pK_1(p\rho),$$

and its temporal Fourier component is equal to

$$E_\phi(\rho, z, \omega) = -\frac{\sqrt{2}q}{\sqrt{\pi}c} \exp(i\omega z/v)pK_1(p\rho) \tag{2.95}$$

A similar calculation with the magnetic field of (2.93) yields $B_\phi = 0$ and

$$B_\rho(\vec{r}, \omega) = \frac{\sqrt{2}q}{\sqrt{\pi}v} \exp(i\omega z/v)pK_1(p\rho), \tag{2.96}$$

$$B_z(\vec{r}, \omega) = \frac{iq\sqrt{2}\omega}{\sqrt{\pi}c^2} \left(\epsilon - \frac{c^2}{v^2}\right) \exp(i\omega z/v)K_0(p\rho), \tag{2.97}$$

where $J_0$ and $K_0$ are, respectively, the Bessel function and the modified Bessel function of the zero kind. As seen from equations (2.95-2.97), the electromagnetic field of a moving magnetic monopole possesses the azimuthal symmetry about the velocity vector $\vec{v}$. Its behavior at large distances from the monopole's trajectory, i.e. at $\rho \to \infty$, crucially depends on whether the velocity of the monopole moving in the medium is superluminal, $v > c/\sqrt{\epsilon}$, or subluminal, $v < c/\sqrt{\epsilon}$. In the latter case the parameter $p^2 = (\frac{\omega^2}{v^2} - \frac{\epsilon(\omega)\omega^2}{c^2}) > 0$, the functions $K_{0,1}(p\rho)$ at $p\rho \gg 1$ are proportional to $\exp(-p\rho)$; therefore, at large distances from the source the electromagnetic field is exponentially small, and, hence, there is no radiation. On the contrary, if $v > c/\sqrt{\epsilon}$,

$$K_{0,1}(s\rho) \approx \sqrt{\frac{\pi}{2s\rho}} \exp(is\rho), \; s\rho \gg 1, \; s^2 \equiv -p^2 > 0,$$

which corresponds to a cylindrical electromagnetic wave propagating apart from the monopole's trajectory. Indeed, in this case the electromagnetic field at large distances varies, according to equations (2.95-2.97), as

$$\exp\left[i\left(s\rho + \frac{\omega}{v}z\right)\right] = \exp\left[i\left(\frac{\omega}{v}z + \rho\sqrt{\frac{\omega^2}{c^2}\epsilon - \frac{\omega^2}{v^2}}\right)\right],$$

which is indicative of the electromagnetic wave emitted in the direction specified by the Cherenokov resonance condition:

$$\cos\theta = \frac{k_z}{k} = \frac{\omega/v}{\sqrt{\epsilon}\omega/c} = \frac{c}{v\sqrt{\epsilon}} \tag{2.98}$$

The radiation power can be derived as the total flux of the electromagnetic energy through the cylindrical surface of a large radius, $\rho \to \infty$, and the length equal to $v$, the path travelled by the particle per unit time:

$$I = v \lim_{\rho \to \infty} \int_{-\infty}^{+\infty} dt 2\pi \rho \frac{c}{4\pi} (\vec{E} \times \vec{B})_\rho = \frac{1}{2} vc \lim_{\rho \to \infty} \int_{-\infty}^{+\infty} dt \rho E_\phi B_z \qquad (2.99)$$

By representing $E_\phi(t)$ and $B_z(t)$ as the Fourier integrals:

$$[E_\phi(t), \, B_z(t)] = \frac{1}{\sqrt{2\pi}} \int_{-\infty}^{+\infty} d\omega \exp(-i\omega t)[E_\phi(\omega), \, B_z(\omega)],$$

one gets that

$$\int_{-\infty}^{+\infty} dt E_\phi(t) B_z(t) = \frac{1}{2\pi} \int_{-\infty}^{+\infty} d\omega \int_{-\infty}^{+\infty} d\omega' \int_{-\infty}^{+\infty} dt E_\phi(\omega) B_z(\omega') \exp[-i(\omega + \omega')t]$$

$$= \int_{-\infty}^{+\infty} d\omega E_\phi(\omega) B_z(-\omega) = \int_{-\infty}^{+\infty} d\omega E_\phi(\omega) B_z^*(\omega) = 2Re \int_0^{+\infty} d\omega E_\phi(\omega) B_z^*(\omega)$$

It has been used here that

$$\int_{-\infty}^{+\infty} dt \exp[-i(\omega + \omega')t] = 2\pi\delta(\omega + \omega'),$$

and that for any real function $f(t)$ its Fourier component $f(-\omega) = f^*(\omega)$. Clearly, in the limit $\rho \to \infty$ only frequencies for which $\sqrt{\epsilon(\omega)} > c/v$ make contribution to the power (2.99). Thus, by using the above-given asymptotics of the Bessel functions, one gets for large $\rho$ that

$$E_\phi(\rho, \omega) \approx iq\sqrt{\frac{s}{\rho}} \exp[i(\omega z/v + s\rho)],$$

$$B_z(\rho, \omega) \approx \frac{iq\omega}{c^2\sqrt{s\rho}} \left(\epsilon - \frac{c^2}{v^2}\right) \exp[i(\omega z/v + s\rho)]$$

Insertion of these expressions into (2.99) yields

$$I = \frac{q^2}{v} \int d\omega \omega \left[\frac{\epsilon(\omega)v^2}{c^2} - 1\right], \qquad (2.100)$$

with the integration in (2.100) carried out over the frequency range where $\sqrt{\epsilon(\omega)} > c/v$.

## Problem 2.4.2

Re-derive the power of Cherenkov radiation obtained in Problem 2.4.1 by considering the electromagnetic drag force exerted on the monopole due to this radiation.

It follows from the energy balance consideration that Cherenkov radiation in a transparent medium should result in a drag force $\vec{F}$ exerted on the monopole, so that

$$I = -\vec{F} \cdot \vec{v} = -q\vec{v} \cdot \vec{B}(\vec{r} = \vec{v}t) \tag{2.101}$$

(it is clear from the symmetry consideration that the electric field makes no contribution to the drag force). In order to determine the magnetic field that features in (2.101), one starts with its Fourier transform (2.93), which yields

$$\vec{B}(\vec{r}, t) = \frac{1}{(2\pi)^2} \int d\vec{k} d\omega \vec{B}(\vec{k}, \omega) \exp[i(\vec{k} \cdot \vec{r} - \omega t)] =$$

$$\frac{iq}{2\pi^2} \int d\vec{k} \frac{\exp[i(\vec{k} \cdot \vec{r} - \vec{k} \cdot \vec{v}t)]}{[k^2 - \epsilon(\vec{k} \cdot \vec{v})^2/c^2]} \left[ \frac{\epsilon\vec{v}(\vec{k} \cdot \vec{v})}{c^2} - \vec{k} \right]$$

Then, by inserting the field $\vec{B}(\vec{r} = \vec{v}t)$ that follows from the above expression into equation (2.101), one gets

$$I = -\frac{iq^2}{2\pi^2} \int d\vec{k} \frac{(\vec{k} \cdot \vec{v})\left(\frac{\epsilon v^2}{c^2} - 1\right)}{[k^2 - \epsilon(\vec{k} \cdot \vec{v})^2/c^2]} =$$

$$-\frac{iq^2}{2\pi^2 v} \int d\vec{k}_\perp \int_{-\infty}^{+\infty} d\omega \frac{\omega \left[\frac{\epsilon(\omega)v^2}{c^2} - 1\right]}{\left[k_\perp^2 + \frac{\omega^2}{v^2} - \frac{\epsilon(\omega)\omega^2}{c^2}\right]}$$

In this integral the new variables are introduced: $\vec{k}_\perp$, the component of $\vec{k}$ which is perpendicular to the velocity vector $\vec{v}$, and the frequency $\omega = (\vec{k} \cdot \vec{v})$, so that $d\vec{k} = d\vec{k}_\perp d\omega/v$. Since the frequencies $\pm\omega$ correspond, by their physical meaning, to the same wave (only a positive frequency makes physical sense), the spectral power of the emitted radiation, if the latter is present, can be written in the following form:

$$\frac{dI}{d\omega} = -\frac{iq^2}{2\pi^2 v} \int d\vec{k}_\perp \Sigma_\pm \frac{\omega \left[\frac{\epsilon(\omega)v^2}{c^2} - 1\right]}{\left[k_\perp^2 + \frac{\omega^2}{v^2} - \frac{\epsilon(\omega)\omega^2}{c^2}\right]}, \tag{2.102}$$

where the symbol $\Sigma_\pm$ means the sum of the contributions of the positive and the negative $\omega$. It will be seen from what follows, that the very existence of Cherenkov radiation is associated with the singularity in the integrand of (2.102), when its denominator turns into zero. Therefore, in order to make correct derivation of the radiated power, it is helpful to resolve this singularity by introducing an infinitely weak dissipation in the medium, i.e., a small imaginary part, $\epsilon''(\omega)$, in the permittivity. By taking now into account that $\epsilon''(\omega)$ is an odd function of $\omega$, while the real part of the permittivity, $\epsilon'(\omega)$, is an even one (these properties of $\epsilon(\omega)$ follow from equation (2.73)), the expression (2.102) can be written as

$$\frac{dI}{d\omega} = -\frac{iq^2\omega}{\pi v} \int\limits_0^\infty k_\perp dk_\perp \left(\frac{\epsilon' v^2}{c^2} - 1\right) \frac{2i\omega^2\epsilon''/c^2}{\left(k_\perp^2 + \frac{\omega^2}{v^2} - \epsilon'\frac{\omega^2}{c^2}\right)^2 + \left(\epsilon''\frac{\omega^2}{c^2}\right)^2}$$

Since

$$\lim_{\epsilon''\to 0} \frac{2i\omega^2\epsilon''/c^2}{\left(k_\perp^2 + \frac{\omega^2}{v^2} - \epsilon'\frac{\omega^2}{c^2}\right)^2 + \left(\epsilon''\frac{\omega^2}{c^2}\right)^2} = \pi\delta\left(k_\perp^2 + \frac{\omega^2}{v^2} - \epsilon\frac{\omega^2}{c^2}\right),$$

in a transparent medium

$$\frac{dI}{d\omega} = \frac{q^2\omega}{v}\left(\frac{\epsilon(\omega)v^2}{c^2} - 1\right)\int\limits_0^\infty dk_\perp^2\,\delta\left(k_\perp^2 + \frac{\omega^2}{v^2} - \epsilon(\omega)\frac{\omega^2}{c^2}\right)$$

As seen, the radiation with the frequency $\omega$ is possible only for a superluminal source with $v > c/\sqrt{\epsilon(\omega)}$, as otherwise the argument of the delta function cannot be made equal to zero at any frequency $\omega$. If, however, $v > c/\sqrt{\epsilon(\omega)}$, the radiation is emitted in the Cherenkov cone defined by equation (2.98), with the spectral power

$$\frac{dI}{d\omega} = \frac{q^2\omega}{v}\left(\frac{\epsilon(\omega)v^2}{c^2} - 1\right),$$

in accordance with equation (2.100).

## Problem 2.4.3

Determine the power of Cherenkov radiation for a relativistic neutron, which is moving with a constant velocity $\vec{v}$ in a transparent medium of dielectric permittivity $\epsilon(\omega)$.

In the reference frame where the neutron is at rest, it possesses its own magnetic dipole moment equal to $\vec{m}_0$ (the own electric dipole moment of a neutron is absent). However, in the laboratory frame the neutron may have both the electric, $\vec{d}$, and the magnetic, $\vec{m}$, dipole moments, which depend on the orientation of the neutron's spin. Therefore, consider first the simplest case, when $\vec{m}_0$ is directed along the velocity $\vec{v}$. Then, the transformation rules (1.12-1.13) for the dipole moments yield

$$\vec{d} = 0, \quad \vec{m} = \vec{m}_0/\gamma,$$

where $\gamma = 1/\sqrt{1 - v^2/c^2}$ is the relativistic factor of the neutron. Consequently, the external current due to the moving neutron is equal to $\vec{j}_{ext} = c(\vec{\nabla} \times \vec{M})$, with $\vec{M}(\vec{r}, t) = \vec{m}\delta(\vec{r} - \vec{v}t)$. Its Fourier transform reads

$$\vec{j}_{ext}(\vec{k}, \omega) = \frac{ic(\vec{k} \times \vec{m})}{2\pi}\delta(\omega - \vec{k} \cdot \vec{v}),$$

which being inserted into Maxwell's equations yields the following Fourier components of the electromagnetic field in the medium:

$$\vec{E}(\vec{k}, \omega) = -\frac{2\omega/c}{(k^2 - \frac{\epsilon\omega^2}{c^2})}(\vec{k} \times \vec{m})\delta(\omega - \vec{k} \cdot \vec{v}),$$

$$\vec{B}(\vec{k}, \omega) = \frac{2}{(k^2 - \frac{\epsilon\omega^2}{c^2})}[k^2\vec{m} - \vec{k}(\vec{k} \cdot \vec{m})]\delta(\omega - \vec{k} \cdot \vec{v}) \qquad (2.103)$$

The next step is to calculate the electromagnetic drag force, $\vec{F}$, exerted on the moving neutron. Thus, according to the results obtained in Problem 1.0.4, in this case $\vec{F} = \vec{\nabla}(\vec{m} \cdot \vec{B})$, where the spatial derivatives of $\vec{B}$ should be obtained at the neutron's location $\vec{r} = \vec{v}t$. Thus, by using expression (2.103), this force takes the form

$$\vec{F}(\vec{r} = \vec{v}t) = \frac{im^2}{2\pi^2} \int \frac{d\vec{k}k_\perp^2 \vec{k}}{k^2 - \epsilon\omega^2/c^2},$$

where $\omega = \vec{k} \cdot \vec{v}$. Then, the power of the drag force, $I = -\vec{v} \cdot \vec{F}$, which determines the sought after intensity of Cherenkov radiation, reads

$$I = \frac{-im^2}{2\pi^2} \int \frac{d\vec{k}k_\perp^2 \omega}{k^2 - \epsilon\omega^2/c^2}$$

Since $d\vec{k} = (d\omega/v)2\pi d(k_\perp^2/2)$, the calculations, similar to those performed in Problem 2.4.2, yield the following spectral power of Cherenkov radiation:

$$\frac{dI}{d\omega} = \frac{\omega m^2}{v}\left(\frac{\epsilon\omega^2}{c^2} - \frac{\omega^2}{v^2}\right) = \frac{m_0^2\omega^3\epsilon}{vc^2\gamma^2}\left(1 - \frac{c^2}{\epsilon v^2}\right),$$

which holds in the frequency range where $\epsilon(\omega) > c^2/v^2$.

Consider now the situation when the magnetic moment of the neutron is directed perpendicular to its velocity. In this particular case the transformation relations (1.12-1.13) yield:

$$\vec{d} = \frac{1}{c}(\vec{v} \times \vec{m}_0), \quad \vec{m} = \vec{m}_0 \qquad (2.104)$$

Therefore, the external current is

$$\vec{j}_{ext} = \frac{\partial \vec{P}}{\partial t} + c(\vec{\nabla} \times \vec{M}),$$

with

$$\vec{P}(\vec{r}, t) = \vec{d}\delta(\vec{r} - \vec{v}t), \quad \vec{M}(\vec{r}, t) = \vec{m}\delta(\vec{r} - \vec{v}t)$$

Thus, the Fourier transform of the external current reads

$$\vec{j}_{ext}(\vec{k}, \omega) = \frac{i[-\omega\vec{d} + c(\vec{k} \times \vec{m})]}{2\pi}\delta(\omega - \vec{k} \cdot \vec{v})$$

Its insertion into Maxwell's equations yields the following Fourier components of the electromagnetic field in a medium:

$$\vec{E}(\vec{k}, \omega) = 2\left(k^2 - \frac{\epsilon\omega^2}{c^2}\right)^{-1} \times$$

$$\left[\frac{\omega^2}{c^2}\vec{d} - \frac{\omega}{c}(\vec{k} \times \vec{m}) - \frac{(\vec{k} \cdot \vec{d})}{\epsilon}\vec{k}\right]\delta(\omega - \vec{k} \cdot \vec{v}), \qquad (2.105)$$

$$\vec{B}(\vec{k}, \omega) = 2\left(k^2 - \frac{\epsilon\omega^2}{c^2}\right)^{-1} \times$$

$$\left[\frac{\omega}{c}(\vec{k} \times \vec{d}) - \vec{k} \times (\vec{k} \times \vec{m})\right]\delta(\omega - \vec{k} \cdot \vec{v}) \qquad (2.106)$$

The electromagnetic drag force, $\vec{F}$, exerted on the moving neutron, according to the result obtained in Problem 1.0.4, is now equal to

$$\vec{F} = (\vec{d} \cdot \vec{\nabla})\vec{E} + \vec{\nabla}(\vec{m} \cdot \vec{B}) + \frac{1}{c}\vec{d} \times (\vec{v} \cdot \vec{\nabla})\vec{B}$$

By using expressions (2.105-2.106), this force takes the form

$$\vec{F}(\vec{r} = \vec{v}t) = \frac{2i}{(2\pi)^2}\int \frac{d\vec{k}}{(k^2 - \epsilon\omega^2/c^2)} \times$$

$$\left\{(\vec{k} \cdot \vec{d})\left[\frac{\omega^2}{c^2}\vec{d} - \frac{\omega}{c}(\vec{k} \times \vec{m} - \frac{(\vec{k} \cdot \vec{d})}{\epsilon}\vec{k}\right]\right.$$

$$+\vec{k}\left[\frac{\omega}{c}\vec{m} \cdot (\vec{k} \times \vec{d}) + k^2 m^2 - (\vec{k} \cdot \vec{m})^2\right]$$

$$\left. +\frac{\omega}{c}\left[\frac{\omega}{c}\vec{d} \times (\vec{k} \times \vec{d}) + k^2(\vec{d} \times \vec{m}) - (\vec{k} \cdot \vec{m})(\vec{d} \times \vec{k})\right]\right\},$$

where $\omega = \vec{k} \cdot \vec{v}$. Then, the power of the drag force, $I = -\vec{v} \cdot \vec{F}(\vec{r} = \vec{v}t)$, can be written (with the help of relation (2.104)) as

$$I = -\frac{i}{2\pi^2}\int d\vec{k}\frac{\omega}{k^2 - \epsilon\omega^2/c^2} \times$$

$$\left\{\frac{1}{c^2}[\vec{v} \cdot (\vec{k} \times \vec{m})]^2\left(1 - \frac{1}{\epsilon}\right) + \left(1 - \frac{v^2}{c^2}\right)[m^2(k^2 - \omega^2/c^2) - (\vec{k} \cdot \vec{m})^2]\right\}$$

(it has been used here that vector $\vec{m}$ is perpendicular to the velocity $\vec{v}$). By putting there $d\vec{k} = (d\omega/v)d(k_\perp^2/2)d\phi$, where $\phi$ is the angle between the vectors $\vec{k}_\perp$ and $\vec{m}$ in the plane perpendicular to $\vec{v}$. Then, the standard calculation (see Problem 2.4.2) yields the following spectral power

$$\frac{dI}{d\omega d\phi} = \frac{\omega m^2}{2\pi v}\left[\frac{v^2}{c^2}k_\perp^2\left(1-\frac{1}{\epsilon}\right)\sin^2\phi\right.$$
$$\left. -k_\perp^2\left(1-\frac{v^2}{c^2}\right)\cos^2\phi + \left(1-\frac{v^2}{c^2}\right)(k^2-\omega^2/c^2)\right],$$

where $k^2 = \epsilon\omega^2/c^2$ and $k_\perp^2 = k^2 - \omega^2/v^2$. Therefore, unlike the previous case of the parallel $\vec{m}$ and $\vec{v}$, the above expression, which depends on the angle $\phi$, reveals no azimuthal symmetry of the radiated power in the Cherenkov cone $\cos\theta = c/v\sqrt{\epsilon}$. Finally, the total (integrated over the angle $\phi$) spectral power can be written as

$$\frac{dI}{d\omega} = \frac{\omega^3\epsilon m_0^2}{2vc^2}\left[2\left(1-\frac{1}{\epsilon}\right)^2 - \left(1-\frac{v^2}{\epsilon c^2}\right)\left(1-\frac{c^2}{\epsilon v^2}\right)\right],$$

which, as in all other examples of Cherenkov radiation, is valid only in the "superluminal" frequency range, where $v > c/\sqrt{\epsilon(\omega)}$.

## Problem 2.4.4

Determine the amplitude of the Langmuir wave (see Problem 2.1.3) induced in a cold electron plasma by a uniformly charged plane (surface charge $\sigma$), which is moving normally to its surface with a constant velocity $u$.

If the plane is moving along the $z$-axis, the electric field $E_z \equiv E$, the plasma space charge $\rho$, and the electric current $j_z \equiv j$ depend only on $z$ and time $t$, and satisfy the following equations:

$$\frac{\partial E}{\partial z} = 4\pi(\rho + \rho_{ext}), \qquad \frac{\partial\rho}{\partial t} + \frac{\partial j}{\partial z} = 0, \qquad (2.107)$$

with $\rho_{ext} = \sigma\delta(z - ut)$. By differentiating the second of these equations with respect to time:

$$\frac{\partial^2\rho}{\partial^2 t} + \frac{\partial^2 j}{\partial z\partial t} = 0,$$

and taking into account that in the linear approximation

$$j = -nev, \qquad \frac{\partial j}{\partial t} = -ne\frac{\partial v}{\partial t} = \frac{ne^2 E}{m} = \omega_{pe}^2\frac{E}{4\pi}$$

(here $n$, $-e$, $m$, $v$ are, respectively, the density of electrons, their charge, mass, and velocity), one gets

$$\frac{\partial^2 \rho}{\partial^2 t} + \frac{\omega_{pe}^2}{4\pi}\frac{\partial E}{\partial z} = 0$$

Together with the first equation in (2.107), it yields

$$\frac{\partial^2 \rho}{\partial^2 t} + \omega_{pe}^2 \rho = -\omega_{pe}^2 \rho_{ext} = -\sigma \omega_{pe}^2 \delta(z - ut) \qquad (2.108)$$

This equation is identical to the equation of motion of an oscillator, which is forced into motion by an instant kick at $t = z/u$. Thus, by integrating equation (2.108) over an infinitesimal interval of time around this moment, one gets $(\partial \rho/\partial t)_{t=(z/u)+\epsilon} = -\sigma \omega_{pe}^2/u$. Then, by solving equation (2.108) as the one for the free oscillator with this initial condition (and the second one which is $\rho_{t=z/u} = 0$), one finds the following perturbation of the plasma space charge:

$$\rho(z, t) = \begin{cases} -(\sigma \omega_{pe}/u)\sin[\omega_{pe}(t - z/u)], & t > z/u \\ 0, & t < z/u \end{cases} \qquad (2.109)$$

Thus, the space charge perturbations associated with the Langmuir wave forms a wake behind the moving charged plane. The resulting electric field, $E(z, t)$, that follows from equation (2.109) and Poisson equation, is equal to

$$E(z, t) = \begin{cases} -4\pi\sigma \cos[\omega_{pe}(t - z/u)], & t > z/u \\ 0, & t < z/u \end{cases} \qquad (2.110)$$

As seen from (2.110), the wave vector of the induced Langmuir wave is equal to $k = \omega_{pe}/u$, which indicates the expected Cherenkov resonance condition $\omega = ku$.

Consider now the energy aspect of this process. Since the wake of Langmuir oscillations is continuously generated behind the moving plane, their total energy per unit surface area of the plane, $U$, is increasing in time with the rate equal to

$$\frac{dU}{dt} = Wu = \frac{E_0^2}{8\pi}u = \frac{(4\pi\sigma)^2}{8\pi}u = 2\pi\sigma^2 u,$$

where $W = E_0^2/8\pi$ is the Langmuir wave energy per unit volume (see Problem 2.1.14). Therefore, according to the energy conservation law, the moving charged plane must experience a drag force, $F$, which is equal (per unit area) to $F = u^{-1}dU/dt = 2\pi\sigma^2$. Indeed, the drag force, if present, should be equal to $\sigma E_p$, where $E_p$ is the electric field produced by the plasma itself at the location of the moving charged plane. Since the electric field of the charged plane is equal to

$$E_{ext} = \begin{cases} -2\pi\sigma, & z < ut \\ 2\pi\sigma, & z > ut \end{cases}$$

it follows from equation (2.110) that $E_p(z = ut) = E(z = ut) - E_{ext} = -2\pi\sigma$, so the required energy balance is fulfilled.

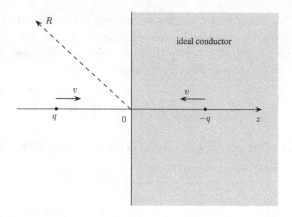

**FIGURE 2.12**

A simple model of the transition radiation

## Problem 2.4.5

A non-relativistic point charge $q$ is moving in free space with a constant velocity $v \ll c$, which is directed normally towards the plane boundary surface of a perfectly conducting material (see Figure 2.12). Determine the spectral and angular distribution, and the polarization of the resulting transition radiation.

A perfectly conducting medium annihilates any electromagnetic field inside it. Therefore, in the present case its polarization is equivalent to the appearance of a fictitious "mirror" charge $-q$ located at the symmetrical point respective to the boundary plane as shown in Figure 2.12. Thus, the electric dipole moment of the system, $\vec{d}$, is equal to

$$\vec{d} = \begin{cases} -2q\vec{v}t, & t \leq 0 \\ 0, & t \geq 0 \end{cases} \qquad (2.111)$$

(it is assumed that the charge reaches the boundary surface at $t = 0$). Since vector $\ddot{\vec{d}} = 2q\vec{v}\delta(t)$, a burst of the so-called transition electromagnetic radiation occurs at the moment $t = 0$, when the charge "transits" the boundary surface. According to the well-known electrodynamic relations (see, e.g., L. D. Landau and E. M. Lifshitz, *The Classical Theory of Fields*, Ch. 9, Pergamon Press, 1981) for the dipole radiation, at large distances (in the wave zone), the electromagnetic field is as follows:

$$\vec{E} = \frac{[(\ddot{\vec{d}} \times \vec{n}) \times \vec{n}]}{c^2 R}, \quad \vec{B} = \frac{(\ddot{\vec{d}} \times \vec{n})}{c^2 R},$$

where $\vec{n} = \vec{R}/R$. Thus, this radiation is linearly polarized with the vector $\vec{B}$ being perpendicular to the plane formed by the wave vector $\vec{k}$ of the radiated wave and the vector $\vec{v}$ (note that in the wave zone $\vec{k}$ is directed along $\vec{n}$). The spectral and angular distribution power reads

$$\frac{dI}{d\omega d\Omega} = \frac{c}{4\pi}|\vec{B}_\omega|^2 R^2 = \frac{[\ddot{\vec{d}}_\omega \times \vec{n}]^2}{2\pi c^3}$$

The Fourier transform of expression (2.111) is $\ddot{\vec{d}}_\omega = 2q\vec{v}/\sqrt{2\pi}$, which yields

$$\frac{dI}{d\omega d\Omega} = \frac{q^2(\vec{v} \times \vec{n})^2}{\pi^2 c^3} \tag{2.112}$$

As seen from (2.112), the resulting spectral power does not depend on the frequency $\omega$, which makes the total radiation power divergent at high frequencies. The physical reason of this divergence is the upfront assumption of the perfect conductivity of a medium, which is clearly not applicable for a sufficiently high frequency. Indeed, for any medium its permittivity $\epsilon(\omega) \to 1$ at the high-frequency limit. Therefore, the expression (2.112) holds only in the frequency range where $\epsilon(\omega) \gg 1$. For a conducting medium with a static conductivity equal to $\sigma_0$ it implies that $\omega \ll \sigma_0$.

## Problem 2.4.6

A non-relativistic point charge $q$ moves in free space towards the plane boundary with a transparent dielectric of permittivity $\epsilon(\omega)$. Its velocity $\vec{v}$ is constant and directed normally to the boundary surface. Determine the energy of the surface electromagnetic waves (see Problem 2.1.12), which are induced by the charge penetrating into the dielectric.

Since the charge is non-relativistic, $v \ll c$, one can assume that the induced electric field $\vec{E}$ is approximately a potential one: $\vec{E} = -\vec{\nabla}\phi$. Then, the Poisson equation reads

$$\vec{\nabla} \cdot (\hat{\epsilon}\vec{E}) = -\vec{\nabla} \cdot (\hat{\epsilon}\vec{\nabla}\phi) = 4\pi\rho_{ext},$$

where $\hat{\epsilon}$ is the permittivity operator, and $\rho_{ext} = q\delta(\vec{r} - \vec{v}t)$. Then, by denoting the boundary plane as $z = 0$, and by making the Fourier transform with respect to the transverse coordinates $(x, y)$ and time $t$, one can solve the resulting equation for $\phi(\vec{k}_\perp, \omega, z)$ separately in the free space (domain 1, $z < 0$), and in the dielectric (domain 2, $z > 0$). By taking into account that

$$\rho_{ext}(\vec{k}_\perp, \omega, z) = \frac{1}{(2\pi)^{3/2}} \int d\vec{r}_\perp dt q\delta(\vec{r} - \vec{v}t) \exp[-i(\vec{k}_\perp \cdot \vec{r}_\perp - \omega t)] =$$

$$\frac{q \exp(i\omega z/v)}{(2\pi)^{3/2}v},$$

one arrives to the following equation for the free space potential $\phi_1$:

$$\vec{\nabla}^2 \phi_1 = -k^2 \phi_1 + \frac{d^2\phi}{d^2 z} = -4\pi \rho_{ext} = -\frac{2q \exp(i\omega z/v)}{(2\pi)^{1/2} v},$$

with the solution

$$\phi_1 = A \exp(kz) + \frac{2q \exp(i\omega z/v)}{(2\pi)^{1/2} v (k^2 + \omega^2/v^2)} \qquad (2.113)$$

Here and in what follows $k$ means the modulus of the vector $\vec{k}_\perp$, and only the solution decreasing at $z \to -\infty$ is retained in equation (2.113). In a similar way, the solution in the dielectric, (at $z > 0$), reads

$$\phi_2 = B \exp(-kz) + \frac{2q \exp(i\omega z/v)}{(2\pi)^{1/2} v \epsilon(\omega) (k^2 + \omega^2/v^2)} \qquad (2.114)$$

The yet unknown constants $A$ and $B$ in expressions (2.113-2.114) have to be determined from the boundary conditions at the interface $z = 0$: the continuity of the potential $\phi$ and of the normal component of the electric induction $\vec{D} = -\epsilon\vec{\nabla}\phi$. Then, a straightforward calculation yields:

$$A = -\frac{2q(\epsilon - 1)}{(2\pi)^{1/2} v (\epsilon(\omega) + 1)(k^2 + \omega^2/v^2)},$$

$$B = \frac{2q(\epsilon - 1)}{(2\pi)^{1/2} v \epsilon(\epsilon(\omega) + 1)(k^2 + \omega^2/v^2)}$$

By knowing the potentials (2.113-2.114), one can find the electric field and, therefore, the drag electric force exerted on the moving charge. The next task is to derive the work of the drag force, and, then, to retrieve from it the contribution associated with the excitation of surface waves. Clearly, this contribution is due to polarization of the dielectric, while the potentials (2.113-2.114) comprise also the ones created by the external charge itself. In order to subtract the latter from expressions (2.113-2.114) and, hence, to determine the potential $\tilde{\phi}$, which is due entirely to the polarization of the dielectric, it is helpful to note that the potential created by the external charge alone is given by the same expressions (2.113-2.114) but with $\epsilon(\omega) \equiv 1$ (the charge moving in free space). Therefore, the required subtraction yields

$$\tilde{\phi}_1 = A \exp(kz),$$

$$\tilde{\phi}_2 = B \exp(-kz) - \frac{2q[1 - \epsilon^{-1}(\omega)] \exp(i\omega z/v)}{\sqrt{2\pi} v (k^2 + \omega^2/v^2)}$$

Thus, the Fourier transform of the electric field component responsible for the drag force, $\tilde{E}_z$, takes the following form:

$$\tilde{E}_{z1}(k, \omega, z) = -kA \exp(kz),$$

$$\tilde{E}_{z2}(k, \omega, z) = kB \exp(-kz) +$$

$$\frac{2qi\omega(1 - \epsilon^{-1}(\omega)) \exp(i\omega z/v)}{v\sqrt{2\pi} v (k^2 + \omega^2/v^2)}$$

Then, the inverse Fourier transformation brings about the sought after electric field at the instantaneous location of the moving charge:

$$\tilde{E}_{z1}(\vec{r}_\perp = 0, t = z/v) = -\frac{1}{\sqrt{2\pi}} \int_0^\infty dk k^2 \int_{-\infty}^{+\infty} d\omega A \exp(k - i\omega/v)z,$$

$$\tilde{E}_{z2}(\vec{r}_\perp = 0, t = z/v) = -\frac{1}{\sqrt{2\pi}} \int_0^\infty dk k \int_{-\infty}^{+\infty} d\omega \times \qquad (2.115)$$

$$\left( kB \exp -(k + i\omega/v)z + \frac{i\omega}{v} \frac{2q\omega(1 - \epsilon^{-1}(\omega))}{v\sqrt{2\pi}v(k^2 + \omega^2/v^2)} \right),$$

with the total energy losses equal to

$$W = -\int_{-\infty}^{+\infty} dz q \tilde{E}_z(\vec{r}_\perp = 0, t = z/v)$$

The last term in expression (2.115) for $\tilde{E}_{z2}$ does not depend on $z$, and, therefore, it corresponds to the energy losses, if any, of a charge moving in a uniform dielectric. Thus, it is irrelevant to the "transition" part of the energy losses under discussion, which are described by the two other terms in (2.115), each of which tends to zero away from the boundary surface, i.e., at $|z| \to \infty$. Thus, after the integration over $z$ in (2.115) one gets

$$W_{tr} = -\frac{q^2}{\pi v} \int_0^\infty k^2 dk \int_{-\infty}^{+\infty} d\omega \frac{(\epsilon - 1)}{(\epsilon + 1)(k^2 + \omega^2/v^2)} \times$$

$$\left( \frac{1}{k - i\frac{\omega}{v}} + \frac{1}{\epsilon(\omega)(k + i\frac{\omega}{v})} \right) \qquad (2.116)$$

In order to find out the contribution of induced electromagnetic surface waves, one should recall that, according to Problem 2.1.12, in the nonrelativistic limit the frequency of these waves, $\omega_0$, is specified by the condition $\epsilon(\omega_0) = -1$. Therefore, the sought after part of the energy losses is associated with the poles of the integrand in (2.116) at $\omega = \pm\omega_0$, where $(\epsilon+1) = 0$. Thus, one can put $\epsilon = -1$ everywhere in (2.116) except the poles, which yields

$$W_{sw} = \frac{4iq^2}{\pi v^2} \int_0^\infty dk k^2 \int_{-\infty}^{+\infty} \frac{d\omega\omega}{[\epsilon(\omega) + 1](k^2 + \omega^2/v^2)^2} \qquad (2.117)$$

It is seen now that $W_{sw}$, the energy of the induced surface waves, is determined by the imaginary part of the integral in equation (2.117). In a transparent medium it can be derived in the way similar to the one already used in Problem 2.4.2 for the Cherenkov radiation. Thus, assume that an infinitesimal imaginary part of the permittivity $\epsilon(\omega)$ is present. Then the poles of the

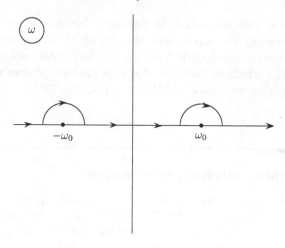

**FIGURE 2.13**
The "passing rule" for integration in (2.117)

integrand in equation (2.117) become slightly shifted down from the real axis in the plane of the complex variable $\omega$, which reduces (2.117) to calculation of the two half-residues around the points $\omega = \pm\omega_0$ as shown in Figure 2.13. Indeed, in the vicinity of $\omega = \omega_0$ the factor $\epsilon(\omega) + 1$ can be approximated as

$$\epsilon(\omega) + 1 \approx \epsilon(\omega_0) + \left(\frac{\partial\epsilon}{\partial\omega}\right)_{\omega_0}(\omega - \omega_0) + i\epsilon''(\omega_0) + 1 =$$

$$\left(\frac{\partial\epsilon}{\partial\omega}\right)_{\omega_0}(\omega - \omega_0) + i\epsilon''(\omega_0)$$

Thus, the account of $\epsilon''$ moves the pole to the point $\omega = \omega_0 - i\gamma$, where $\gamma = \epsilon''(\omega_0)/(\partial\epsilon/\partial\omega)_{\omega_0}$. Under a state of thermodynamic equilibrium both $\epsilon''(\omega_0)$ and $(\partial\epsilon/\partial\omega)_{\omega_0}$ are positive for $\omega_0 > 0$. Indeed the former, according to equation (2.12), determines the dissipation power in a medium, while the latter, as shown in Problem 2.1.16, is associated with the energy of the surface electromagnetic wave. Hence, $\gamma > 0$, and the pole is shifted to the lower half-plane, and, therefore, should be passed above in the integral in equation (2.117). Furthermore, since the real, $\epsilon'$, and the imaginary, $\epsilon''$, parts of the dielectric permittivity are, respectively, even and odd functions of $\omega$ on the real axis, a similar procedure at $\omega = -\omega_0$ yields the same result: shifting the pole down. This proves that the only contribution to the imaginary part of the integral in (2.117) arises from these two half-residues, yielding

$$W_{sw} = \frac{8q^2\omega_0}{v^2(\partial\epsilon/\partial\omega)_{\omega_0}}\int\limits_{0}^{\infty} dk\frac{k^2}{(k^2 + \omega_0^2/v^2)^2} \tag{2.118}$$

The integrand in expression (2.118) indicates distribution of the induced surface waves against the wave vector $\vec{k}_\perp$, which is, as expected, isotropic with a characteristic wavelength $\lambda \sim v/\omega_0$. Their total energy is $W_{sw} = 2\pi q^2/v(\partial\epsilon/\partial\omega)_{\omega_0}$, which in the case of a cold electron plasma with $\epsilon(\omega) = 1 - \omega_{pe}^2/\omega^2$ yields $\omega_0 = \omega_{pe}/\sqrt{2}$ and $W_{sw} = \pi q^2\omega_{pe}/2\sqrt{2}v$.

## 2.5　Non-linear interaction of waves

Electrodynamic equations in a medium are, generally speaking, non-linear. Therefore, the expression (2.1) for the electric current in terms of the linear operator of conductivity is, in fact, only the first term of its presentation as a power series of the electromagnetic field amplitude. In this approximation (linear electrodynamics) any electromagnetic field can be viewed as a superposition of the eigenmodes (the electromagnetic waves) described by equations (2.8-2.9), when each of them evolves independently of the others. If non-linearity (the terms in the series after the first) is taken into account, a weak interaction between the modes occurs, which brings about the exchange of the energy and the momentum between the waves, while the total energy and momentum remain conserved in a transparent medium.

　　In the case of a weak non-linearity the main non-linear process is the **resonant three-wave interaction**. This process is possible if the wave vectors and frequencies of the three waves $(1, 2, 3)$ involved in the interaction satisfy the following conditions:

$$\vec{k}_1 = \vec{k}_2 + \vec{k}_3, \ \omega(\vec{k}_1) = \omega(\vec{k}_2) + \omega(\vec{k}_3), \tag{2.119}$$

where the frequencies $\omega(\vec{k})$ are determined by the linear dispersion equation (2.9) (here and in what follows it is assumed that all wave frequencies are positive). The evolution equations for the wave amplitudes, $C_{1,2,3}$, take the form:

$$\dot{C}_1 = V_1 C_2 C_3, \ \dot{C}_2 = V_2 C_1 C_3^*, \ \dot{C}_3 = V_3 C_1 C_2^* \tag{2.120}$$

It is convenient to normalize the amplitudes $C(\vec{k})$ in such a way that the energy of the wave, $W(\vec{k})$, given by equation (2.13), becomes equal to $W(\vec{k}) = \pm|C(\vec{k})|^2\hbar\omega(\vec{k})$, where the signs $\pm$ correspond to a wave of the positive/negative energy, and $\hbar$ is the Planck constant. Although all the processes under consideration are entirely classical, the introduction of the quantum constant $\hbar$ helps to make a simple visual interpretation of the non-linear interaction of waves. Indeed, if $\hbar\omega$ and $\hbar\vec{k}$ are the energy and the momentum of a single quantum, the relations (2.119) can be viewed as the energy and momentum conservation conditions at the elementary act of the three-wave interaction: the decay of the quantum (1) into quanta (2) and (3) (and the reverse process of the merging of quanta (2) and (3) into (1)). It also implies that $|C(\vec{k})|^2 \equiv N(\vec{k})$, the number density of quanta (quasiparticles) with the momentum $\hbar\vec{k}$ and the energy $\hbar\omega$. The quantities $V$ in equations (2.120) are called the **matrix elements** of the three-wave interaction. If all of the three interacting waves have a positive energy, the following symmetry relations hold for $V$:

$$V_2 = V_3 = -V_1^*, \tag{2.121}$$

which ensure (it can be easily varified with the help of equations (2.120)) conservation of the total energy (and momentum): $W = \Sigma_k W(\vec{k}) = \Sigma_k N(\vec{k})\hbar\omega(\vec{k}) = const$. If one of the interacting waves (say, wave (1)) has a negative energy, one gets, instead of (2.121), that

$$V_2 = V_3 = V_1^*, \tag{2.122}$$

and the conserved total energy is $W = \Sigma_k N(\vec{k})\hbar\omega(\vec{k}) - \Sigma'_k N(\vec{k})\hbar\omega(\vec{k})$, where the waves with a negative energy are summed up with negative sign.

On top of the conservation of total energy and momentum, the three-wave interaction also possesses the conservation laws which involve the number of quasipartcles $N(\vec{k})$. Thus, it follows from equations (2.120) and (2.121) that

$$N_1 + N_2 = const, \ N_1 + N_3 = const, \ N_2 - N_3 = const \tag{2.123}$$

These are the so-called **Manley-Rowe relations**, the meaning of which becomes obvious from the quantum viewpoint. Indeed, at the elementary act of the three-wave interaction a quasiparticle of type (1) decays into two quasiparticles of types (2) and (3); therefore a decrease of $N_1$ is accompanied by the equal increase of $N_2$ and $N_3$ (a similar consideration applies to the reverse process — the merging of the quanta (2) and (3) into the quantum(1)). If, however, one of the interacting waves has a negative energy, one gets, instead of (2.123), that

$$N_1 - N_2 = const, \ N_1 - N_3 = const, \ N_2 - N_3 = const \tag{2.124}$$

In this case an elementary act of the interaction is simultaneous creation (or annihilation) of all three quasiparticles.

## Problem 2.5.1

By knowing the wave dispersion relation $\omega(\vec{k})$ in an isotropic medium, determine whether the resonant three-wave interaction is possible for such waves.

The three-wave interaction is possible if the resonant conditions (2.119) can be satisfied. In an isotropic medium the wave frequency, $\omega(\vec{k})$, does not depend on the direction of the wave vector, i.e., $\omega(\vec{k}) \equiv \omega(k)$, and the sought

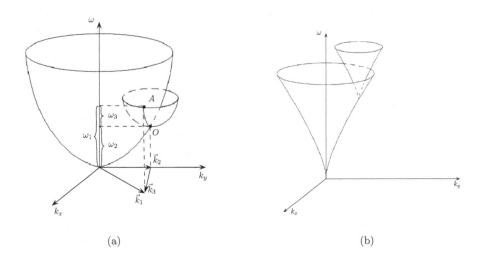

**FIGURE 2.14**
Graphical illustration for decay and non-decay spectra

after condition can be easily found by graphical means as shown in Figure 2.14. It follows from relations (2.119) that all three wave vectors of the interacting waves lie in the same plane, say $(k_x, k_y)$. Then the function $\omega(k)$ defines in the $(k_x, k_y, \omega)$ space a surface of revolution about the $\omega$-axis. Consider now some point $O$ lying on this surface, and by using it as the origin, draw another surface $\omega(k)$ (see Figure 2.14(a)). If these two surfaces intersect, the resonant conditions (2.119) can be satisfied, and such a spectrum $\omega(k)$ is called a decay spectrum. Indeed, take any point $A$ lying on the intersection line, and consider projections of the points $O$ and $A$ on the plane $(k_x, k_y)$ and the $\omega$-axis. As seen from the figure, the three wave vectors obtained in this way, $\vec{k}_{1,2,3}$, together with the respective frequencies $\omega_{1,2,3}$, do satisfy the resonant conditions (2.119). An example of a non-decay spectrum is shown in Figure 2.14(b). In this case the two above constructed surfaces do not intersect, and, hence, the conditions (2.119) cannot be satisfied.

## Problem 2.5.2

Determine possibility of the resonant three-wave interaction for the following types of waves: (a) $\omega = \sqrt{\alpha k^3/\rho}$ — capillary waves on the surface of a deep fluid, where $\rho$ is the fluid density, and $\alpha$ is the capillary coefficient; (b) $\omega = \sqrt{gk}$ — gravity waves on the surface of a deep

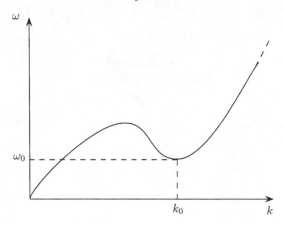

**FIGURE 2.15**

Spectrum of excitations in liquid helium

fluid; (c) $\omega = \sqrt{\omega_{pe}^2 + k^2 c^2}$ — electromagnetic waves in a plasma; (d) $\omega = kc_s/\sqrt{1 + k^2 r_d^2}$ — ion sound waves in a plasma, where $c_s$ is the speed of sound, and $r_d$ is the Debye radius; (e) collective excitation in liquid helium with the spectrum shown in Figure 2.15 (the Landau spectrum): $\omega \approx ku$ for $k \ll k_0$- phonons, $\omega \approx \omega_0 + (k - k_0)^2/2\mu$ for $|k - k_0| \ll k_0$ - rotons.

Here one can apply the method described in Problem 2.5.1. Thus, the spectra (a) and (b) are of the type shown in Figure 2.14; hence, the capillary waves have a decay spectrum, while the gravity waves correspond to a non-decay one. By plotting the respective figures for the rest of the waves (see Figure 2.16), one concludes that electromagnetic and ion sound waves in a plasma have non-decay spectra. In the liquid helium a decay of a roton into a phonon and another roton is possible. In this case the intersection of the dispersion surfaces takes place in the plane of the figure, which indicates that all three interacting waves have collinear wave vectors. Such a one-dimensional decay process is, however, not possible, for example, for capillary waves.

## Problem 2.5.3

Find the minimum frequency of the wave that can decay into two other waves, if the dispersion law is $\omega(k) = \omega_0 + \alpha k^2$.

The correct answer is not $2\omega_0$, as it may look at first glance, because

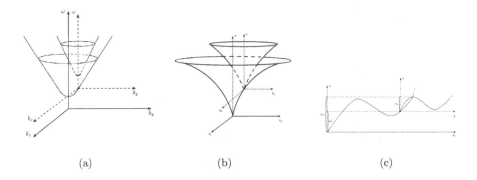

(a)                    (b)                    (c)

**FIGURE 2.16**
Examples of decay and non-decay spectra: (a) electromagnetic waves in a
plasma; (b) ion-sound waves in a plasma; (c) collective excitations in liquid
helium

not only the energy, but also the linear momentum of the three involved
waves must be conserved. Thus, it follows from the conditions (2.119) and the
dispersion relation $\omega(k) = \omega_0 + \alpha k^2$ that $(\vec{k}_2 \cdot \vec{k}_3) = \omega_0/2\alpha$. Therefore, the
problem is reduced to finding a minimum of the sum $(k_2^2 + k_3^2)$ under a fixed
scalar product of these two vectors. Clearly, the minimum is achieved when $\vec{k}_2$
and $\vec{k}_3$ are parallel and equal to each other, which yields $k_2 = k_3 = \sqrt{\omega_0/2\alpha}$.
Thus, the minimum frequency of the decaying wave is equal to $3\omega_0$.

### Problem 2.5.4

Consider a medium that supports two types of waves: waves of type
(l), with the spectrum $\omega_l = \omega_0 + \alpha k^2$, and sound waves, (s), with the
spectrum $\omega_s = kc_s$. Find the minimum frequency of the wave, which can
decay according to the scheme $l \rightarrow l' + s$.

By plotting the respective graphical diagram as discussed in Problem 2.5.1,
one concludes that two different cases of the decay under discussion are possible (see Figure 2.17). In the case (a), when the group velocity of the wave
(l') is less than the speed of sound $c_s$, the decaying wave (l) has the group
velocity bigger than $c_s$. Otherwise, if the frequency of the wave (l') is high
enough so that its group velocity exceeds $c_s$, the case (b), the decaying wave
(l) may have the frequency which is very close to that of the wave (l'). Therefore, since for a wave of type (l) the group velocity increases together with its
frequency, the conclusion is that the minimum $\omega_l$ is achieved when the group
velocity of the wave (l') is equal to $c_s$. In this case $\omega_l \rightarrow \omega_{l'}$, $\omega_s \rightarrow 0$, and since

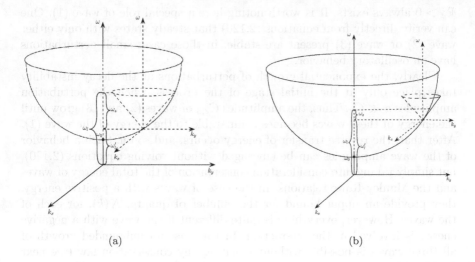

(a)                                           (b)

**FIGURE 2.17**
Decay of wave $(l)$ into waves $(l')$ and $(s)$

$(\partial\omega_l/\partial k) = 2\alpha k = c_s$, the respective wave vector is equal to $k = k_0 = c_s/2\alpha$, which yields the minimum frequency $\omega_{min} = \omega_0 + \alpha k_0^2 = \omega_0 + c_s^2/4\alpha$.

## Problem 2.5.5

Determine the growth rate of the **decay instability** that occurs in the three-wave interaction of waves with a positive energy.

Consider a steady solution of equations (2.120), when the amplitudes of any of the two waves involved in the three wave interaction are equal to zero, while the amplitude of the third one remains, therefore, a constant. It turns out, however, that this steady state is unstable, if the initially present wave is the one with the highest frequency (the wave (1) in equations (2.119-2.120)). In this case, under a small perturbation the energy of the wave (1) becomes transferred to waves (2) and (3). Thus, consider the initial state $C_1(0) = C_0, C_2(0) = C_3(0) = 0$, subjected to a small perturbation of the form: $C_1 = C_0 + a_1 \exp(\gamma t)$, $C_2 = a_2 \exp(\gamma t)$, $C_3 = a_3 \exp(\gamma t)$. By substituting these expressions into equations (2.120), and taking into account relations (2.121), one gets in the linear approximation that

$$a_1 = 0, \ \gamma a_2 = -V_1^* C_0 a_3, \ \gamma a_3 = -V_1^* C_0 a_2^*$$

It yields $|\gamma|^2 = |V_1|^2 |C_0|^2$, so that the exponentially growing perturbation with

$\Re\gamma > 0$ always exists. It is worth noting here a special role of wave (1). One can verify directly from equations (2.120) that steady states with only either wave (2) or wave (3) present are stable: in those cases small perturbations have an oscillatory behavior.

Clearly, the exponential growth of perturbations at the decay instability takes place only at the initial stage of the process, when the pertubation amplitudes $a \ll C_0$. Thus, the amplitudes $C_{2,3}$ of waves (2) and (3) grow until the energy of these waves becomes comparable to the energy of the wave (1). After that, the reverse transfer of energy occurs, and so on. Such a behavior of the wave amplitudes can be envisaged without solving equations (2.120), but simply taking into consideration conservation of the total energy of waves and the Manley-Rowe relations. In the case of waves with a positive energy, they provide an upper bound for the number of quanta, $N(\vec{k})$, for each of the waves. However, everything is quite different if the wave with a negative energy is involved in the interaction. In this case an unbounded growth of all three waves is possible without violating any conservation law (see next Problem).

## Problem 2.5.6

Show that if a wave with a negative energy takes part in the three-wave interaction, the so-called **"explosive" instability** is possible, when the amplitudes of all three waves formally diverge at a finite time interval.

Consider the simplest case, when wave (1) with the maximum frequency has a negative energy, and at the initial moment $t = 0$ only waves (2) and (3) are present: $C_1(0) = 0$, $C_2(0) = C_3(0) = C_0$. Then, if one also assumes that the initial amplitude $C_0$ and the matrix elements $V$ are real, it follows from equations (2.120) and relation (2.122) that all the amplitudes remain real at any time, and it follows then from the Manley-Rowe relations (2.124) that

$$C_2(t) = C_3(t) \equiv C(t), \quad C_1^2(t) - C_2^2(t) = const = -C_0^2, \quad C^2(t) = C_1^2(t) + C_0^2$$

After that, the first of equations (2.120) takes the form

$$\frac{dC_1}{dt} = V_1 C_2 C_3 = V_1 C^2 = V_1(C_1^2 + C_0^2),$$

the integration of which yields $C_1(t) = C_0 \tan(V_1 C_0 t)$. Thus, the amplitudes of all three waves become infinite at the instant $t = t_0 = \pi/2 V_1 C_0$, and at its vicinity they explode as $C_{1,2,3} \propto (t_0 - t)^{-1}$.

## Problem 2.5.7

Derive equations describing non-linear interaction of the electromagnetic and Langmuir waves in a cold electron plasma (see Problem 2.1.3).

The dielectric permittivity of such a plasma is $\epsilon(\omega) = 1 - \omega_{pe}^2/\omega^2$, which yields the dispersion law $\omega_t(k) = \sqrt{\omega_{pe}^2 + k^2c^2}$ for transverse electromagnetic waves, and $\omega_l(k) = \omega_{pe}$ for longitudinal Langmuir waves. Although each of these spectra is of a non-decay type (see Problem 2.5.2), the three-wave interaction is possible if, for example, the waves (1) and (2) are the electromagnetic ones, and the wave (3) is the Langmuir wave. In the simplest case all three waves propagate along the $z$-axis, and the electromagnetic waves are linearly polarized with the electric field along the $x$-axis (so that only $E_x$ and $B_y$ are present). The resonant conditions $k_1 = k_2 + k_3$, $\omega_t(k_1) = \omega_t(k_2) + \omega_l(k_3)$ can be satisfied if the frequency of the decaying electromagnetic wave, $\omega_1 \equiv \omega_t(k_1)$, exceeds $2\omega_{pe}$, with $k_1$, $k_2$ being the wave vectors of the electromagnetic waves, and $k_3$ the wave vector of the Langmuir wave. The system evolution is governed by Maxwell's equations and the equation of motion for electrons:

$$\vec{\nabla} \times \vec{E} = -\frac{1}{c}\frac{\partial \vec{B}}{\partial t},$$

$$\vec{\nabla} \times \vec{B} = \frac{1}{c}\frac{\partial \vec{E}}{\partial t} - \frac{4\pi e}{c}(n + \delta n)\vec{v} =$$

$$\frac{1}{c}\frac{\partial \vec{E}}{\partial t} - \frac{4\pi n e}{c}\vec{v} + \frac{\vec{\nabla} \cdot \vec{E}}{c}\vec{v}, \qquad (2.125)$$

$$\frac{\partial \vec{v}}{\partial t} + (\vec{v} \cdot \vec{\nabla})\vec{v} = -\frac{e}{m}\left(\vec{E} + \frac{1}{c}(\vec{v} \times \vec{B})\right),$$

where $\delta n = -\vec{\nabla} \cdot \vec{E}/4\pi e$ is the electron density perturbation expressed here with the help of the Poisson equation. In the case under consideration, the system is described by the following quantities: $E_x$, $E_z$, $B_y$, $v_x$, $v_z$, and by representing each of them as a Fourier series like, say,

$$E_x(z, t) = \Sigma_k E_{xk}(t)\exp(ikz)$$

Equations (2.125) can be written as

$$\frac{\partial E_{xk}}{\partial t} = -ickB_{yk} + 4\pi n e v_{xk} - i\Sigma_{k'}k'E_{zk'}v_{x(k-k')},$$

$$\frac{\partial B_{yk}}{\partial t} = -ickE_{xk},$$

$$\frac{\partial v_{xk}}{\partial t} = -\frac{e}{m}E_{xk} - i\Sigma_{k'}k'v_{xk'}v_{z(k-k')} + \frac{e}{mc}\Sigma_{k'}B_{yk'}v_{z(k-k')}, \quad (2.126)$$

$$\frac{\partial E_{zk}}{\partial t} = 4\pi nev_{zk} - i\Sigma_{k'}k'E_{zk'}v_{z(k-k')},$$

$$\frac{\partial v_{zk}}{\partial t} = -\frac{e}{m}E_{zk} - i\Sigma_{k'}k'v_{zk'}v_{z(k-k')} - \frac{e}{mc}\Sigma_{k'}B_{yk'}v_{x(k-k')}$$

The non-linearity of these equations is exhibited by the sums over $k'$ in the right-hand sides of equations (2.126). In the linear approximation, when all these non-linear terms are omitted, the Langmuir wave, which is associated with $(E_z, v_z)$, and the electromagnetic wave, involving $(E_x, B_y, v_x)$, evolve independently of each other.

## Problem 2.5.8

Derive the evolution equations for the amplitudes of waves involved in the three-wave interaction by starting from a set of non-linear equations of the type of (2.126).

The equations (2.126) for the Fourier components of the wave perturbations in a medium can be symbolically written in the following form:

$$\frac{\partial A_{\alpha\vec{k}}}{\partial t} - L_{\alpha\beta}(\vec{k})A_{\beta\vec{k}} = \Sigma_{\vec{k}'}T_{\alpha\beta\gamma}(\vec{k}, \vec{k}')A_{\beta\vec{k}'}A_{\gamma(\vec{k}-\vec{k}')}, \quad (2.127)$$

where $\vec{A}$ is a "vector of state" of a medium (for example, in Problem 2.5.7, $\vec{A} = (E_x, B_y, v_x, E_z, v_z)$). Such a general representation allows one to consider the non-linear interaction of not only the electromagnetic waves, but also of waves of any other origin. The matrix $L_{\alpha\beta}(\vec{k})$ in equation (2.127) describes the waves in the linear approximation, while the matrix $T_{\alpha\beta\gamma}(\vec{k}, \vec{k}')$ there is due to the nonlinearity of a system. The goal is to find the evolution equations for the amplitudes of the waves, which result from their non-linear interaction, by treating the non-linear term in equation (2.127) as a weak perturbation. To begin with, consider the eigenvectors of the linearized equations:

$$\phi_{\alpha\vec{k}}(t) = C_{\vec{k}}\exp(-i\omega_{\vec{k}}t)\psi_{\alpha\vec{k}},$$

where $(-i\omega_{\vec{k}})$ is the eigenvalue of the matrix $L_{\alpha\beta}(\vec{k})$. In the absence of dissipation the frequencies $\omega_{\vec{k}}$ must be real; therefore, the matrix $L_{\alpha\beta}$ is an antihermitian one: $L_{\beta\alpha} = -L_{\alpha\beta}^*$. It is convenient to normalize the vectors $\vec{\phi}_{\vec{k}}$ and $\vec{\psi}_{\vec{k}}$ in such a way that $\psi_{\alpha\vec{k}}\psi_{\alpha\vec{k}}^* = \hbar\omega(\vec{k})$, and $\phi_{\alpha\vec{k}}\phi_{\alpha\vec{k}} = W_{\vec{k}}$, the volumetric energy of the wave. Thus, $W_{\vec{k}} = |C_{\vec{k}}|^2\hbar\omega_{\vec{k}}$, so that $|C_{\vec{k}}|^2 \equiv N_{\vec{k}}$ is the number density of quasiparticles with the given linear momentum $\hbar\vec{k}$ and energy $\hbar\omega_{\vec{k}}$. Since the components of the vector $\vec{A}$ are physical quantities which must be real, the Fourier amplitudes should satisfy the condition $C_{-\vec{k}} = C_{\vec{k}}^*$, while $\omega_{-\vec{k}} = -\omega_{\vec{k}}$. Therefore, a pair of harmonics with $(\vec{k}, \omega_{\vec{k}})$ and $(-\vec{k}, -\omega_{\vec{k}})$ correspond to the same physical wave (in what follows the frequency of a wave is assumed to be positive). For a wave with a negative energy the normalization $W_{\vec{k}} = -|C_{\vec{k}}|^2 \hbar \omega_{\vec{k}} = -N_{\vec{k}} \hbar \omega_{\vec{k}}$ will be used.

The non-linearity brings about variation of the amplitudes $C_{\vec{k}}$ with time, as well as a small correction to the "unit" vector $\psi_{\vec{k}}$:

$$\phi_{\vec{k}} = C_{\vec{k}}(t) \exp(-i\omega_{\vec{k}} t)[\psi_{\vec{k}} + \delta\psi\vec{k}(t)],$$

with $\psi_{\alpha\vec{k}} \delta\psi_{\alpha\vec{k}}^* = 0$ due to the imposed normalization of the vector $\psi_{\vec{k}}$. By inserting the above expression into equation (2.127), where a weak non-linearity is assumed, one finds with the required degree of accuracy:

$$\dot{C}_{\vec{k}} \exp(-i\omega_{\vec{k}} t)\vec{\psi}_{\alpha\vec{k}} - i\omega_{\vec{k}} C_{\vec{k}} \exp(-i\omega_{\vec{k}} t)(\psi_{\alpha\vec{k}} + \delta\psi_{\alpha\vec{k}}) -$$
$$C_{\vec{k}} \exp(-i\omega_{\vec{k}} t) L_{\alpha\beta}(\vec{k})(\psi_{\alpha\vec{k}} + \delta\psi_{\alpha\vec{k}}) =$$
$$\Sigma_{\vec{k}'} T_{\alpha\beta\gamma}(\vec{k}, \vec{k}') C_{\vec{k}'} C_{(\vec{k}-\vec{k}')} \psi_{\beta\vec{k}'} \psi_{\gamma(\vec{k}-\vec{k}')} \exp[-i(\omega_{\vec{k}'} + \omega_{(\vec{k}-\vec{k}')})t]$$

Since $\vec{\psi}_{\vec{k}}$ is an eigenvector of the $L_{\alpha\beta}(\vec{k})$, it follows then that

$$\dot{C}_{\vec{k}}\vec{\psi}_{\alpha\vec{k}} - i\omega_{\vec{k}} C_{\vec{k}}\delta\psi_{\alpha\vec{k}} - C_{\vec{k}} L_{\alpha\beta}(\vec{k})\delta\psi_{\beta\vec{k}} =$$
$$\Sigma_{\vec{k}'} T_{\alpha\beta\gamma}(\vec{k}, \vec{k}') C_{\vec{k}'} C_{(\vec{k}-\vec{k}')} \psi_{\beta\vec{k}'} \psi_{\gamma(\vec{k}-\vec{k}')} \times$$
$$\exp[i(\omega_{\vec{k}} - \omega_{\vec{k}'} - \omega_{(\vec{k}-\vec{k}')})t] \tag{2.128}$$

From a vector equation (2.128), together with the orthogonality condition $\psi_{\alpha\vec{k}} \delta\psi_{\alpha\vec{k}}^* = 0$, one can derive, in principle, both the amplitude time derivative $\dot{C}_{\vec{k}}$ and the correction $\delta\vec{\psi}_{\vec{k}}$ to the "unit" eigenvector $\vec{\psi}_{\vec{k}}$. However, in order to derive only $\dot{C}_{\vec{k}}$, which is of interest here, it is helpful to get a single scalar equation by taking the scalar product of equation (2.128) with the vector $\vec{\psi}_{\alpha\vec{k}}^*$. Then, the second term on the left-hand side of equation (2.128) vanishes because of the required orthogonality, while the third one can be transformed as follows:

$$L_{\alpha\beta}\delta\psi_{\beta}\psi_{\alpha}^* = -L_{\beta\alpha}^*\psi_{\alpha}^*\delta\psi_{\beta} = -i\omega\psi_{\beta}^*\delta\psi_{\beta} = 0$$

(recall that $L_{\alpha\beta}$ is an antihermitian matrix, and $\psi_{\alpha}$ is its eiegenvector), which eliminates $\delta\psi_{\vec{k}}$. The only significant contribution to the right-hand side of equation (2.128) is made by the resonant terms, for which $(\omega_{\vec{k}} - \omega_{\vec{k}'} - \omega_{(\vec{k}-\vec{k}')}) = 0$ (otherwise the exponent in equation (2.128) oscillates very rapidly, which prevents effective interaction of waves). Therefore, by leaving there only the resonant terms (the respective sum is denoted as $\tilde{\Sigma}$), one finally gets the sought after equation for $\dot{C}_{\vec{k}}$:

$$\dot{C}_{\vec{k}} = \tilde{\Sigma}_{\vec{k}'} V(\vec{k}, \vec{k}') C_{\vec{k}'} C_{(\vec{k}-\vec{k}')},$$
$$V(\vec{k}, \vec{k}') = (\hbar\omega_{\vec{k}})^{-1} T_{\alpha\beta\gamma}(\vec{k}, \vec{k}')\psi_{\alpha\vec{k}}^*\psi_{\beta\vec{k}'}\psi_{\gamma(\vec{k}-\vec{k}')} \tag{2.129}$$

## Problem 2.5.9

Derive the matrix elements for the resonant three-wave interaction of the electromagnetic and Langmuir waves in a cold electron plasma (see Problem 2.5.7).

Here the scheme developed in Problem 2.5.8 can be applied to equations (2.126), which describe decay of the electromagnetic wave into the two other waves: the Langmuir wave and another electromagnetic wave. Thus, by introducing the "vector of state"

$$\vec{A} = [E_x,\ B_y,\ (4\pi n m)^{1/2} v_x,\ E_z,\ (4\pi n m)^{1/2} v_z],$$

the system (2.126) can be re-written as

$$\frac{\partial A_{1k}}{\partial t} = -ick A_{2k} + \omega_{pe} A_{3k} - \frac{2i\sqrt{\pi} e}{m\omega_{pe}} \Sigma_{k'} (k - k') A_{3k'} A_{4(k-k')},$$

$$\frac{\partial A_{2k}}{\partial t} = -ick A_{1k},$$

$$\frac{\partial A_{3k}}{\partial t} = -\omega_{pe} A_{1k} - \frac{2i\sqrt{\pi} e}{m\omega_{pe}} \Sigma_{k'} k' A_{3k'} A_{5(k-k')} +$$

$$\frac{2\sqrt{\pi} e}{mc} \Sigma_{k'} A_{2k'} A_{5(k-k')}, \qquad (2.130)$$

$$\frac{\partial A_{4k}}{\partial t} = \omega_{pe} A_{5k} - \frac{2i\sqrt{\pi} e}{m\omega_{pe}} \Sigma_{k'} k' A_{4k'} A_{5(k-k')},$$

$$\frac{\partial A_{5k}}{\partial t} = -\omega_{pe} A_{4k} - \frac{2i\sqrt{\pi} e}{m\omega_{pe}} \Sigma_{k'} k' A_{5k'} A_{5(k-k')} - \frac{2\sqrt{\pi} e}{mc} \Sigma_{k'} k' A_{2k'} A_{3(k-k')}$$

In accordance with the general rule discussed in Problem 2.5.8, the linear matrix $L_{\alpha\beta}$ is the antihermitian one with the following non-zero components:

$$L_{12} = -L_{21}^* = -ick, \quad L_{13} = -L_{31}^* = \omega_{pe}, \quad L_{45} = -L_{54}^* = \omega_{pe}$$

Its eigenvectors correspond to the transverse electromagnetic wave, (t), and to the longitudinal Langmuir wave, (l):

$$\vec{\phi}^{(l)} = C^{(l)} \exp(-i\omega_{pe} t) \vec{\psi}^{(l)}, \quad \vec{\phi}^{(t)} = C^{(t)} \exp(-i\omega^{(t)} t) \vec{\psi}^{(t)},$$

where $\omega^{(t)} = \sqrt{\omega_{pe}^2 + c^2 k^2}$, and the properly normalized "unit" vectors $\vec{\psi}^{(l),(t)}$ read

$$\vec{\psi}^{(l)} = \sqrt{\hbar\omega_{pe}} \left(0,\ 0,\ 0,\ \frac{1}{\sqrt{2}},\ -\frac{1}{\sqrt{2}}\right),$$

$$\vec{\psi}^{(t)} = \sqrt{\hbar\omega^{(t)}} \left(\frac{1}{\sqrt{2}},\ \frac{\sqrt{\epsilon(\omega^{(t)})}}{\sqrt{2}},\ -\frac{i\omega_{pe}}{\sqrt{2}\omega^{(t)}},\ 0,\ 0\right)$$

Assume now that the electromagnetic wave with the wave vector $k_1$, the frequency $\omega_1 = \omega^{(t)}(k_1)$, and the amplitude $C_1$ decays into another electromagnetic wave, $(k_2, \omega_2 = \omega^{(t)}(k_2), C_2)$, and the Langmuir wave, $(k_3 = k_1 - k_2, \omega_3 = \omega_{pe} = \omega_1 - \omega_2, C_3)$. In this case, according to equations (2.130), the non-linear interaction matrix $T_{\alpha\beta\gamma}$ has the following non-zero components:

$$T_{134} = -i\frac{2\sqrt{\pi}e}{m\omega_{pe}}(k_1 - k_2), \quad T_{325} = -T_{523} = -\frac{2\sqrt{\pi}e}{mc},$$

$$T_{335} = T_{445} = T_{555} == i\frac{2\sqrt{\pi}e}{m\omega_{pe}}k_2$$

Thus, it follows from equation (2.130) and the above given expressions for the "unit" vectors $\vec{\psi}^{(l,t)}$ that $\dot{C}_1 = V_1 C_2 C_3$, where

$$V_1 = \frac{1}{\hbar\omega_1}[T_{134}\psi^*_{1k_1}\psi_{3k_2}\psi_{4k_3} + T_{325}\psi^*_{3k_1}\psi_{2k_2}\psi_{5k_3} +$$

$$T_{335}\psi^*_{3k_1}\psi_{3k_2}\psi_{5k_3}] =$$

$$-\frac{\sqrt{\pi/2}e}{\hbar m\omega_1}\sqrt{\hbar^3\omega_1\omega_2\omega_{pe}}\left(-\frac{k_3}{\omega_2} + \frac{\omega_{pe}\sqrt{\epsilon(\omega_2)}}{c\omega_1} - \frac{k_2\omega_{pe}}{\omega_1\omega_2}\right) =$$

$$= \frac{e(k_1 - k_2)}{m}\left(\frac{\pi\hbar\omega_{pe}}{2\omega_1\omega_2}\right)^{1/2}$$

In order to derive similar equations for the amplitudes $C_2$ and $C_3$, one can note that the resonant conditions for these three waves can be formally rewritten as $k_2 = k_1 + (-k_3)$, $\omega_2 = \omega_1 + (-\omega_3)$. Therefore, since changing the sign of a wave vector and frequency of a wave is equivalent to transition to the complex conjugate values of its amplitude and the "unit" vector, one gets that $\dot{C}_2 = V_2 C_1 C_3^*$, with

$$V_2 = \frac{1}{\hbar\omega_2}[T_{134}\psi^*_{1k_2}\psi_{3k_1}\psi^*_{4k_3} + T_{325}\psi^*_{3k_2}\psi_{2k_1}\psi^*_{5k_3} +$$

$$T_{335}\psi^*_{3k_2}\psi_{3k_1}\psi^*_{5k_3}] =$$

$$= -\frac{e(k_1 - k_2)}{m}\left(\frac{\pi\hbar\omega_{pe}}{2\omega_1\omega_2}\right)^{1/2}$$

Similarly, the equation for the amplitude $C_3$ reads $\dot{C}_3 = V_3 C_1 C_2^*$, with

$$V_3 = \frac{1}{\hbar\omega_{pe}}[T_{523}\psi^*_{5k_3}\psi_{2k_1}\psi^*_{3k_2} + T_{523}\psi^*_{5k_3}\psi^*_{2k_2}\psi_{3k_1}] = V_2$$

Altogether, the obtained equation can be written in a "standard" form (2.120) for the interaction of waves with the positive energies:

$$\dot{C}_1 = VC_2C_3, \quad \dot{C}_2 = -V^*C_1C_3^*, \quad \dot{C}_3 = -V^*C_1C_2^*,$$

$$V = \frac{e(k_1 - k_2)}{m}\left(\frac{\pi\hbar\omega_{pe}}{2\omega_1\omega_2}\right)^{1/2} \tag{2.131}$$

Although the so derived interaction matrix element $V$ contains the Planck constant $\hbar$, the resonant three-wave interaction is an entirely classical process. Clearly, the appearance of $\hbar$ in equation (2.131) is due to the imposed normalization of the amplitudes, with the occupation numbers $N_k = |C_k|^2 \propto \hbar^{-1}$. The advantage of such a normalization is that it provides a simple interpretation of the three-wave interaction as a decay or merging of quasiparticles. From the quantum viewpoint equations (2.131) correspond to the classical limit of the large occupation numbers, when the induced processes (rather than the spontaneous ones) dominate.

## Problem 2.5.10

Derive a kinetic equation for the spectral occupation numbers in an ensemble of resonantly interacting waves.

If a broad spectrum of waves is present in a medium, the resonant conditions (2.119) could be satisfied for many different sets of three waves. In this case each wave interacts simultaneously with a large number of other waves, which makes applicable the statistical description of a system. Thus, the phases of waves may be considered as chaotically distributed (the so-called **random phases approximation**), and the system can be described in terms of the spectral occupation numbers $N_{\vec{k}}$. In order to derive the respective kinetic equation for $\dot{N}_{\vec{k}}$, consider a generalization of equations (2.128-2.130) for the case of a continuous spectrum of waves (so that the integration over $\vec{k}'$ must be carried out):

$$\frac{dC_{\vec{k}}}{dt} = \frac{1}{2} \int d\vec{k}' V_{\vec{k}\vec{k}'} C_{\vec{k}'} C_{\vec{k}''} \exp[i(\omega_{\vec{k}} - \omega_{\vec{k}'} - \omega_{\vec{k}''})t], \qquad (2.132)$$

where $\vec{k}'' = \vec{k} - \vec{k}'$, and the factor of a half arises because an interchange of the vectors $\vec{k}'$ and $\vec{k}''$ correspond to the same process. Thus, the rate of the temporal variation of the occupation number is equal to

$$\frac{dN_{\vec{k}}}{dt} = C_{\vec{k}}^* \frac{dC_{\vec{k}}}{dt} + c.c. =$$

$$\frac{1}{2} \int d\vec{k}' V_{\vec{k}\vec{k}'} C_{\vec{k}'} C_{\vec{k}''} C_{\vec{k}}^* \exp[i(\Delta\omega)t] + c.c., \qquad (2.133)$$

$$\Delta\omega = \omega_{\vec{k}} - \omega_{\vec{k}'} - \omega_{\vec{k}''}$$

The next step is to apply the random phases approximation for the amplitudes $C_{\vec{k}}$, $C_{\vec{k}'}$, and $C_{\vec{k}''}$ in equation (2.133). To do so, one can explore the perturbation method, when the amplitude is represented as a series of the interaction matrix element $V$: $C_{\vec{k}} = C_{\vec{k}}^{(0)} + C_{\vec{k}}^{(1)} + ....$ Then, the random phases state means that there are no phase correlations in the zeroth-order approximation, that is

$$\langle C_{\vec{k}}^{(0)} \rangle = 0, \quad \langle C_{\vec{k}}^{(0)} C_{\vec{k}'}^{(0)*} \rangle = N_{\vec{k}} \delta(\vec{k} - \vec{k}'),$$

$$\langle C_{\vec{k}}^{(0)} C_{\vec{k}'}^{(0)} \rangle = N_{\vec{k}} \delta(\vec{k} + \vec{k}'),$$

where the symbol $\langle \rangle$ means averaging over the phases, and the relation $C_{-\vec{k}} = C_{\vec{k}}^*$ has been used. Therefore, in the zeroth-order approximation the right-hand side of equation (2.133) becomes equal to zero, and, hence, the next term in the series, $C_{\vec{k}}^{(1)}$, should be used in the derivation of $\dot{N}_{\vec{k}}$. Thus, according to equation (2.132),

$$\frac{dC_{\vec{k}}^{(1)}}{dt} \approx \frac{1}{2} \int d\vec{k}' V_{\vec{k}\vec{k}'} C_{\vec{k}'}^{(0)} C_{\vec{k}''}^{(0)} \exp[i(\Delta\omega)t] \qquad (2.134)$$

It is convenient to assume here that the interaction is switched-on adiabatically at $t \to -\infty$, so that $V \propto \exp(-\gamma^2 t^2)$ and $C_{\vec{k}}^{(1)}(t \to -\infty) = 0$. Then, integration in equation (2.134) yields

$$C_{\vec{k}}^{(1)} = \frac{1}{2} \int d\vec{k}' V_{\vec{k}\vec{k}'} C_{\vec{k}'}^{(0)} C_{\vec{k}''}^{(0)} \int_{-\infty}^{t} dt' \exp[-i(\Delta\omega)t' - \gamma^2 t'^2] =$$

$$\int d\vec{k}' V_{\vec{k}\vec{k}'} C_{\vec{k}'}^{(0)} C_{\vec{k}''}^{(0)} \frac{\sqrt{\pi}}{2\gamma} \exp[-(\Delta\omega)^2/4\gamma^2], \qquad (2.135)$$

where the amplitudes $C_{\vec{k}', \vec{k}''}^{(0)}$ are considered constants (taking into account their variation with time leads to the higher-order corrections). Since

$$\lim_{\gamma \to 0} \frac{\exp[-(\Delta\omega)^2/4\gamma^2]}{2\gamma\sqrt{\pi}} = \delta(\Delta\omega),$$

it follows from equation (2.135) that in this limit

$$C_{\vec{k}}^{(1)} \approx \frac{\pi}{2} \int d\vec{q} V_{\vec{k}\vec{q}} C_{\vec{q}}^{(0)} C_{\vec{k}-\vec{q}}^{(0)} \delta(\Delta\omega) \qquad (2.136)$$

The required corrections $C_{\vec{k}'}^{(1)}$ and $C_{\vec{k}''}^{(1)}$ can be obtained in a similar way. Thus, by using equations (2.120) together with the symmetry relations (2.121) (it is assumed that the interacting waves have a positive energy), one gets

$$C_{\vec{k}'}^{(1)} \approx -\frac{\pi}{2} \int d\vec{q} V_{\vec{k}\vec{q}}^* C_{\vec{q}}^{(0)} C_{\vec{q}-\vec{k}'}^{(0)*} \delta(\Delta\omega),$$

$$C_{\vec{k}''}^{(1)} \approx -\frac{\pi}{2} \int d\vec{q} V_{\vec{k}\vec{q}}^* C_{\vec{q}}^{(0)} C_{\vec{q}-\vec{k}''}^{(0)*} \delta(\Delta\omega)$$

By substituting these expressions together with (2.136) into equation (2.133), one gets

$$\frac{dN_{\vec{k}}}{dt} = \frac{1}{2}\int d\vec{k}' V_{\vec{k}\vec{k}'} \exp[i(\Delta\omega)t]\langle C_{\vec{k}'}^{(1)} C_{\vec{k}''}^{(0)} C_{\vec{k}}^{(0)*} + C_{\vec{k}'}^{(0)} C_{\vec{k}''}^{(1)} C_{\vec{k}}^{(0)*} +$$

$$C_{\vec{k}'}^{(0)} C_{\vec{k}''}^{(0)} C_{\vec{k}}^{(1)*}\rangle + c.c. = \frac{\pi}{2}\int d\vec{k}' V_{\vec{k}\vec{k}'} \left\{\int d\vec{q} V_{\vec{k}\vec{q}}^* \langle C_{\vec{k}''}(0) C_{\vec{k}''}^{(0)} C_{\vec{q}}^{(0)*} C_{\vec{k}-\vec{q}'}^{(0)*}\rangle \right.$$

$$\left. - \langle C_{\vec{k}''}(0) C_{\vec{k}}^{(0)*} C_{\vec{q}}^{(0)} C_{\vec{q}-\vec{k}'}^{(0)}\rangle - \langle C_{\vec{k}'}(0) C_{\vec{k}}^{(0)*} C_{\vec{q}}^{(0)} C_{\vec{q}-\vec{k}'}^{(0)*}\rangle\right\}\delta(\Delta\omega)$$

Since in the zeroth-order approximation there are no phase correlations in the amplitudes, the phase averaging in the products of four of them in equation (2.133) brings about a product of the two occupation numbers. For example, in the first term on the right-hand side of equation (2.133) a non-zero result arises only in two cases: $\vec{q} = \vec{k}'$ or $\vec{q} = \vec{k}''$. By proceeding with a similar "pairing" in the other two terms, one arrives to the sought after kinetic equation:

$$\frac{dN_{\vec{k}}}{dt} = \pi\int d\vec{k}' |V_{\vec{k}\vec{k}'}|^2 (N_{\vec{k}'} N_{\vec{k}-\vec{k}'} -$$

$$N_{\vec{k}} N_{\vec{k}'} - N_{\vec{k}} N_{\vec{k}-\vec{k}'})\delta(\omega_{\vec{k}} - \omega_{\vec{k}'} - \omega_{\vec{k}-\vec{k}'}) \tag{2.137}$$

The equations for $N_{\vec{k}'}$ and $N_{\vec{k}-\vec{k}'}$ immediately follow from the Manley-Rowe relations (2.123):

$$\frac{dN_{\vec{k}'}}{dt} = \frac{dN_{\vec{k}-\vec{k}'}}{dt} = -\frac{dN_{\vec{k}}}{dt} \tag{2.138}$$

It is noteworthy that the kinetic equation (2.137) can also be derived using the quantum mechanical correspondence principle. Indeed, consider the elementary act of the three-wave interaction. If $w(\vec{k}, \vec{k}')d\vec{k}'$ is the probability of the decay for a quasiparticle with momentum $\hbar\vec{k}$ into two other quasiparticles with their momenta in the intervals $[\hbar\vec{k}', \hbar(\vec{k}' + d\vec{k}')]$ and $[\hbar(\vec{k} - \vec{k}'), \hbar(\vec{k} - \vec{k}' - d\vec{k}')]$, it follows from the principle of detailed balance (see, e.g., E. M. Lifshitz and L. P. Pitaevskii, *Physical Kinetics*, §2, Butterworth-Heinemann, 1995) that a reverse process (in this case the merging of two quasiparticles into the one) has the same probability. Therefore, the balance equation for the occupation number $N_{\vec{k}}$ reads:

$$\frac{dN_{\vec{k}}}{dt} = \frac{1}{2}\int d\vec{k}' w(\vec{k}, \vec{k}')[(N_{\vec{k}}+1)N_{\vec{k}'} N_{\vec{k}-\vec{k}'} - N_{\vec{k}}(N_{\vec{k}'}+1)(N_{\vec{k}-\vec{k}'}+1)] \tag{2.139}$$

A factor of a half appears for the very same reason as in equation (2.132), while the combination of the occupation numbers is determined here by the eigenvalues of the operators of creation and annihilation of quasiparticles (see, e.g., L. D. Landau and E. M. Lifshitz, *Quantum Mechanics*, §64, Pergamon, 1984). In the classical limit, when $N_{\vec{k}}$, $N_{\vec{k}'}$, $N_{\vec{k}-\vec{k}'} \gg 1$, the kinetic equation (2.137) follows from (2.139) with the probability

$$w(\vec{k}, \vec{k}') = 2\pi|V_{\vec{k}\vec{k}'}|^2\delta(\omega_{\vec{k}} - \omega_{\vec{k}'} - \omega_{\vec{k}-\vec{k}'}),$$

which is in accordance with the quantum-mechanical "golden rule" for calculation of probabilities (see, e.g., L. D. Landau and E. M. Lifshitz, *Quantum Mechanics*, §43, Pergamon, 1984). Unlike the dynamic equations (2.120) for the amplitudes of the interacting waves, the kinetic equation (2.137) is the statistical one, and, therefore, it does not possess the property of the time reversibility. Thus, the kinetic equation describes an irreversible process with a positive production of the entropy (see next Problem).

## Problem 2.5.11

Prove the second law of thermodynamics (the increase of the total entropy) for a gas of quasiparticles whose occupation numbers are governed by the kinetic equation (2.137).

In the classical limit of large occupation numbers, the entropy of a unit volume of the Bose gas is equal to $S = \int d\vec{k} \ln N_{\vec{k}}$ (see, e.g., L. D. Landau and E. M. Lifshitz, *Statistical Physics*, Part 1, §55, Pergamon, 1981). Thus, the rate of change of the entropy is equal to

$$\frac{dS}{dt} = \int \frac{d\vec{k}}{N_{\vec{k}}} \frac{dN_{\vec{k}}}{dt} \qquad (2.140)$$

In the calculation of the integral in (2.140) one should take into account the simultaneous change of the occupation numbers for all three of the interacting quasiparticles. Therefore, according to equation (2.140) taken together with equations (2.137) and (2.138), one gets:

$$\frac{dS}{dt} = \int d\vec{k} \frac{dN_{\vec{k}}}{dt} \left( \frac{1}{N_{\vec{k}}} - \frac{1}{N_{\vec{k}'}} - \frac{1}{N_{\vec{k}-\vec{k}'}} \right) =$$

$$\pi \int d\vec{k} d\vec{k}' |V|^2 \delta(\Delta\omega) \frac{(N_{\vec{k}'} N_{\vec{k}-\vec{k}'} - N_{\vec{k}} N_{\vec{k}'} - N_{\vec{k}} N_{\vec{k}-\vec{k}'})^2}{N_{\vec{k}} N_{\vec{k}'} N_{\vec{k}-\vec{k}'}} \geq 0$$

Thus, the total entropy of the gas does not increase only when

$$\frac{1}{N_{\vec{k}}} = \frac{1}{N_{\vec{k}'}} + \frac{1}{N_{\vec{k}-\vec{k}'}} \qquad (2.141)$$

It is easy to verify that such a relation corresponds to the state of thermodynamic equilibrium in a gas of quasiparticles. Indeed, in the classical limit the thermodynamic equilibrium is described by the Rayleigh-Jeans formula with $N_{\vec{k}} \propto T/\omega_{\vec{k}}$, in which case relation (2.141) becomes satisfied identically due to the resonant condition (2.119) for the frequencies of the interacting waves.

However, the kinetic equations of the type of (2.137) could also possess steady solutions which correspond to the states that are far away from the thermodynamic equilibrium. These non-equilibrium solutions usually occur in situations when quasiparticles are generated by some external source, which operates in the frequency/wavelength band that lies well apart from the range where dissipative processes take place. Therefore, they describe the flux of energy over the spectrum of waves from the generation range to the dissipative one (see, e.g., V. E. Zakharov, Kolmogorov spectra in weak turbulence problems, in *Handbook of Plasma Physics*, vol. 2, North-Holland, 1983).

# 3

## Fluid dynamics

## 3.1 Dynamics of an ideal fluid

The motion of an ideal fluid is described by the following set of equations.
The continuity equation

$$\frac{\partial \rho}{\partial t} + \vec{\nabla} \cdot (\rho \vec{v}) = 0 \tag{3.1}$$

The equation of motion (**Euler's equation**)

$$\frac{\partial \vec{v}}{\partial t} + (\vec{v} \cdot \vec{\nabla})\vec{v} = -\frac{1}{\rho}\vec{\nabla}p \tag{3.2}$$

The adiabatic condition

$$\frac{ds}{dt} = \frac{\partial s}{\partial t} + (\vec{v} \cdot \vec{\nabla})s = 0 \tag{3.3}$$

where $\rho(\vec{r}, t)$, $p(\vec{r}, t)$, and $\vec{v}(\vec{r}, t)$ are, respectively, the density, pressure, and velocity of the fluid, and $s(p, \rho)$ is the entropy per unit mass.

For a steady flow **Bernoulli's equation** follows from equations (3.2) and (3.3):

$$\vec{v} \cdot \vec{\nabla}(\frac{v^2}{2} + w) = 0,$$

which indicates conservation of the quantity $(\frac{v^2}{2} + w)$ along a streamline of the flow. Here $w$ is the heat function (enthalpy) per unit mass.

Conservation of velocity circulation along any closed "liquid contour" that moves together with a fluid is known as **Kelvin's circulation theorem**:

$$\frac{d}{dt} \oint_C \vec{v} \cdot d\vec{l} = 0$$

The equation of motion (3.2) is equivalent to the conservation law for the linear momentum of a fluid, which can be written in the "divergent" form as

$$\frac{\partial(\rho v_i)}{\partial t} = -\frac{\partial \Pi_{ik}}{\partial x_k}, \tag{3.4}$$

where $\Pi_{ik} = p\delta_{ik} + \rho v_i v_k$ is the **linear momentum flux tensor**.

In a similar way, the energy conservation law takes the following form:

$$\frac{\partial}{\partial t}\left(\frac{1}{2}\rho v^2 + \rho\epsilon\right) = -\vec{\nabla}\cdot\vec{q},$$

where $\vec{q} = \rho\vec{v}(v^2/2 + w)$ is the **energy flux vector**, and $\epsilon$ is the internal energy per unit mass.

## Problem 3.1.1

Derive the evolution equation for the flow vorticity $\vec{\Omega} = \vec{\nabla}\times\vec{v}$ in an ideal fluid. Prove that vector lines of vorticity are "frozen" into the flow.

Euler's equation (3.2) can be written as

$$\frac{\partial\vec{v}}{\partial t} + (\vec{v}\cdot\vec{\nabla})\vec{v} = -\frac{1}{\rho}\vec{\nabla}p = -\vec{\nabla}w,$$

or

$$\frac{\partial\vec{v}}{\partial t} - \vec{v}\times(\vec{\nabla}\times\vec{v}) = -\vec{\nabla}\left(w + \frac{v^2}{2}\right)$$

By taking the curl of this equation, one gets the following equation for the flow vorticity $\vec{\Omega} = \vec{\nabla}\times\vec{v}$:

$$\frac{\partial\vec{\Omega}}{\partial t} = \vec{\nabla}\times(\vec{v}\times\vec{\Omega}),$$

which can be re-written as

$$\frac{\partial\vec{\Omega}}{\partial t} = (\vec{\Omega}\cdot\vec{\nabla})\vec{v} - (\vec{v}\cdot\vec{\nabla})\vec{\Omega} - \vec{\Omega}\vec{\nabla}\cdot\vec{v}$$

By substituting here the expression for $\vec{\nabla}\cdot\vec{v}$ that follows from the continuity equation (3.1):

$$\vec{\nabla}\cdot\vec{v} = -\frac{1}{\rho}\frac{\partial\rho}{\partial t} - \frac{1}{\rho}\vec{v}\cdot(\vec{\nabla}\rho),$$

and making a simple re-arrangement of terms, one gets

$$\left(\frac{\partial}{\partial t} + \vec{v}\cdot\vec{\nabla}\right)\frac{\vec{\Omega}}{\rho} \equiv \frac{d}{dt}\frac{\vec{\Omega}}{\rho} = \left(\frac{\vec{\Omega}}{\rho}\cdot\vec{\nabla}\right)\vec{v} \tag{3.5}$$

Consider now a "fluid line," which is a line that moves together with fluid particles that constitute it. If $\delta\vec{l}$ is an element of length along this line, its variation with time is due to the velocity difference, $\delta\vec{v}$, at the two ends of the element: $\delta \vec{v} = (\delta \vec{l} \cdot \vec{\nabla})\vec{v}$, thus $d(\delta \vec{l})/dt = (\delta \vec{l} \cdot \vec{\nabla})\vec{v}$. It shows that temporal variations of the two vectors, $\vec{\Omega}/\rho$ and $\delta \vec{l}$, are described by identical equations. Therefore, if these two vectors are parallel at some initial moment, they will remain parallel at any time, and their ratio is the integral of motion. In other words, if $\delta \vec{l}$ between the two fluid particles is initially parallel to the vector $\vec{\Omega}$, i.e., the particles are located on the same vector line of the vorticity field, they will always remain on this vorticity line in the course of the fluid flow. By extending this feature from the neighboring fluid particles to those separated by any distance along a vorticity line, one concludes that a field line of the vorticity field moves together with the fluid particles that lie on it, i.e., the lines of vorticity are "frozen" into an ideal (inviscid) fluid.

It is easy to see that Kelvin's circulation theorem is a direct consequence of the "frozen-in" vorticity. Indeed, any closed fluid contour in the course of its motion does not cut vorticity lines. Therefore, the flux of the vorticity vector field through any surface spanning the fluid contour does not vary with time. Since, according to Stokes' theorem, this flux is equal to the velocity circulation along the fluid contour, the latter also remains constant. Another consequence of this property of an ideal fluid flow is as follows. If initially a flow is vortex-free, it will remain so at any time, and, hence, it can be described in terms of the velocity potential $\phi$ as $\vec{v} = -\vec{\nabla}\phi$.

## Problem 3.1.2

Write down the one-dimensional ideal fluid dynamics equations in terms of the Lagrangian variables.

The Lagrangian description of a fluid flow is made in terms of the functions $x(x_0, t)$, $\rho(x_0, t)$, $p(x_0, t)$, where $x_0$ is the initial position of a fluid element (which in the one-dimensional case is a layer). The mass conservation condition, which reads $\rho dx = \rho_0 dx_0$, yields

$$\left(\frac{\partial x}{\partial x_0}\right)_t = \frac{\rho_0(x_0)}{\rho(x_0, t)} \tag{3.6}$$

The equation of motion for a layer of a width $dx$:

$$\rho dx \left(\frac{\partial^2 x}{\partial^2 t}\right)_{x_0} = -p(x + dx) + p(x) + \rho dx f(x_0, t),$$

where $f$ is the external force per unit mass, yields

$$\left(\frac{\partial^2 x}{\partial^2 t}\right)_{x_0} = -\frac{1}{\rho_0(x_0)}\left(\frac{\partial p}{\partial x_0}\right)_t + f(x_0, t) \tag{3.7}$$

Finally, the entropy conservation equation reads $(\partial s/\partial t)_{x_0} = 0$, where $s(p, \rho)$ is the entropy per unit mass.

## Problem 3.1.3

In a uniform medium of density $\rho_0$ and pressure equal to zero (a dust) a non-uniform flow with velocity $v(x) = v_0 \sin(\pi x/l)$ is instantly created at $t = 0$ in the domain $0 \leq x \leq l$. Determine the resulting distribution of the density $\rho(x, t)$.

This problem can be easily solved by using the Lagrangian coordinates, where all governing equations are linear (which is not the case for the Euler's description with the non-linear equation of motion). Thus, in the absence of pressure and external force, it follows from the equation of motion (3.7) that each element of the dust continues to move with the initially acquired velocity, hence

$$x = x_0 + v(x_0)t = x_0 + v_0 t \sin(\pi x_0/l)$$

Then, the continuity equation (3.6) yields

$$\rho(x_0, t) = \frac{\rho_0}{(\partial x/\partial x_0)_t} = \rho_0 \left[1 + \frac{\pi v_0 t}{l} \cos\left(\frac{\pi x_0}{l}\right)\right]^{-1} \qquad (3.8)$$

Together with the function $x(x_0, t)$ given above, this equation determines, though implicitly, the density profile $\rho(x, t)$. An important point here is that at some locations the density tends to infinity in the course of the flow. Indeed, as seen from equation (3.6), such a singularity occurs when the derivative $(\partial x/\partial x_0)_t$ becomes equal to zero. Thus, according to equation (3.8), it first takes place at $x_0 = l$ at the instant $t = t_* = l/\pi v_0$. This phenomenon is called "breaking" of the velocity profile, since the respective graph $v(x, t)$ acquires at this moment an infinite derivative at the point $x_0 = l$. At $t > t_*$ an "intersection of trajectories" occurs, when different elements of a fluid end up at the same location. Therefore, it makes the Lagrangian description invalid.

## Problem 3.1.4

Describe expansion of a uniformly charged spherical cloud of dust that is initially at rest. The initial mass density is equal to $\rho_0$, and has charge per unit mass $\alpha$.

In the Lagrangian description the interest is in finding the functions $r(r_0, t)$ and $v(r_0, t)$. If there is no profile "breaking," which will be confirmed in what follows, the net charge inside each expanding spherical surface remains constant: $q(r) = q(r_0) = 4\alpha\rho_0\pi r_0^3/3$. Thus, the equation of motion takes the form

$$\left(\frac{\partial^2 r}{\partial t^2}\right)_{r_0} = f(r_0, t) = \alpha E(r_0, t) = \frac{4\pi\alpha^2 \rho_0 r_0^3}{3r^2}$$

By integrating this equation along with the initial conditions $r(0) = r_0$, $\dot{r}(0) = 0$, one gets

$$\omega_0 t = \int_1^{r/r_0} dx \frac{\sqrt{x}}{\sqrt{x-1}}, \qquad (3.9)$$

where the parameter $\omega_0$, which has a dimension of frequency, is equal to $(8\pi\alpha^2\rho_0/3)^{1/2}$. This equation implicitly determines the expansion radius $r(r_0, t)$, which can be represented as $r = r_0 F(\omega_0 t)$. Thus, the absence of the profile "breaking" is now evident. Such a form of the solution also demonstrates that the cloud undergoes a uniform expansion, with density $\rho(t) = \rho_0 F^{-3}(\omega_0 t)$. At long times, when $\omega_0 t \gg 1$, the function $F(\omega_0 t) \approx \omega_0 t$, which indicates expansion with a constant velocity $v(r_0) = \omega_0 r_0$ (at this stage of expansion the electric field becomes weak and, hence, plays no role).

## Problem 3.1.5

Find criterion for the profile "breaking" in a spherical cloud of charged dust (see Problem 3.1.4) with a non-uniform initial density.

In the case of a non-uniform density the basic equation (3.9) can be generalized to the following form:

$$\omega(r_0)t = \int_1^{r/r_0} dx \frac{\sqrt{x}}{\sqrt{x-1}}, \quad \omega(r_0) = \left(\frac{8\pi\alpha^2}{r_0^3}\int_0^{r_0} \rho_0(x)x^2 dx\right)^{1/2} \qquad (3.10)$$

The absence of "breaking" requires a positive derivative $(\partial r/\partial r_0)_t$, which, according to equation (3.10), is equal to

$$\left(\frac{\partial r}{\partial r_0}\right)_t = tr_0 \frac{d\omega}{dr_0}\left(1 - \frac{r_0}{r}\right)^{1/2} + \frac{r}{r_0}$$

Thus, by eliminating $t$ from this expression with the help of (3.10), one arrives to the following condition of no "breaking":

$$\frac{r_0}{\omega}\frac{d\omega}{dr_0} > -\frac{x^{3/2}/\sqrt{x-1}}{\int_1^x \sqrt{z}dz/\sqrt{z-1}} \equiv -P(x), \qquad (3.11)$$

where $x = r/r_0$, and the inequality (3.11) must be satisfield in the entire interval of $1 \leq x \leq \infty$. Thus, the next step is to consider the minimum of the function $P(x)$ in this domain. A simple integration yields that

$$P(x) = \frac{x^{3/2}}{\sqrt{(x-1)}} \left[ \sqrt{x(x-1)} - \frac{1}{2} \ln \left| \frac{\sqrt{x} - \sqrt{(x-1)}}{\sqrt{x} + \sqrt{(x-1)}} \right| \right]^{-1},$$

which is a positive function, with $P(x) \to \infty$ at $x \to 1$, and $P(x) \to 1$ at $x \to \infty$. It can be obtained numerically that its minimum, $P_{min} \approx 0.93$, is achieved at $x \approx 8$. Therefore, the condition (3.11) takes the form

$$\rho_0(r_0)r_0^3 > 1.14 \int_0^{r_0} \rho_0(x)x^2 dx,$$

which must be satisfied for each $r_0$ inside the initial cloud. For example, if the initial density has a power law distribution: $\rho_0(r_0) \propto r_0^{-\kappa}$ with $\kappa < 3$ (so that the total mass of the cloud is finite), the profile "breaking" occurs if $\kappa > 1.86$.

## Problem 3.1.6

In a cold uniform electron plasma (see Problem 2.1.2) the electron velocity distribution $v(x) = v_0 \sin(\pi x/l)$ is instantaneously created at the moment $t = 0$. Determine a critical value of the velocity amplitude $v_0$, exceeding which leads to "breaking" in the flow of electrons.

Since in this case the only force exerted on electrons is the electric field $E_x \equiv E$, the equation of motion (3.7) takes the form $(\partial^2 x/\partial^2 t)_{x_0} = -eE/m$, where the electric field appears due to space charge caused by the electron flow (recall that ions are assumed immobile). Thus, if there is no "intersection" of electron trajectories, a displacement of the electron layer with initial coordinate $x_0$ to a new position with coordinate $x$ brings about a surplus charge (per unit area) equal to $\sigma = ne(x - x_0)$ on the one side of the layer, and the equal charge of the opposite sign on its other side. The resulting electric field is equal to $E = 4\pi\sigma = 4\pi ne(x - x_0)$, which being inserted into the equation of motion for electrons yields

$$(\partial^2 x/\partial^2 t)_{x_0} = -eE/m = -\omega_{pe}^2(x - x_0), \quad \omega_{pe}^2 = \frac{4\pi ne^2}{m},$$

with a general solution $(x - x_0) = A \sin(\omega_{pe}t) + B \cos(\omega_{pe}t)$. The constants $A$ and $B$ are specified by the initial conditions, namely

$$(x - x_0)_{t=0} = 0, \quad \left( \frac{\partial x}{\partial t} \right)_{t=0} = v_0 \sin(\pi x_0/l)$$

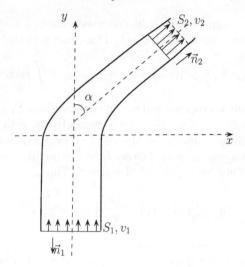

**FIGURE 3.1**
Ejection of an incompressible fluid

Thus, one gets

$$x(x_0, t) = x_0 + \frac{v_0}{\omega_{pe}} \sin(\pi x_0/l) \sin(\omega_{pe}t),$$

so that

$$\left(\frac{\partial x}{\partial x_0}\right)_t = 1 + \frac{\pi v_0}{\omega_{pe}l} \cos(\pi x_0/l) \sin(\omega_{pe}t)$$

As seen from this expression, if $\pi v_0/\omega_{pe}l < 1$, the derivative $(\partial x/\partial x_0)_t$ is positive in any location at any time, and, hence, there is no breaking in the electron flow. It also shows that the most "dangerous" location for the occurrence of breaking is $x_0 = l$, where it does occur if $v_0 > \omega_{pe}l/\pi$.

## Problem 3.1.7

An ideal incompressible fluid is ejected into a free space of negligibly small pressure with the help of a tube, shown in Figure 3.1. Derive the force exerted on the bent part of the tube, by assuming that the flow is uniform at the cross sections $S_1$, $S_2 < S_1$, and the tube discharge is equal to $Q$.

For a steady flow the net flux of the linear momentum through any closed surface is equal to zero: $\int \Pi_{ik} dS_k = 0$, where $\Pi_{ik} = p\delta_{ik} + \rho v_i v_k$ is the momentum flux tensor. In this particular case one can apply this relation to the closed surface that comprises the tube cross sections $S_{1,2}$ and the part of the lateral surface of the tube between them. Thus, one gets that

$$n_{1k}S_1(p_1\delta_{ik} + \rho v_{1i}v_{1k}) + n_{2k}S_2\rho v_{2i}v_{2k} + \int p\,dS_i = 0 \qquad (3.12)$$

Here $\vec{n}_{1,2}$ are the unit vectors normal to the cross sections $S_{1,2}$, where the fluid velocity is equal to $\vec{v}_1 = -v_1\vec{n}_1$, $\vec{v}_2 = v_2\vec{n}_2$, respectively, with $v_{1,2} = Q/\rho S_{1,2}$. A value for the fluid pressure $p_1$ at the cross section $S_1$, that is required to provide for the given discharge $Q$, follows from Bernoulli's equation, which in an incompressible fluid reads $p + \rho v^2/2 = const$. Thus,

$$p_1 = \frac{\rho}{2}(v_2^2 - v_1^2) = \frac{Q^2}{2\rho}\left(\frac{1}{S_2^2} - \frac{1}{S_1^2}\right)$$

(recall that there is no pressure at the free cross section $S_2$). The sought after force, which is due to the fluid pressure acting at the lateral surface of the tube, is expressed by the respective integral term in equation (3.12), hence

$$F_i = -n_{1i}S_1(p_1 + \rho v_1^2) - n_{2i}S_2\rho v_2^2 \qquad (3.13)$$

Then, by introducing the coordinate system as shown in Figure 3.1, one gets that

$$F_x = -n_{2x}S_2\rho v_2^2 = -\frac{Q^2}{\rho S_2}\sin\alpha,$$

$$F_y = S_1(p_1 + \rho v_1^2) - S_2\rho v_2^2\cos\alpha = \frac{Q^2}{2\rho}\left(\frac{S_1}{S_2^2} + \frac{1}{S_1}\right) - \frac{Q^2}{\rho S_2}\cos\alpha$$

## Problem 3.1.8

Derive the dispersion relation for a surface wave propagating along a horizontal interface between two ideal incompressible fluids of densities $\rho_1$ and $\rho_2$ in the gravity field $\vec{g}$ (see Figure 3.2). The interface surface tension coefficient is $\alpha$.

For the wave of a small amplitude, the fluid motion can be considered as a potential one (see, e.g., L. D. Landau and E. M. Lifshitz, *Fluid Dynamics*, §9, Pergamon Press, 1987); hence, the linearized equation of motion and the continuity equation take the following form:

$$\rho\frac{\partial\vec{v}}{\partial t} = -\vec{\nabla}\delta p, \ \vec{\nabla}\cdot\vec{v} = 0, \ \vec{v} = -\vec{\nabla}\phi, \qquad (3.14)$$

where $\delta p$ is the pressure perturbation due to the wave. Without it the pressure profile in the fluids is determined by the gravity force: $p_1 = p_0 - \rho_1 gz$, $p_2 =$

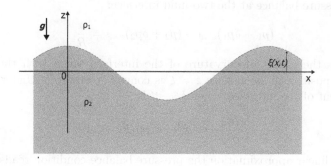

**FIGURE 3.2**
Surface wave in a fluid

$p_0 - \rho_2 g z$, where $p_0$ is the pressure at the interface $z = 0$ (see Figure 3.2). If the wave propagates along, say, the $x$-axis, the perturbed quantities can be written as follows:

$$\phi(x, z, t) = f(z) \exp[i(kx - \omega t)], \quad \delta p(x, z, t) = \psi(z) \exp[i(kx - \omega t)],$$
$$\xi(x, t) = \xi_0 \exp[i(kx - \omega t)], \tag{3.15}$$

where $\xi(x, t)$ is vertical displacement of the fluid's interface. Thus, the functions $f(z)$ and $\psi(z)$ should be determined separately in the lower and the upper fluids. It follows from equations (3.14) that the velocity potential $\phi$ satisfies the Laplace equation $\vec{\nabla}^2 \phi = 0$. Since the interest is in a surface wave, all perturbations must vanish at $z \to \pm\infty$, therefore

$$\phi_1 = A \exp(-kz) \exp[i(kx - \omega t)],$$
$$\phi_2 = B \exp(kz) \exp[i(kx - \omega t)] \tag{3.16}$$

The constants $A$, $B$ and $\xi_0$ in equations (3.15) and (3.16) are not independent from each other but relate by the requirement of continuity at the interface of the normal component of the fluid velocity, which in the linear approximation yields

$$-\left(\frac{\partial \phi_1}{\partial z}\right)_{z=0} = -\left(\frac{\partial \phi_2}{\partial z}\right)_{z=0} = \frac{\partial \xi}{\partial t},$$

so that $B = -A = i\omega\xi_0/k$. By substituting this into the linearized equation of motion in (3.14), one gets the following expressions for the pressure perturbations for both fluids:

$$\delta p_1 = -i\omega\rho_1\phi_1 = -\frac{\omega^2 \rho_1}{k}\xi_0 \exp(-kz) \exp[i(kx - \omega t)],$$
$$\delta p_2 = -i\omega\rho_2\phi_2 = \frac{\omega^2 \rho_2}{k}\xi_0 \exp(kz) \exp[i(kx - \omega t)],$$

Then, the sought after dispersion relation $\omega(k)$ follows from Laplace's formula for the pressure balance at the two-fluid interface:

$$(p_1 + \delta p_1)_{z=\xi} - (p_2 + \delta p_2)_{z=\xi} = \frac{\alpha}{R},$$

where $R$ is the radius of curvature of the interface taken with the proper sign ($R$ is positive if the surface $z = \xi$ is convex downward). Under a small displacement of the interface, when $k\xi_0 \ll 1$,

$$\frac{1}{R} \approx \frac{\partial^2 \xi}{\partial^2 x} = -k^2 \xi_0 \exp[i(kx - \omega t)],$$

and in the linear approximation the pressure balance condition reads:

$$\xi_0 \exp[i(kx - \omega t)] \left( -\rho_1 g - \frac{\omega^2 \rho_1}{k} + \rho_2 g - \frac{\omega^2 \rho_2}{k} \right) = -\alpha k^2 \xi_0 \exp[i(kx - \omega t)],$$

which finally yields

$$\omega(k) = \left[ \frac{gk(\rho_2 - \rho_1) + \alpha k^3}{(\rho_1 + \rho_2)} \right]^{1/2} \tag{3.17}$$

Consider now some particular cases of the general dispersion equation (3.17).

a) A free boundary of a heavy fluid, when $\rho_1 = 0$, $\rho_2 = \rho$. In this case $\omega(k) = \sqrt{gk + \alpha k^3/\rho}$, so that in a long wavelength limit, $k \ll \sqrt{\rho g/\alpha}$, one gets $\omega(k) \approx \sqrt{gk}$ — gravity waves on a surface of a deep fluid. In the opposite limit of a short wavelength it yields $\omega(k) \approx \sqrt{\alpha k^3/\rho}$ — capillary surface waves. If $\rho_1 \neq 0$, but $\rho_1 < \rho_2$ (a light fluid above a heavy one), all qualitative properties of the surface waves remain unchanged.

b) A heavy fluid above a light one: $\rho_1 > \rho_2$. In this case the interface becomes unstable, and, in the absence of capillarity, the instability growth rate is equal to

$$\gamma = i\omega = \left[ \frac{gk(\rho_1 - \rho_2)}{(\rho_1 + \rho_2)} \right]^{1/2}$$

This is the **Rayleigh-Taylor instability**. A non-zero surface tension brings about stabilization of short wavelength perturbations with $k > k_* = \sqrt{g(\rho_1 - \rho_2)/\alpha}$, but the instability survives at long wavelength, when $k < k_*$.

## Problem 3.1.9

Derive the energy and linear momentum of a gravity-capillary wave that propagates on the free surface of a deep ideal incompressible fluid.

According to derivations given in the previous Problem, the vertical displacement $\xi(x, t) = \xi_0 \cos(kx - \omega t)$ of the fluid surface corresponds to the velocity potential

$$\phi(x, z, t) = -\frac{\omega}{k}\xi_0 \sin(kx - \omega t)\exp(kz)$$

Thus, the fluid velocity reads

$$v_x = -\frac{\partial \phi}{\partial x} = \omega \xi_0 \exp(kz)\cos(kx - \omega t), \quad v_z = -\frac{\partial \phi}{\partial z} = \omega \xi_0 \exp(kz)\sin(kx - \omega t),$$

and the fluid kinetic energy per unit surface area is equal to

$$E_{kin} = \int\limits_{-\infty}^{0} dz\frac{\rho <v^2>}{2} = \frac{\rho \omega^2}{4k}\xi_0^2,$$

where the symbol $<>$ means averaging over time. Since for a linear harmonic oscillation the potential energy (which in this case comprises the gravitational and the capillary energy) is, on average, equal to the kinetic energy, the total energy per unit surface area for such a wave is

$$E = 2E_{kin} = \frac{\rho \omega^2}{2k}\xi_0^2 \tag{3.18}$$

As far as the linear momentum is concerned, it may look at first glance that its total value is equal to zero because the time average of the fluid velocity components vanish, and the density of an incompressible fluid is a constant. Nevertheless, the wave possesses the total linear momentum, which is brought about by the correlation between the vertical displacement, $\xi(x, t)$, of the fluid surface, and the horizontal velocity component there, $v_x(x, z = 0, t)$. The latter is positive at the wave crests, where $\xi > 0$, and negative at its troughs, where $\xi < 0$. As a result, at any instant in time a slightly larger volume of the fluid is moving to the right than to the left, which indicates the linear momentum in the direction of the wave vector $k_x$. In order to make an explicit calculation, it is convenient to introduce the **stream function** $\psi(x, z, t)$ defined as

$$\vec{v} = (\vec{\nabla}\psi \times \vec{e}_y), \quad v_x = -\frac{\partial \psi}{\partial z}, \quad v_z = \frac{\partial \psi}{\partial x}$$

In the present case

$$\psi = -\frac{\omega}{k}\xi_0 \exp(kz)\cos(kx - \omega t);$$

therefore

$$P_x = \left\langle \int_{-\infty}^{\xi} dz \rho v_x \right\rangle = -\rho \left\langle \int_{-\infty}^{\xi} dz \frac{\partial \psi}{\partial z} \right\rangle =$$

$$-\rho \langle \psi(z = \xi) \rangle = \rho \frac{\omega}{k} \xi_0 \langle \exp[k\xi_0 \cos(kx - \omega t)] \cos(kx - \omega t) \rangle \approx$$

$$\rho \frac{\omega}{k} \xi_0 \langle [1 + k\xi_0 \cos(kx - \omega t)] \cos(kx - \omega t) \rangle = \frac{\rho \omega}{2} \xi_0^2$$

By comparing this expression with equation (3.18), one confirms the anticipated relation between the energy and the linear momentum of the wave: $\vec{P} = E\vec{k}/\omega$.

## Problem 3.1.10

Investigate the Rayleigh-Taylor instability for the case of a smooth transition between the two incompressible fluids, when the density variation with height is as follows:

$$\rho(z) = \begin{cases} \rho_2, & z \leq 0 \\ \rho_2 \exp(z/h), & 0 \leq z \leq h \\ \rho_1 = e\rho_2, & z \geq h \end{cases} \tag{3.19}$$

Due to the symmetry in the horizontal $(x, y)$ plane, without loss of generality all perturbations, $(\delta p, \delta \rho, \vec{v})$, can be written in the form $f(z) \exp(ikx + \gamma t)$. Thus, for an incompressible fluid, where $\vec{\nabla} \cdot \vec{v} = 0$, one gets

$$ikv_x + \frac{dv_z}{dz} = 0, \rightarrow v_x = \frac{i}{k}\frac{dv_z}{dz}$$

Furthermore, the linearized continuity equation

$$\frac{\partial \delta \rho}{\partial t} + v_z \frac{d\rho}{dz} = 0$$

yields

$$\delta \rho = -\frac{v_z}{\gamma}\frac{d\rho}{dz}$$

The linearized equation of motion reads $\gamma \rho \vec{v} = -\vec{\nabla}\delta p + \vec{g}\delta\rho$, so its $x$-component, $\gamma \rho v_x = -ik\delta p$, yields

$$\delta p = \frac{i\gamma \rho}{k}v_x = -\frac{\gamma \rho}{k^2}\frac{dv_z}{dz}$$

By substituting now the above expressions for $\delta\rho$ and $\delta p$ into the $z$-component of the linearized equation of motion, one arrives to the following equation for the vertical velocity component $v_z$:

$$\frac{1}{\rho}\frac{d}{dz}\left(\rho\frac{dv_z}{dz}\right) + \left(\frac{gk^2}{\gamma^2\rho}\frac{d\rho}{dz} - k^2\right)v_z = 0$$

It can be written in a more convenient form by using the new unknown function $q(z) = \rho^{1/2}(z)v_z$:

$$\frac{d^2q}{d^2z} - k^2\left[1 - \frac{g}{\gamma^2\rho}\frac{d\rho}{dz} + \frac{1}{k^2\rho^{1/2}}\frac{d^2(\rho^{1/2})}{d^2z}\right]q = 0 \qquad (3.20)$$

For the equilibrium density profile of (3.19) the solution of this equation, which is localized at the transition region, takes the form:

$$q(z) = \begin{cases} \exp(kz), & z \leq 0 \\ A_2\exp(\kappa z) + B_2\exp(-\kappa z), & 0 \leq z \leq h \\ A_3\exp(-kz), & z \geq h, \end{cases}$$

where

$$\kappa^2 = k^2\left(1 - \frac{g}{\gamma^2 h} + \frac{1}{4k^2h^2}\right),$$

and the constants $A_2$, $B_2$, $A_3$ have to be found from the matching conditions at $z = 0$ and $z = h$. The latter follow from equation (3.20) and are comprised of continuity of $q$, and an appropriate discontinuity of its derivative at these points. Indeed, by integrating this equation over an infinitesimal interval around the point $z = 0$, one gets:

$$\int_{0-\epsilon}^{0+\epsilon}\frac{d^2q}{d^2z} - \int_{0-\epsilon}^{0+\epsilon}\frac{q}{\sqrt{\rho}}\frac{d^2(\sqrt{\rho})}{d^2z} = 0,$$

which yields

$$\left(\frac{dq}{dz}\right)_{0+\epsilon} - \left(\frac{dq}{dz}\right)_{0-\epsilon} = \frac{q(0)}{\sqrt{\rho(0)}}\left(\frac{d\sqrt{\rho}}{dz}\right)_{0+\epsilon} = \frac{q(0)}{2h} = \frac{1}{2h}$$

Similarly,

$$\left(\frac{dq}{dz}\right)_{h+\epsilon} - \left(\frac{dq}{dz}\right)_{h-\epsilon} = -\frac{q(h)}{\sqrt{\rho(h)}}\left(\frac{d\sqrt{\rho}}{dz}\right)_{h-\epsilon} = -\frac{q(h)}{2h} = -\frac{A_3\exp(-kh)}{2h}$$

These result in the following relations:

$$A_2 + B_2 = 1, \quad \kappa(A_2 - B_2) - k = \frac{1}{2h}, \quad A_2\exp(\kappa h) + B_2\exp(-\kappa h) = A_3\exp(-kh),$$

$$-kA_3\exp(-kh) - \kappa A_2\exp(-\kappa h) + \kappa B_2\exp(-\kappa h) = -\frac{A_3\exp(-kh)}{2h},$$

which yield

$$A_2 = \frac{(k + \kappa + 1/2h)}{2\kappa}, \quad B_2 = \frac{(-k + \kappa - 1/2h)}{2\kappa},$$
$$A_3 = \exp(kh)(A_2 \exp(\kappa h) + B_2 \exp(-\kappa h))$$

Then, the dispersion relation, which determines the instability growth rate $\gamma$, reads

$$\exp(2\kappa h) = \frac{[(4h^2)^{-1} - (k - \kappa)^2]}{[(4h^2)^{-1} - (k + \kappa)^2]},$$

or, after a simple transformation,

$$\frac{\tanh(\kappa h)}{(\kappa h)} = \frac{2kh}{\left(\frac{1}{4} - k^2 h^2 - \kappa^2 h^2\right)} \tag{3.21}$$

Two different cases, depending on the sign of the parameter $\kappa^2$, have to be considered. Thus, Figure 3.3 depicts functions $f_1(\kappa h)$ and $f_2(\kappa h)$, which represent, respectively, the left- and the right-hand sides of equation (3.21) for the case of $\kappa^2 > 0$. As seen, the sought after solution, $f_1 = f_2$, is possible only if $f_2(0) = 2kh/(1/4 - k^2 h^2) \leq 1$, i.e., $kh \leq (\sqrt{5}/2 - 1) \approx 0.12$. For a long wavelength perturbation with $kh \ll 1$ and, hence, $f_2(0) \ll 1$, the two curves, $f_1$ and $f_2$, intersect close to $(\kappa h) = 1/2$. Therefore, by putting $\kappa h = 1/2 - \epsilon$ with $\epsilon \ll 1$, one gets from equation (3.21) that $\epsilon \approx kh/\tanh(1/2)$. According to the given above definition of $\kappa$, it yields the following instability growth rate:

$$\gamma^2 \approx \frac{gk^2 h}{\epsilon} = gk \tanh(1/2) = gk\frac{(e - 1)}{(e + 1)} = gk\frac{(\rho_1 - \rho_2)}{(\rho_1 + \rho_2)}, \tag{3.22}$$

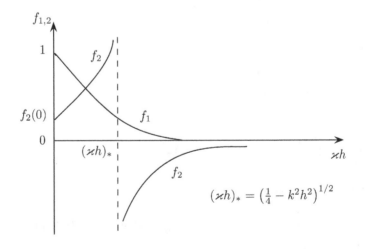

**FIGURE 3.3**
Solution of the dispersion equation in the case of $\kappa^2 > 0$

which is the result obtained in Problem 3.1.8 for the sharp boundary model of the Rayleigh-Taylor instability. When the perturbation wavelength becomes comparable to the width of the transition layer, i.e., for $kh \sim 1$, the instability growth rate saturates at $\gamma \sim \sqrt{g/h}$.

In the case of $\kappa^2 < 0$ the dispersion equation (3.21) can be written as

$$\frac{\tan(\alpha h)}{(\alpha h)} = \frac{2kh}{\left(\frac{1}{4} - k^2h^2 + \alpha^2h^2\right)},$$

where $\alpha = i\kappa$, and the respective functions, $f_1(\alpha h)$ and $f_2(\alpha h)$, are shown in Figure 3.4. It is seen that for a fixed perturbation wavelength, i.e., the parameter $kh$, there is the infinite number of discrete intersections. It means that, unlike the sharp boundary model, in the case of a smooth density variation an infinite set of eigenmodes corresponds to each perturbation wavelength. For a long wavelength one, $(kh \ll 1)$, so that $f_2(0) \approx 8kh \ll 1$, the first root of the equation $f_1 = f_2$ occurs at $(\alpha h)_1 \approx \pi$, which yields the instability increment $\gamma_1 \sim \sqrt{gk}\sqrt{kh}$, which is significantly smaller than that of equation (3.22). The higher order modes have the increment $\gamma_n \sim \gamma_1$, which is gradually decreasing for large $n$. The short wavelength perturbations with $kh \gg 1$ are localized inside the transition region and have a growth rate $\gamma \sim \sqrt{g/h}$. Therefore, the overall conclusion is that the sharp boundary model is adequate for long wavelength perturbations with $kh \ll 1$, while behavior of the short wavelength ones, with $kh \geq 1$, is sensitive to the particular details of the density transition profile.

## Problem 3.1.11

A quasimonochromatic packet of gravity waves, consisting of a large number, $N \gg 1$, of crests and troughs, propagates on a fluid surface toward a light float. How many "ups" and "downs" will the float undergo while the packet passes through?

The vertical displacement of the fluid surface due to this wave packet can be written as

$$\xi(x, t) = F(x - v_g t) \cos(kx - \omega t),$$

where $F(x, t)$ is the packet envelope (see Problem 2.3.4). Since the length of the packet is equal to $L = N\lambda = 2\pi N/k$, and it propagates with the group velocity $v_g$, the float will be forced to move by the passing wave packet during the time interval $\Delta t = L/v_g = 2\pi N/kv_g$. For a quasimonochromatic packet the period of the induced oscillation of the fluid surface is equal to $T = 2\pi/\omega(k)$; therefore, the number of times the float bounces back and forth

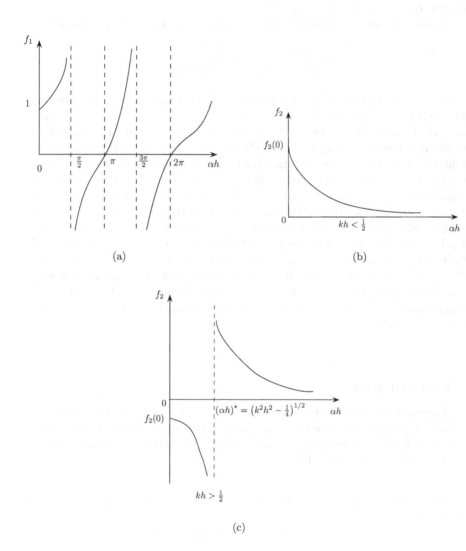

(a)

(b)

(c)

**FIGURE 3.4**
Solution of the dispersion equation in the case of $\alpha^2 \equiv -\kappa^2 > 0$

is equal to

$$n = \frac{\Delta t}{T} = \frac{2\pi N \omega(k)}{2\pi k v_g} = N\frac{\omega(k)}{k v_g} = N\frac{v_{ph}}{v_g}$$

For the gravity wave on the surface of a deep fluid $\omega(k) = \sqrt{gk}$; so its phase velocity $v_{ph} = \omega/k$ is twice as large as its group velocity $v_g = \partial\omega/\partial k$. Hence, $n = 2N$.

## Problem 3.1.12

When the horizontal surface of a deep water undergoes an instant point-like perturbation (for example, the one made by a falling stone), the gravity surface waves start to propagate outwards. As a result, the vertical displacement of the surface represents a set of alternating crests and troughs, which are confined inside an expanding boundary circle of radius $R(t)$. Estimate the value of $R(t)$.

The crests and troughs on the water surface are lines of constant phase of the oscillation, with the phase difference between the adjacent crest and trough being equal to $\pi$. In order to explain the surface pattern under discussion, consider how the phase of the surface oscillation varies with a distance $r$ from the source at a given instant of time $t$. The external pertubation, that occurred at $t = 0$, is a source of gravity waves, which propagate from the origin with the group velocity $v_g = d\omega(k)/dk = \sqrt{g/k}/2$. Therefore, at the instant $t$, the distance $r$ is reached by such a wave, whose group velocity is equal to $r/t$; hence, its wave vector $k(r, t) = gt^2/4r^2$, and the phase of the oscillation is given by

$$\phi(r,\, t) = \omega(k)t - kr = \frac{gt^2}{4r} \tag{3.23}$$

This expression implies that at a given moment $t$ the phase varies very sharply at small distances, which indicates a large number of crests and troughs there. On the other hand, at large $r$ its variation is smooth: $\phi \to 0$ at $r \to \infty$, and it acquires the value of $\pi$ at $r = r_* = gt^2/4\pi$. This means that the variation of the phase from $r > r_*$ to $r \to \infty$ is less than $\pi$ and, therefore, there are no crests and troughs in this domain. Hence, the sought after $R(t) \sim r_* \sim gt^2$.

## Problem 3.1.13

Determine the wave pattern formed on the surface of a deep water by a point source moving there with a constant speed $V$ (the **Kelvin wedge**.)

Let the water surface be the $(x, y)$ plane, with the source moving to the left along the $x$-axis and located at the origin of this coordinate system at $t = 0$. Then, the perturbation caused by the moving source in the point with the surface coordinates $(x, y)$ at $t = 0$ can be viewed as a superposition of instantaneous point-like pulses (see the previous Problem), which originate from each point of the source's preceding trajectory, a straight line $y = 0$, $0 \le x < +\infty$. Thus, by introducing the "retardation" time $\tau$, which corresponds to the source location at $x = V\tau$, one gets, according to equation (3.23), that this elementary perturbation has the phase (see Figure 3.5):

$$\phi(x, y, t = 0)_\tau = \frac{g\tau^2}{4r_*} = \frac{g\tau^2}{4\sqrt{y^2 + (V\tau - x)^2}} \tag{3.24}$$

As seen from the above expression, this phase depends on $\tau$, which means that, in a general case, the perturbations originating from the neighboring points of the trajectory interfere destructively, i.e., cancel each other. Therefore, the only significant perturbation is associated with the pulse that comes from the point of a "stationary phase," where $d\phi/d\tau = 0$. According to equation (3.24), this condition reads

$$\frac{g\tau}{2r_*} = \frac{g\tau^2 V(V\tau - x)}{4r_*^3},$$

which allows a simple physical interpretation. Indeed, since $r_*/\tau = v_{gr} = v_{ph}/2$, the stationary phase requirement implies that

$$\frac{(V\tau - x)}{r_*} = \cos\theta = \frac{2r_*}{V\tau} = \frac{v_{ph}}{V},$$

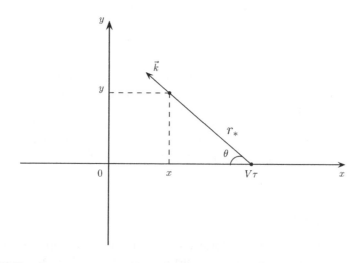

**FIGURE 3.5**
Kinematics of the point-like perturbation

which means that the wave vector $\vec{k}$ of the incoming perturbation is directed in such a way that $\cos\theta = -k_x/k = v_{ph}/V$, the usual Cherenkov resonance condition (see Problem 2.4.1). However, since the gravity waves are dispersive, i.e., their phase velocity depends on the wavelength, the angle $\theta$ varies along a line of the constant phase (such as a crest or a trough line), because the waves of different wavelengths constructively interfere at different points along such a line.

By using the derived above expressions for the phase $\phi$ and $\cos\theta$, one can represent the lines of constant phase in parametric form as follows:

$$x = V\tau - r_* \cos\theta = \frac{V^2\phi}{g} \cos\theta(2 - \cos^2\theta),$$

$$y = r_* \sin\theta = \frac{V^2\phi}{g} \sin\theta \cos^2\theta \qquad (3.25)$$

The resulting pattern is depicted in Figure 3.6. A wake just behind the source corresponds to the angle $\theta$ close to $\pi/2$, which means waves with a small phase velocity ($v_{ph} \ll V$, the short wavelength perturbations). On the other hand, at the point $A$ the angle $\theta = 0$, so there $v_{ph} = V$, and $k = g/V^2$. All lines of crests and troughs are geometrically similar, and they differ from each other only by changing the phase $\phi$ in equations (3.25) by $2\pi$. The entire wave pattern is confined inside the wedge of angle $2\alpha_0$, which can be determined by requiring the maxima of $x$ and $y$ at the point $B$ on Figure 3.6. It follows

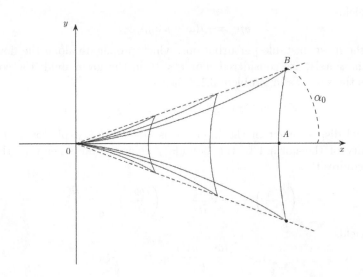

**FIGURE 3.6**
The Kelvin wedge pattern

then from relation (3.25), that $\sin \theta_B = 1/\sqrt{3}$, therefore

$$\tan \alpha_0 = \frac{y_{max}}{x_{max}} = \frac{\sin \theta_B \cos \theta_B}{(2 - \cos^2 \theta_B)} = \frac{1}{2\sqrt{2}},$$

which yields $\alpha_0 \approx 19.5$ degrees.

## Problem 3.1.14

Investigate the stability of a horizontal interface between two ideal incompressible fluids in a gravity field $\vec{g}$. The upper fluid is much lighter than the lower one, $\rho_1 \ll \rho_2$, and flows with a uniform horizontal velocity equal to $V$. The surface tension coefficient of the interface is $\alpha$.

In the absence of flow, the interface perturbations represent the gravity-capillary surface waves with the dispersion law $\omega(k) = \sqrt{gk + \alpha k^3 / \rho}$. Thus, the question of interest is when the flow of the upper light fluid can make these perturbations unstable. Thus, by following Problem 3.1.8, in the linear approximation one gets for $z > 0$, the upper fluid,

$$\phi_1 = A \exp(-kz) \exp[i(kx - \omega t)],$$

$$\rho_1 \left( \frac{\partial \vec{v}}{\partial t} + V \frac{\partial \vec{v}}{\partial x} \right) = -\vec{\nabla} \delta p_1,$$

which yields

$$\delta p_1 = -i(\omega - kV) \rho_1 \phi_1,$$

where the most unstable perturbations, which propagate along the flow, i.e., along the $x$-axis, are considered. For $z < 0$, in the lower fluid, the solution remains the same as in Problem 3.1.8, hence

$$\phi_2 = B \exp(kz) \exp[i(kx - \omega t)], \quad \delta p_2 = -i\omega \rho_2 \phi_2$$

If vertical displacement of the interface is $\xi(x, t) = \xi_0 \exp[i(kx - \omega t)]$, the continuity of the normal to the boundary velocity components of the two fluids requires that

$$-\left( \frac{\partial \phi_1}{\partial z} \right)_{z=0} - V \frac{\partial \xi}{\partial x} = -\left( \frac{\partial \phi_2}{\partial z} \right)_{z=0} = \frac{\partial \xi}{\partial t},$$

which yields

$$A = -i \frac{(\omega - kV)}{k} \xi_0, \quad B = i \frac{\omega}{k} \xi_0$$

Then, the sought after dispersion relation, which is modified by the flow, follows from Laplace's formula for the interface pressure balance:

$$(p_1 + \delta p_1)_{z=\xi} - (p_2 + \delta p_2)_{z=\xi} = \frac{\alpha}{R},$$

which yields the following dispersion law (see Problem 3.1.8 for more details):

$$\omega^2 - 2\epsilon kV\omega - \left(gk + \frac{\alpha k^3}{\rho_2} - \epsilon k^2 V^2\right) = 0,$$

where $\epsilon \equiv \rho_1/\rho_2 \ll 1$. Its solution, which is

$$\omega_{1,2} \approx \epsilon kV \pm \left(gk + \frac{\alpha k^3}{\rho_2} - \epsilon k^2 V^2\right)^{1/2},$$

shows that the instability, which requires $\Im\omega > 0$, occurs when

$$V > \epsilon^{-1/2}\left(\frac{g}{k} + \frac{\alpha k}{\rho_2}\right)^{1/2} = \epsilon^{-1/2}v_{ph}(k),$$

where $v_{ph}(k)$ is the phase velocity for the gravity-capillary waves on the free surface of the lower fluid. This phase velocity depends on the wave vector $k$ and acquires the minimum at $k = k_* = \sqrt{g\rho_2/\alpha}$. Thus, the perturbation with this wave vector is the most unstable one, and the respective critical velocity $V_{cr}$ is equal to $V_{cr} = \epsilon^{-1/2}v_{ph}^{(min)} = \epsilon^{-1/2}(4\alpha g/\rho_2)^{1/4}$.

However, it is worth noting that such an instability, which is the particular example of the hydrodynamic **Kelvin-Helmholtz instability**, has, actually, nothing to do with the generation of the surface gravity-capillary waves. Indeed, the derived above critical velocity depends on the density of the light upper fluid; therefore, this instability requires a strong coupling of the two fluids, which, in fact, leaves no room for the gravity-capillary surface waves. In simple physical terms it can be explained as follows. In the case of the free surface of the lower fluid, when $\rho_1 = p_1 = \delta p_1 = 0$, the pressure perturbation at its surface is equal, according to Laplace's formula, to $\delta p_2 = \rho_2 g\xi + \alpha k^2\xi$; therefore, the $x$-component of the equation of motion yields

$$\rho_2\frac{\partial v_x}{\partial t} \sim \omega v_x = -(\vec{\nabla}\delta p_2)_x \sim k(\rho_2 g + \alpha k^2)\xi$$

Since, according to equation (3.16), $v_x \sim (\partial\phi_2/\partial x) \sim (\partial\phi_2/\partial z) \sim v_z \sim \omega\xi$, it leads to $\omega^2\xi \approx k(\rho_2 g + \alpha k^2)\xi$, i.e., to the above derived dispersion law for the gravity-capillarly waves. When the upper fluid with $\rho_1 \ll \rho_2$ is moving, the main contribution to $\delta p_1$ comes from the variation of the flow velocity due to the surface wave. The point is that since the upper fluid should flow along the interface, which is deformed by the surface wave, its otherwise horizontal flow with velocity $V_x = V$ acquires the vertical component $v_z \sim V\xi/\lambda \sim V(k\xi)$. It follows then from expression (3.16) for the flow potential that the horizontal velocity component is similarly perturbed: $v_x \sim v_z \sim V(k\xi)$. Then, according to Bernoulli's equation, the respective pressure perturbation in the moving light upper fluid is $\delta p_1 = \Delta(\rho_1 v_1^2/2) \sim \rho_1 V v_x \approx -\rho_1 V^2(k\xi)$. Consequently, the equal additional pressure perturbation appears also in the lower fluid, giving the following evolution equation for $\xi$: $\omega^2\xi \approx k(\rho_2 g + \alpha k^2 - k\rho_1 V^2)\xi$.

Thus, the flow of the upper fluid leads to the reduction of the combined (gravity+capillarity) restoring force, which makes these surface waves. In other words, this flow results in the "stiffness" reduction of the interface: as its speed increases, it first reduces the wave frequency, and eventually brings about a "negative stiffness," i.e., instability, when $V^2 > (\rho_2 g + \alpha k^2)/k\rho_1$.

The irrelevance of the Kelvin-Helmholtz instability to the generation of surface waves can be also demonstrated by the following quantitative example. Consider the water-air interface, for which $\rho_1 \approx 1.2 kg/m^3, \rho_2 \approx 10^3 kg/m^3, \alpha \approx 7.3 \times 10^{-2} N/m, g \approx 9.8 m/s^2$. In this case the minimum phase velocity of the surface waves occurs at $k_* \approx 3.7 \times 10^2 m^{-1}$, which corresponds to a ripple with wavelength $\lambda_* \approx 1.7 cm$. Then, the density ratio $\epsilon \approx 1.2 \times 10^{-3}$ yields the critical air flow velocity of $V_{cr} \approx 6.6 m/s$, which is quite a strong wind by itself. However, it requires almost a gale force of $V \approx 36 m/s$ to generate a gravity wave with $\lambda \sim 1m$, which is in apparent contradiction with the fact that even a light breeze makes such a wave. Thus, another mechanism, which is the resonant interaction between the air flow and the surface waves (see next Problem), is actually behind this natural phenomenon.

## Problem 3.1.15

Investigate the resonant interaction of air flow and gravity-capillary waves on the water surface.

This process cannot be understood in the framework of the sharp-boundary model, which assumes a uniform airflow above the water surface. In reality, a finite width boundary layer with a sheared air flow is always present, and this is at the heart of the resonant mechanism under discussion. Inside this layer the air velocity along, say, the $x$-axis, $V(z)$, is equal to zero at the water surface and gradually increases with height. If at some elevation, $z = z_r$, the velocity of air, $V(z_r)$, becomes equal to the phase velocity of a surface wave propagating in the same direction, the resonant exchange of energy and linear momentum between the air and the wave occurs, which can cause the amplitude of the wave to grow exponentially in time (the surface wave instability). Since the air flow velocity required for the resonant interaction is small compared to the one necessary for the Kelvin-Helmholtz instability (see Problem 3.1.14), such a wind has no effect on the gravity-capillarity restoring force, so the surface wave's dispersion law $\omega(k) = \sqrt{gk + \alpha k^3/\rho_w}$ remains unchanged. In other words, unlike the Kelvin-Helmholtz instability, this wave generation mechanism represents a weak wind-wave coupling, when a wave is amplified by a slow transfer of energy and momentum from the airflow.

The essential role here belongs to the vorticity of the airflow $\vec{\Omega} = \vec{\nabla} \times \vec{v}$, which is non-zero due to the initial sheared velocity profile $V(z)$ and, according to Problem 3.1.1, is conserved in the course of its subsequent evolution.

The velocity of air can be considered as a superposition of the initial shear component, $V_x(z)$, and the perturbation caused by the surface wave:

$$\vec{v} = V(z)\vec{e}_x + \vec{u}(z)\sin(kx - \omega t) \tag{3.26}$$

Then, by representing the equation of motion for an incompressible ideal fluid as

$$\frac{\partial \vec{v}}{\partial t} = (\vec{v} \times \vec{\Omega}) - \vec{\nabla}\left(\frac{p}{\rho} + \frac{v^2}{2} + gz\right), \tag{3.27}$$

the rate of change of the $x$-component of the linear momentum of air per unit horizontal area can be evaluated as

$$\frac{dP_{ax}}{dt} = \int_0^\infty dz \rho_a \left\langle \frac{\partial v_x}{\partial t} \right\rangle =$$

$$\int_0^\infty dz \rho_a \left\langle (\vec{v} \times \vec{\Omega})_x \right\rangle = -\rho_a \int_0^\infty dz \left\langle v_z \Omega \right\rangle, \tag{3.28}$$

where $\Omega_y = \Omega(x, z)$ is the only non-zero component of the vorticity in this case, and the symbol $\langle \rangle$ means averaging over the $x$-coordinate (note that the last term in equation (3.27) is the gradient of a periodic function and makes no averaged contribution). Since, according to expression (3.26), $< v_z >= 0$, in order to get a non-zero result in equation (3.28), one should take into account the variations of vorticity, which are present because $\Omega$ is conserved not locally but for a moving fluid element. Thus, it is convenient to consider the trajectory of such an air element, say, the one with initial, at $t = 0$, location $x = x_0$, $z = z_0$. Due to the initial flow with velocity $V(z_0)$, its $x$-coordinate is then evolving as $x(t) = x_0 + V(z_0)t$. Therefore, the vertical velocity acquired by this element due to the surface wave is equal, according to expression (3.26), to

$$\frac{dz}{dt} = u_z(z_0)\sin[k(x_0 + V(z_0)t) - \omega t] = u_z(z_0)\sin[kx_0 - (\omega - kV(z_0))t]$$

Therefore, it also undergoes a vertical displacement $\Delta z$, so its elevation varies with time as

$$z = z_0 + \Delta z = z_0 + \frac{u_z(z_0)}{[\omega - kV(z_0)]} \times$$

$$\{\cos[kx_0 - (\omega - kV(z_0))t] - \cos(kx_0)\} \tag{3.29}$$

It results in variations of the local vorticity because air elements with different initial location $z_0$ reach a fixed elevation $z$ and, hence, bring there their "own" conserved vorticity $\Omega_0(z_0) = V'(z_0)$, which is determined by the initial background flow $V(z)$. Thus, it follows from (3.29) that

$$\Omega(z) = \Omega_0(z_0) = V'(z - \Delta z) \approx V'(z) - V''(z)\Delta z =$$

$$V'(z) - V''(z)\frac{u_z(z)}{[\omega - kV(z)]} \times \{\cos[kx_0 - (\omega - kV(z))t] - \cos(kx_0)\};$$

so the integrand in equation (3.28) can be written as

$$\langle v_z \Omega \rangle = -\langle u_z(z) \sin[kx_0 - (\omega - kV(z))t] \times$$

$$V''(z)\frac{u_z(z)}{[\omega - kV(z)]} \times \{\cos[kx_0 - (\omega - kV(z))t] - \cos(kx_0)\}\rangle$$

After averaging the above expression over $x_0$ (using $< \sin^2(kx_0) > = < \cos^2(kx_0) > = 1/2$, $< \sin(kx_0)\cos(kx_0) > = 0$) and inserting the result into (3.28), one gets the following rate of change for the linear momentum of air:

$$\frac{dP_{ax}}{dt} = \frac{\rho_a}{2} \int\limits_0^\infty dz u_z^2 V''(z)\frac{\sin([\omega - kV(z)]t)}{[\omega - kV(z)]} \tag{3.30}$$

Although the integration in (3.30) is formally carried out over the entire half-space $z > 0$, the main contribution comes from a narrow layer around the elevation $z = z_r$, where the resonant condition $\omega - kV(z_r) = 0$ is satisfied. The reason is that the factor

$$f(z) = \frac{\sin([\omega - kV(z)]t)}{[\omega - kV(z)]}$$

in the integrand of equation (3.30) is sharply peaked at $z = z_r$, with amplitude increasing with time and decreasing width. Therefore, after several wave periods, it can be approximated by the $\delta$-function as

$$f(z) \approx \pi\delta[\omega - kV(z)] = \frac{\pi}{k|V'(z_r)|}\delta(z - z_r),$$

which yields the momentum transfer rate equal to

$$\frac{dP_{ax}}{dt} = \frac{\pi\rho_a}{2k}\frac{u_z^2(z_r)}{|V'(z_r)|}V''(z_r) \tag{3.31}$$

(compare it with the linear momentum exchange between the Langmuir plasma wave and the resonant electrons, considered in Problem 2.1.17 for derivation of the Landau damping of the wave).

It follows then from this expression that, if $V''(z_r) < 0$, the momentum of the airflow is transferred to the surface wave, thus providing the mechanism for the wave generation. If, however, $V''(z_r) > 0$, the momentum transfer is in the opposite direction, which leads to the wave damping. It may seem at first glance that whether or not surface waves are generated by wind is decided more or less by chance, depending on the particular details of the air velocity profile $V(z)$. However, nature is strongly in favor of the wave generation. The reason is that viscosity of air is so small ($\eta_a \approx 1.8 \times 10^{-5} N/s/m^2$) that even a weak wind has a high Reynolds number (for example, a speed of $V \sim 1m/s$ with a scale length $L \sim 10m$ yields the Reynolds number $R = LV\rho_a/\eta_a \sim 10^6$). Therefore, a real airflow is usually turbulent, in which case the above function $V(z)$ should be understood as a mean velocity profile that is averaged over numerous turbulent eddies. It is well known that such a velocity profile possesses a universal feature: the "logarithmic law" (see, e.g., L. D. Landau and E. M. Lifshitz, *Fluid Dynamics*, §42, Pergamon, 1987), for which $V(z) \propto \ln z$, and hence $V''(z) < 0$. Thus, a wind above a water surface does make waves, provided that the resonant condition $V = v_{ph}$ holds at some elevation, which is not too far from the water surface (see below). In reality, this condition is easily achieved, as, for example, a wind of $V \geq 0.23 m/s$ is sufficient to ripple a water surface, while $V \geq 1.25 m/s$ can generate gravity waves with $\lambda \sim 1m$. So far, only waves propagating parallel to the wind were considered. However, the same generation mechanism also works for waves propagating at a non-zero angle with respect to the wind. In this case the resonant condition reads $\omega - \vec{k} \cdot \vec{V}(z_r) = 0$, where both $\vec{k}$ and $\vec{V}$ are two-dimensional vectors in the horizontal plane.

Thus, the important conclusion is that the critical speed of wind required for the resonant generation of surface waves is determined only by the wave phase velocity and, therefore, does not depend at all on the density of air. Where $\rho_a$ does matter is the rate of this generation: according to equation (3.31), the rate of the momentum transfer is proportional to $\rho_a$, and, therefore, so is the surface wave growth rate $\gamma$. In order to derive it, one should compare expression (3.31) with the linear momentum carried by the surface wave itself, which, according to Problem 3.1.9, is equal to $P_x = \rho_w \omega \xi_0^2 / 2$. Thus, the sought after growth rate is

$$\gamma = -\frac{1}{2P_x}\frac{dP_{ax}}{dt} = -\omega\pi\frac{\rho_a}{\rho_w}\frac{u_z^2(z_r)}{\omega^2\xi_0^2}\frac{V''(z_r)}{k|V'(z_r)|}$$

Consider now its simple quantitative estimate. Spatial derivatives of the background air flow can be estimated as $V'(z_r) \sim V/z_r, V'' \sim V/z_r^2$, while $\omega\xi_0$ is $u_z(0)$, the vertical velocity amplitude at the water surface. Thus,

$$\frac{\gamma}{\omega} \sim \frac{\pi}{kz_r}\frac{\rho_a}{\rho_w}\frac{u_z^2(z_r)}{u_z^2(0)}$$

Since the perturbation of the air flow due to the surface wave decays exponentially with height, the resonant mechanism under discussion is effective only if $kz_r \lesssim 1$ (otherwise the ratio $u_z^2(z_r)/u_z^2(0)$ is very small). Therefore, for $kz_r \sim 1$ one gets $\gamma/\omega \sim \rho_a/\rho_w \sim 10^{-3}$, i.e., the growth time is about a hundred of the periods $T = 2\pi/\omega$.

Two final remarks are due here. Firstly, since $\gamma/\omega \ll 1$, the wave amplitude increases only slightly over the wave period $T$. This justifies the above-given derivation of the momentum transfer, where it was tacitly assumed that the amplitude of the wave is a constant. Secondly, this linear resonant mechanism describes only the primary transfer of energy and momentum from the air flow to surface waves, while saturation of their amplitudes is determined by the non-linear wave-wave interactions (see, e.g., V. E. Zakharov, Kolmogorov spectra in weak turbulence problems, in *Handbook of Plasma Physics*, vol.2, North-Holland, 1983).

## Problem 3.1.16

Derive the force exerted on a rigid sphere of radius $a$, which is moving in an ideal incompressible fluid. Assume that the flow around the sphere is a potential one.

For a potential flow of an incompressible fluid, when $\vec{v} = -\vec{\nabla}\phi$, the flow potential satisfies the Laplace equation $\vec{\nabla}^2\phi = 0$ with the boundary condition

$$v_n|_S = -\vec{\nabla}\phi \cdot \vec{n}|_S = (\dot{\vec{R}} \cdot \vec{n})_S,$$

where $\vec{R}(t)$ is location of the sphere's center, and $\vec{n}$ is a unit vector directed outwards normally to the surface of the sphere. The respective solution (see Problem 1.0.6) reads:

$$\phi(\vec{r}, t) = \frac{\alpha \dot{\vec{R}} \cdot \vec{\rho}}{\rho^3}, \quad \vec{\rho} = \vec{r} - \vec{R},$$

$$\vec{v} = -\vec{\nabla}\phi = \frac{\alpha}{\rho^3}[-\dot{\vec{R}} + 3\vec{n}(\dot{\vec{R}} \cdot \vec{n})], \tag{3.32}$$

$$\alpha = \frac{a^3}{2}$$

The sought after force, $\vec{F}_f$, is determined entirely by the fluid pressure distribution at the surface of the sphere:

$$\vec{F}_f = -\int pd\vec{S} = -\int p\vec{n}dS \tag{3.33}$$

The pressure itself can be obtained from the equation of motion with the known velocity field (3.32):

$$\frac{\partial \vec{v}}{\partial t} + (\vec{v} \cdot \vec{\nabla})\vec{v} = -\frac{1}{\rho_0}\vec{\nabla}p,$$

where $\rho_0$ is the fluid density. Since for a potential flow $(\vec{v} \cdot \vec{\nabla})\vec{v} = \vec{\nabla}(v^2/2)$, we have

$$p = p_0 - \rho_0\frac{v^2}{2} + \rho_0\frac{\partial \phi}{\partial t},$$

where $p_0$ is the pressure at infinity, where the fluid is at rest. Furthermore, since

$$\frac{\partial \phi}{\partial t} = \frac{\alpha}{\rho^2}(\ddot{\vec{R}} \cdot \vec{n}) + \frac{3\alpha}{\rho^3}(\dot{\vec{R}} \cdot \vec{n})^2 - \frac{\alpha}{\rho^3}(\dot{\vec{R}})^2, \tag{3.34}$$

$$v^2 = \frac{\alpha^2}{\rho^6}[(\dot{\vec{R}})^2 + 3(\dot{\vec{R}} \cdot \vec{n})^2], \tag{3.35}$$

it turns out that the only non-zero contribution to the integral in (3.33) comes from the term in Equation (3.34), which is proportional to the sphere acceleration $\ddot{\vec{R}}$:

$$\vec{F}_f = -\frac{\alpha\rho_0}{a^2}\int(\ddot{\vec{R}}\cdot\vec{n})\vec{n}dS = -\frac{2\pi\rho_0 a^3}{3}\ddot{\vec{R}} \tag{3.36}$$

Thus, if the sphere of mass $M$ is moving in a fluid due to some external force $\vec{F}_{ext}$, its equation of motion, which is $M\ddot{\vec{R}} = \vec{F}_{ext} + \vec{F}_f$, can be written as $(M + \tilde{m})\ddot{\vec{R}} = \vec{F}_{ext}$, where $\tilde{m} = 2\pi\rho_0 a^3/3$ is the so-called **induced mass**.

In physical terms, the appearance of such an induced extra mass can be explained as follows. The force (3.36), if non-zero, means that a power equal to $-\vec{F}_f\cdot\dot{\vec{R}}$ is transferred to the fluid. However, in an ideal fluid there is no energy dissipation. Furthermore, since the fluid is incompressible, its internal energy does not change, and there is no emission of sound waves. Thus, the only remaining form of energy that is able to account for the energy conservation law is the kinetic energy of a fluid $E_{kin} = \int\rho_0 v^2 dV/2$, so that

$$\frac{dE_{kin}}{dt} = -\vec{F}_f\cdot\dot{\vec{R}} \tag{3.37}$$

In the case of a sphere this kinetic energy, according to Problem 1.0.6, is equal to $E_{kin} = \pi\rho_0 a^3(\dot{\vec{R}})^2/3$, so that the derived-above force (3.36) indeed satisfies the energy balance requirement (3.37). Clearly, such an energy based consideration is applicable to any body, not only a sphere, moving in an ideal incompressible fluid. Therefore, the relation (3.37), together with the formula (1.22) of Problem 1.0.6 for the kinetic energy of a fluid, enables one to determine the force exerted on the body without a direct calculation of this force (which is hardly possible for a body of arbitrary shape). Thus, since the dipole vector $\vec{A}$ in equation (1.22) must be proportional to the velocity $\vec{u}$ of the body, its most general form is $A_i = -\alpha_{ik}u_k$, and the kinetic energy (1.22) can be written as

$$E_{kin} = \frac{\tilde{m}_{ik}u_i u_k}{2}, \quad \tilde{m}_{ik} = -\rho_0(4\pi\alpha_{ik} + V_0\delta_{ik}) \tag{3.38}$$

It follows then from expression (3.37) that the force exerted on the body is equal to $F_i = -\tilde{m}_{ik}\dot{u}_k$, where $\tilde{m}_{ik}$ is called the **induced mass tensor**. Thus, in a general case for arbitrary shape of the moving body, this force depends not only on the absolute magnitude of the body acceleration, but also on its direction (see the next Problem).

## Problem 3.1.17

Derive the induced mass tensor for the body shown in Figure 3.7: a long dumbbell made by two spheres of radius $a$, which are attached to each other by a long, $l \gg a$, thin rigid wire.

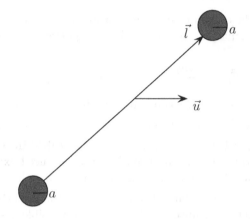

**FIGURE 3.7**

A long, $a \ll l$, dumbbell moving in a fluid

Assume that velocity of the dumbbell is equal to $\vec{u}$. Then, the flow of a fluid around the sphere 1 results in an additional fluid velocity $\vec{v}_{12}$ at the location of the sphere 2, which, according to equation (3.32), is equal to

$$\vec{v}_{12} = \frac{a^3}{2l^3} \left[ -\vec{u} + 3\vec{n}(\vec{n} \cdot \vec{u}) \right], \ \vec{n} = \frac{\vec{l}}{l}$$

This is equivalent to the sphere 2 moving relative to the fluid with the velocity

$$\vec{u}_2 = \vec{u} - \vec{v}_{12} = \vec{u} + \frac{a^3}{2l^3} \left[ \vec{u} - 3\vec{n}(\vec{n} \cdot \vec{u}) \right]$$

Since this additional velocity is the even function of $\vec{n}$, it is the same for the sphere 1, hence $\vec{u}_1 = \vec{u}_2$, and the fluid flow velocity potential can be written as

$$\phi(\vec{r}) \approx \frac{a^3}{2} \frac{\vec{u}_1 \cdot (\vec{r} - \vec{l}/2)}{|\vec{r} - \vec{l}/2|^3} + \frac{a^3}{2} \frac{\vec{u}_2 \cdot (\vec{r} + \vec{l}/2)}{|\vec{r} + \vec{l}/2|^3}, \tag{3.39}$$

with the origin of the coordinate system placed at the center of the dumbbell. Then, according to Problem 3.1.16, in order to determine the induced mass tensor, one needs to find vector $\vec{A}$, which determines the dipole term in the asymptotic expansion of the flow potential (3.39) at large distances, $r \gg l$. Thus, its simple comparison with expression (3.32) yields that in this case

$$\vec{A} = -\frac{a^3}{2}(\vec{u}_1 + \vec{u}_2) = -a^3 \left[ \left( 1 + \frac{a^3}{2l^3} \right) \vec{u} - \frac{3a^3}{2l^3} \vec{n}(\vec{n} \cdot \vec{u}) \right],$$

or, in tensor notation,

$$A_i = -\alpha_{ik} u_k, \ \alpha_{ik} = a^3 \left( 1 + \frac{a^3}{2l^3} \right) \delta_{ik} - \frac{3a^3}{2l^3} a^3 n_i n_k$$

Then, according to relation (3.38), the sought after induced mass tensor reads

$$\tilde{m}_{ik} = -\rho_0(4\pi\alpha_{ik} + V_0\delta_{ik}) =$$
$$\frac{4\pi\rho_0 a^3}{3}\left[\left(1 + \frac{3a^3}{2l^3}\right)\delta_{ik} - \frac{9a^3}{2l^3}n_i n_k\right]$$

As seen from this expression, it is easier to accelerate the immersed dumbbell along its axis than across it, because in the first case, when $\dot{\vec{u}} \parallel \vec{n}$, the respective induced mass $\tilde{m}_\parallel = \frac{4\pi\rho_0 a^3}{3}\left(1 - \frac{3a^3}{l^3}\right)$ is smaller than for $\dot{\vec{u}} \perp \vec{n}$, when $\tilde{m}_\perp = \frac{4\pi\rho_0 a^3}{3}\left(1 + \frac{3a^3}{2l^3}\right)$.

## Problem 3.1.18

A pendulum of length $l$ is made with a small sphere of density $\rho_s$. Find the pendulum frequency, if the sphere is immersed into an ideal incompressible fluid of density $\rho_0 < \rho_s$.

In free space the frequency of the pendulum is equal to $\omega_0 = \sqrt{g/l}$. The effect of the fluid is two-fold. Firstly, the buoyancy force results in a reduction of the effective gravity acceleration by the factor of $(\rho_s - \rho_0)/\rho_s$. Secondly, although in free space the frequency does not depend on the mass of the sphere, it is not so in the fluid. The reason is in the induced mass, which adds only to the inertial mass of a body, but not to its gravitational mass. Hence, another reduction by the factor of $\sqrt{M/(M + \tilde{m})}$, where $M = 4\pi\rho_s a^3/3$ is the mass of the sphere, and $\tilde{m} = 2\pi\rho_0 a^3/3$ is its induced mass in the fluid. Altogether, one gets

$$\omega = \omega_0 \left(1 - \frac{\rho_0}{\rho_s}\right)^{1/2} \left(1 + \frac{\rho_0}{2\rho_s}\right)^{-1/2}$$

## Problem 3.1.19

A very light cylinder of radius $r$ is immersed into a vertical pipe of radius $R$, which is filled with water and plugged at both ends. Determine the vertical acceleration of the cylinder due to gravity, assuming that the cylinder length, $L$, is long compared to $R$.

The cylinder acceleration $a$ should be established at such a value that the buoyancy force becomes balanced by the force which is due to the induced mass $\tilde{m}$ of the fluid involved in the flow around the cylinder (the own weight and inertia of the cylinder are negligibly small). Hence, $\pi r^2 L \rho g = \tilde{m} a$, where $\rho$ is the density of the fluid. According to Problem 3.1.16, to evaluate $\tilde{m}$ one has to calculate the kinetic energy of the fluid, which can be done in the following way. If the cylinder is moving up with a velocity equal to $u$, a counterflow of the fluid occurs due to its incompressibility. In the case of a long cylinder the edge effects are small, and one can assume a uniform counterflow through the annular cross-sectional area of $\pi(R^2 - r^2)$ with a velocity $v = u\pi r^2/\pi(R^2 - r^2) = ur^2/(R^2 - r^2)$. Therefore, the respective kinetic energy, $E_{kin}$, is equal to

$$E_{kin} = \frac{\rho v^2}{2} L\pi(R^2 - r^2) = \frac{\rho u^2}{2} \frac{\pi r^4 L}{(R^2 - r^2)}$$

Since $E_{kin} = \tilde{m} u^2/2$, it yields the induced mass

$$\tilde{m} = \frac{\rho \pi r^4 L}{(R^2 - r^2)},$$

and, hence, the sought after acceleration

$$a = g\frac{\pi r^2 L \rho}{\tilde{m}} = g\frac{(R^2 - r^2)}{r^2}$$

---

## 3.2   Viscous fluids

In a viscous fluid the momentum flux tensor $\Pi_{ik}$ takes the form

$$\Pi_{ik} = p\delta_{ik} + \rho v_i v_k + \pi_{ik}, \tag{3.40}$$

where the term $\pi_{ik}$ accounts for the viscous transfer of linear momentum in a flow with a non-uniform velocity field. If the spatial scale of the flow, $L$, is long compared to the microscopic molecular mean free path $l$, the tensor $\pi_{ik}$ is determined by the first order spatial derivatives of the velocity vector field, and for an isotropic medium its most general form is as follows:

$$\pi_{ik} = -\eta\left(\frac{\partial v_i}{\partial x_k} + \frac{\partial v_k}{\partial x_i} - \frac{2}{3}\delta_{ik}\frac{\partial v_l}{\partial x_l}\right) - \xi\delta_{ik}\frac{\partial v_l}{\partial x_l}, \tag{3.41}$$

where $\eta$ and $\xi$ are the coefficients of viscosity, both of which are positive due to the second law of thermodynamics (see Problem 3.2.1). By substituting expressions (3.40) and (3.41) into the linear momentum conservation equation (3.4), and considering the viscosity coefficients as constants, one arrives to the equation of motion for a viscous fluid:

$$\rho\left[\frac{\partial \vec{v}}{\partial t} + (\vec{v} \cdot \vec{\nabla})\vec{v}\right] = -\vec{\nabla}p + \eta\vec{\nabla}^2\vec{v} + \left(\xi + \frac{1}{3}\eta\right)\vec{\nabla}(\vec{\nabla} \cdot \vec{v})$$

In the particular case of an incompressible fluid, when $\vec{\nabla} \cdot \vec{v} = 0$, it reduces to the **Navier-Stokes equation**

$$\frac{\partial \vec{v}}{\partial t} + (\vec{v} \cdot \vec{\nabla})\vec{v} = -\frac{1}{\rho}\vec{\nabla}p + \nu\vec{\nabla}^2\vec{v}, \tag{3.42}$$

where $\nu \equiv \eta/\rho$ is the kinematic viscosity of a fluid.

## Problem 3.2.1

Derive the energy dissipation rate in a viscous incompressible fluid.

In an incompressible fluid, where there are no compressions or rarefactions in the flow, the internal energy of any fluid element does not vary with time. Therefore, the dissipated energy can be tapped only from the kinetic energy of the flow. Thus, the rate of change for the total kinetic energy within some fixed volume, $E_{kin} = \int_V dV \rho v^2/2$, is equal to

$$\frac{dE_{kin}}{dt} = \int_V dV \rho v_i \frac{\partial v_i}{\partial t} = \int_V dV v_i \left( -\rho v_k \frac{\partial v_i}{\partial x_k} - \frac{\partial p}{\partial x_i} - \frac{\partial \pi_{ik}}{\partial x_k} \right) \tag{3.43}$$

Since

$$v_i \frac{\partial \pi_{ik}}{\partial x_k} = \frac{\partial}{\partial x_k}(v_i \pi_{ik}) - \pi_{ik}\frac{\partial v_i}{\partial x_k},$$

and

$$\rho v_i \left( v_k \frac{\partial v_i}{\partial x_k} + \frac{1}{\rho}\frac{\partial p}{\partial x_i} \right) = \frac{\partial}{\partial x_k}\left[ \rho v_k \left( \frac{v^2}{2} + \frac{p}{\rho} \right) \right]$$

(the condition $\vec{\nabla} \cdot \vec{v} = \partial v_k/\partial x_k = 0$ has been used here), the expression (3.43) can be transformed into

$$\frac{dE_{kin}}{dt} = -\int_S dS_k \left[ \rho v_k \left( \frac{v^2}{2} + \frac{p}{\rho} \right) + v_i \pi_{ik} \right] + \int_V dV \pi_{ik}\frac{\partial v_i}{\partial x_k}$$

The surface integral here accounts for the outflux/influx of the kinetic energy due to the flow (the term $\rho v_k v^2/2$), as well as for the flow power due to the pressure force and the viscous stresses (the terms $p v_k$ and $v_k v_i \pi_{ik}$, respectively). Thus, the sought after dissipation is determined by the integral over the volume, which, by using the symmetry of the tensor $\pi_{ik}$, can be written as

$$\int_V dV \pi_{ik}\frac{\partial v_i}{\partial x_k} = \frac{1}{2}\int_V dV \pi_{ik}\left( \frac{\partial v_i}{\partial x_k} + \frac{\partial v_k}{\partial x_i} \right) = -\frac{\eta}{2}\int_V dV \left( \frac{\partial v_i}{\partial x_k} + \frac{\partial v_k}{\partial x_i} \right)^2$$

Therefore, the viscous dissipation power per unit volume is equal to

$$Q = \frac{\eta}{2}\left( \frac{\partial v_i}{\partial x_k} + \frac{\partial v_k}{\partial x_i} \right)^2 \tag{3.44}$$

## Problem 3.2.2

A plane bottom surface of an infinitely deep viscous fluid of density $\rho$ and kinematic viscosity $\nu$ undergoes horizontal motion with the velocity $V_x = v_0 \cos(\omega t)$. Derive the mean power of an external force that supports this oscillation, and verify by direct calculation that it is equal to the total energy dissipation rate in the fluid.

If the fluid occupies the domain $z > 0$, its velocity can be written as $v_x(z,\, t) \equiv v(z,\, t)$, and the equation of motion takes the form

$$\frac{\partial v}{\partial t} = \nu \frac{\partial^2 v}{\partial^2 z} \tag{3.45}$$

Since the fluid motion should be also an oscillatory one with the same frequency $\omega$, one can represent its velocity as $v(z,\, t) = \Re[U(z) \exp(-i\omega t)]$, which inserting into equation (3.45) yields

$$-i\omega U = \nu \frac{d^2 U}{d^2 z}$$

Its solution, vanishing at $z \to +\infty$ and equal to $v_0$ at $z = 0$, reads

$$U = v_0 \exp\left[-(1-i)z\sqrt{\frac{\omega}{2\nu}}\right];$$

hence, the fluid velocity is

$$v(z,\, t) = v_0 \exp(-z/h) \cos(\omega t - z/h), \tag{3.46}$$

where $h = \sqrt{2\nu/\omega}$ is the so-called viscous skin-depth that determines the penetration depth of the oscillation into the fluid. Thus, the above-derived solution holds if the depth of the fluid exceeds several skin-depths.

The external force per unit area $\vec{F}$, which is required to move the bottom plane, is equal to the flux of the linear momentum at $z = 0$:

$$F_i = (\Pi_{ik} n_k)_{z=0} = \Pi_{iz}(z=0) = \pi_{iz}(z=0)$$

Thus, its only non-zero component, $F_x$, is equal to

$$\pi_{xz}(z=0) = -\rho\nu \left(\frac{\partial v_x}{\partial z}\right)_{z=0} = v_0 \rho \nu \sqrt{\frac{\omega}{2\nu}}[\cos(\omega t) - \sin(\omega t)],$$

which yields the mean power $\dot{W} = \langle F_x v_x(0,\, t)\rangle = \rho v_0^2 \sqrt{\nu\omega/8}$, where the symbol $\langle \rangle$ means averaging over the time. Clearly, this power must be transformed into the heat released in the fluid due to the viscous energy dissipation. Indeed, according to equation (3.44), the dissipation rate per unit volume is equal to $Q = \rho\nu(dv_x/dz)^2$; hence, it follows from expression (3.45) that

$$\langle Q \rangle = \rho\nu\langle(dv_x/dz)^2\rangle = \frac{\rho\nu v_0^2}{h^2}\exp(-2z/h)$$

Therefore, the total mean dissipation rate is equal to

$$\int\limits_0^{+\infty} dz\langle Q \rangle = \frac{\rho\nu v_0^2}{2h} = \dot{W}$$

## Problem 3.2.3

A plate of mass per unit area equal to $\mu$ is lying above the layer of a viscous fluid (density $\rho$, kinematic viscosity $\nu$) with depth equal to $H$. A plane bottom surface of the fluid oscillates with the frequency $\omega$ and the amplitude $A$. Derive the oscillation amplitude of the plate.

By choosing the same coordinate system as in Problem 3.2.2, one gets the same equation (3.45) for the fluid velocity, and its general solution reads

$$v(z, t) = \{v_1 \exp[(1 - i)z/h] + v_2 \exp[-(1 - i)z/h]\}\exp(-i\omega t),$$

where $h = \sqrt{2\nu/\omega}$ is the viscous skin-depth, and the yet unknown constants $v_1$ and $v_2$ have to be determined from the boundary conditions at the bottom of the fluid, at $z = 0$, and at its surface, $z = H$. Thus, if the plate oscillates with amplitute equal to $a$, the boundary conditions yield the following relations:

$$v_1 + v_2 = -i\omega A,$$
$$v_1 \exp[(1 - i)H/h] + v_2 \exp[-(1 - i)H/h] = -i\omega a,$$

so that

$$v_1 = \frac{-i\omega\left(a - A\exp[-(1 - i)H/h]\right)}{\exp[(1 - i)H/h] - \exp[-(1 - i)H/h]},$$
$$v_2 = \frac{i\omega\left(a - A\exp[(1 - i)H/h]\right)}{\exp[(1 - i)H/h] - \exp[-(1 - i)H/h]}$$

The system of equations becomes complete by adding the equation of motion for the plate:

$$-\mu\omega^2 a = F_x = -\rho\nu\left(\frac{\partial v}{\partial z}\right)_{z=H}$$

Finally, one gets the following expression for the sought after amplitude of the plate:

$$a = A \left\{ \frac{\alpha}{(1+i)} \left[ \exp[(1-i)H/h] - \exp[-(1-i)H/h] \right] + \right.$$

$$\left. \frac{1}{2} \left[ \exp[(1-i)H/h] + \exp[-(1-i)H/h] \right] \right\}^{-1}, \tag{3.47}$$

where the dimensionless parameter $\alpha \equiv \mu/\rho h$. Since this general expression is quite cumbersome, consider two limiting cases of a light plate, when $\alpha \ll 1$, and a heavy one, for which $\alpha \gg 1$. In the former case, the plate has a little effect on the motion of the fluid, so the amplitude $a$ is equal to that of the free boundary surface at $z = H$:

$$|a| \approx |A| \left( \frac{2}{\cosh(2H/h) + \cos(2H/h)} \right)^{1/2}$$

In the case of a heavy plate, the answer depends on the relation of the two parameters: $\alpha \gg 1$ and the ratio $H/h$. For example, if the fluid layer is thin, i.e., $H \ll h$, it follows from expression (3.47) that $|a| \approx |A|(1+4\alpha^2 H^2/h^2)^{-1/2}$, and the plate amplitude becomes small if $\alpha \gg h/H \gg 1$.

## Problem 3.2.4

A plane bottom surface of an infinitely deep incompressible viscous fluid (density $\rho$, kinematic viscosity $\nu$) is instantly brought into motion with a constant velocity $V_x = V_0$. Determine the subsequent fluid flow and the viscous friction force exerted per unit area of the bottom surface.

By introducing dimensionless quantity $u(z, t) \equiv v_x(z, t)/V_0$, one concludes that $u(z, t)$ should be a solution of the equation of motion (3.45) with the boundary conditions $u(0, t) = 1, u(z \to +\infty) = 0$. Since the bottom plane starts to move instantly, the problem under consideration does not possess any characteristic temporal and spatial scales, with the kinematic viscosity $\nu$ being the only dimensional parameter involved. Therefore, the respective solution should be a **self-similar** one, which means that the fluid velocity profiles at different moments of time can be transformed into each other by an appropriate change of the spatial scale. Furthermore, since the kinematic viscosity $\nu$ has a dimension of $m^2/s$, the only dimensionless self-similar variable that can be constructed from the available variables $z, t$, and $\nu$ is $\xi = z/\sqrt{\nu t}$. Thus, the solution for $u(z, t)$ should have the form $u(z, t) = f(\xi)$, with $f(0) = 1$, $f(\xi \to +\infty) = 0$, and, by inserting it into equation (3.45), one arrives to the following equation for $f(\xi)$:

$$\frac{\xi}{2} \frac{df}{d\xi} = \frac{d^2 f}{d^2 \xi}$$

Integrating once yields

$$\frac{df}{d\xi} = C \exp\left(-\frac{\xi^2}{4}\right),$$

with the yet unknown constant $C$. Then, since $f(0) = 1$, one gets

$$f(\xi) = 1 + C \int_0^\xi dy \exp\left(-\frac{y^2}{4}\right)$$

The second boundary condition, $f(\xi \to +\infty) = 0$, together with the equality

$$\int_0^{+\infty} dy \exp\left(-\frac{y^2}{4}\right) = \sqrt{\pi},$$

finally yields the sought after solution

$$v(z, t) = V_0 \left[ 1 - \frac{1}{\sqrt{\pi}} \int_0^{z/\sqrt{\nu t}} dy \exp\left(-\frac{y^2}{4}\right) \right] \tag{3.48}$$

Thus, as time progresses, the velocity imposed at the bottom surface diffuses into the fluid, increasing in this way the amount of fluid involved into motion; as follows from the above deduced self-similarity, the total momentum of the fluid, $P_x$, is proportional to $\sqrt{t}$. Accordingly, since this fluid momentum ultimately originates from the friction force $F$ at the bottom surface, $F = \dot{P}_x$ should be proportional to $\sqrt{1/t}$. Indeed, as it follows from expression (3.48),

$$F = -\rho\nu \left(\frac{\partial v}{\partial z}\right)_{z=0} = \rho V_0 \sqrt{\nu/\pi t}$$

## Problem 3.2.5

A vertical tube of radius $R$ is plugged at both ends and filled up with an incompressible viscous fluid of density $\rho$ and viscosity $\eta$. A long, $L \gg R$, light cylinder (its density is small compared to $\rho$) of radius $r = R - h$, $h \ll R$ is immersed coaxially into the tube. Determine the upward velocity of the cylinder due to the gravity $g$.

The velocity under question, $u$, is established at such a value that the cylinder buoyancy force $F_A \approx \pi R^2 L\rho g$ (A here refers to Archimedes) is balanced by the viscous drag force, $F_d$, exerted on the cylinder at its lateral surface. In order to derive the latter, one should consider the fluid flow, which, by neglecting the edge effects, $(L \gg R)$, takes place inside a narrow gap between the cylinder and the tube surface. Thus, it can be approximated as a plane flow with $v_z \equiv v(x)$, where the radial coordinate $x$ varies from $x = 0$ at the cylinder surface to $x = h$ at that of the tube. The equation of motion for such a steady flow reads

$$-\frac{dp_1}{dz} + \eta \frac{d^2 v}{d^2 x} = 0, \tag{3.49}$$

where $p_1(z)$ comes on top of the hydrostatic pressure distribution $p_0 = -\rho g z$. The appearance of this extra pressure gradient becomes evident from the following consideration. If $(dp_1/dz) = 0$, equation (3.49) yields a linear velocity profile inside the gap, and since $v(x = 0) = u$, $v(x = h) = 0$, the fluid would be flowing upward together with the cylinder. However, this is not possible in the plugged tube of a finite length filled with an incompressible fluid. Therefore, the additional pressure gradient, $(dp_1/dz) > 0$, builds up, which pushes the fluid down so that the following volume balance relation holds:

$$\pi R^2 u = -2\pi R \int_0^h dx v(x) \tag{3.50}$$

Thus, the general solution of equation (3.49) reads

$$v(x) = \frac{x^2}{2\eta} \frac{dp_1}{dz} + ax + b$$

The constants $a$ and $b$ here are determined by the boundary conditions for $v(x)$ at $x = 0$ and $x = h$, which yield

$$b = u, \quad a = -\frac{h}{2\eta} \frac{dp_1}{dz} - \frac{u}{h}$$

After that one finds from the condition (3.50) that

$$\frac{dp_1}{dz} = \frac{6uR\eta}{h^3}$$

Then, the viscous drag force exerted on the cylinder becomes equal to

$$F_d = -2\pi R L \eta \left(\frac{dv}{dx}\right)_{x=0} = -2\pi R L \eta a = \frac{6\pi \eta u R^2 L}{h^2}$$

By equating it to the buoyancy force $F_A$, one finds that the cylinder floats up with the velocity $u = gh^2/6\nu$, where $\nu = \eta/\rho$ is the kinematic viscosity of the fluid.

## Problem 3.2.6

Determine frequency and damping rate for a small amplitude radial oscillation of an ideal gas spherical bubble of radius $R$, immersed into an incompressible viscous liquid of density $\rho_0$, pressure $p_0$, and viscosity $\eta$. Assume that the adiabatic ratio of the gas is equal to $\gamma$, and that the capillary effects are negligibly weak.

A small radial oscillation of the bubble implies that its radius, $R_b$, varies with time as $R_b = R + a \exp(-i\omega t)$, with the oscillation amplitude $a \ll R$. The induced flow of the surrounding liquid is in the radial direction with $v_r(r, t) \equiv v(r, t)$. Since the liquid is incompressible, $vr^2 = const$, hence $v(r, t) = A \exp(-i\omega t)/r^2$. At the surface of the bubble the liquid velocity is equal to $\dot{R}_b$, therefore $A = -i\omega a R^2$, so that $v(r, t) = -i\omega a R^2 \exp(-i\omega t)/r^2$. Then, by knowing the liquid velocity, the pressure perturbation there, $p_l$, can be determined from the linearized equation of motion:

$$\rho_0 \frac{\partial \vec{v}}{\partial t} = -\vec{\nabla} p_l + \eta \vec{\nabla}^2 \vec{v} \tag{3.51}$$

Furthermore, since $\vec{\nabla}^2 \vec{v} = \vec{\nabla}(\vec{\nabla} \cdot \vec{v}) - \vec{\nabla} \times (\vec{\nabla} \times \vec{v})$, for such a flow, where $\vec{\nabla} \cdot \vec{v} = 0$ and $\vec{\nabla} \times \vec{v} = 0$, $\vec{\nabla}^2 \vec{v} = 0$, and, hence, the viscous term in equation (3.51) vanishes. Thus, its radial component yields

$$\frac{\partial p_l}{\partial r} = -\rho_0 \frac{\partial v}{\partial t} = i\omega \rho_0 v = \frac{\rho_0 \omega^2 a R^2 \exp(-i\omega t)}{r^2},$$

integration of which brings

$$p_l(r, t) = p_0 - \frac{\rho_0 \omega^2 a R^2}{r} \exp(-i\omega t),$$

where $p_0$ is the unperturbed liquid pressure far away from the bubble. As far as the gas pressure inside the bubble, $p_g$, is concerned, the consecutive compressions and rarefactions due to the oscillation can be considered as quasistatic (see below). Therefore, the pressure there remains almost uniform, and it can be determined from the adiabaticity condition $p_g V_b^\gamma \propto p_g R_b^{3\gamma} = const$, which in the linear approximation yields

$$p_g = p_0 - \frac{3\gamma p_0 a}{R} \exp(-i\omega t)$$

Finally, the sought after frequency of this oscillation follows from the boundary condition at the gas-fluid interface at $r = R$, which requires continuity of the momentum flux. In this particular case it reduces to a single equation, $\Pi_{rr}^{(g)} = \Pi_{rr}^{(l)}$, where $\Pi_{rr}^{(g)} = p_g$, and $\Pi_{rr}^{(l)} = p_l - 2\eta(\partial v/\partial r)$. Thus, by using the above obtained expressions for the pressure $p_{g,l}$ and the liquid velocity $v$, one gets the following result:

$$\omega^2 + \frac{4i\eta}{\rho_0 R^2}\omega - \frac{3\gamma p_0}{\rho_0 R^2} = 0,$$

or

$$\omega = -\frac{2i\eta}{\rho_0 R^2} \pm \left(\frac{3\gamma p_0}{\rho_0 R^2} - \frac{4\eta^2}{\rho_0 R^4}\right)^{1/2} \tag{3.52}$$

Thus, the oscillation frequency is equal to

$$\Omega = \left(\frac{3\gamma p_0}{\rho_0 R^2} - \frac{4\eta^2}{\rho_0 R^4}\right)^{1/2},$$

if $\eta < \sqrt{3\gamma\rho_0 p_0 R^2/4}$, with the damping rate $\Gamma = 2\eta/\rho_0 R^2$. In the case of a stronger viscosity, the bubble radial displacement damps aperiodically, while in an inviscid liquid it oscillates undamped with the frequency $\Omega = \omega_0 = \sqrt{3\gamma p_0/\rho_0 R^2}$.

Consider now the validity of the quasistatic approximation, which has been adopted above for the evolution of the gas pressure inside the bubble. It holds if the oscillation period $T = 2\pi/\Omega$ is long compared to the characteristic dynamical time, which in this case is the sound wave travel time $\tau_s \sim R/c_s$, where $c_s = \sqrt{\gamma p_0/\rho_g}$ is the speed of sound in a gas of density $\rho_g$ (see Section 3.4). According to expression (3.52), the period under consideration is $T \sim \omega_0^{-1} \sim R/(p_0/\rho_0)^{1/2}$, so the ratio $T/\tau_s \sim (\rho_0/\rho_g)^{1/2}$ is large indeed: a liquid is dense compared to a gas.

It is also worth noting that the disappearance of the viscous term in the equation of motion (3.51) clearly does not mean absence of the viscous dissipation in the liquid; otherwise there would be no damping of the oscillation. This point can be verified quantatively by calculating the total viscous dissipation, and then using the result for deriving the oscillation damping rate. Thus, according to expression (3.44), the viscous dissipation rate per unit volume is equal to $Q = \eta(V_{ik})^2/2$, where the tensor $V_{ik} = (\partial v_i/\partial x_k) + (\partial v_k/\partial x_i)$. In the case of the flow under consideration, when the only present velocity component is $v_r = A\cos(\omega t)/r^2$, and, hence the spherical coordinates $(r, \theta, \phi)$ are most appropriate, the non-zero components of the tensor $V_{ik}$, on top of an obvious one $V_{rr} = 2(\partial v_r/\partial r)$, are also $V_{\theta\theta}$ and $V_{\phi\phi}$. A rigorous way forward here is to apply quite cumbersome standard equations for the tensor transformation, but the goal can be achieved much easier by using the following consideration. Firstly, $V_{\theta\theta}$ and $V_{\phi\phi}$ should be equal to each other due to the spherical symmetry of the flow. Secondly, in the case of an incompressible flow, the trace of the tensor $V_{ik}$ is equal to zero, therefore $V_{\theta\theta} = V_{\phi\phi} = -V_{rr}/2$. Then, the time-averaged viscous dissipation rate reads

$$\langle Q \rangle = \frac{\eta}{2}\langle (V_{rr}^2 + V_{\theta\theta}^2 + V_{\phi\phi}^2) \rangle = \frac{3\eta}{4}\langle V_{rr}^2 \rangle = \frac{6\eta A^2}{r^6},$$

which yields the following rate, $\dot{W}$, for the oscillation energy:

$$\dot{W} = -\int dV\langle Q\rangle = -4\pi \int\limits_{R}^{\infty} r^2 dr\langle Q\rangle = -\frac{8\pi\eta}{R^3}A^2$$

On the other hand, the total oscillation energy, $W$, is equal to twice the time-averaged kinetic energy of the liquid, hence

$$W = \rho\int dV\langle v^2\rangle = \frac{2\pi\rho}{R}A^2$$

Finally, since $W \propto \exp(-2\Gamma t)$, the sought after damping rate $\Gamma$ is equal to

$$\Gamma = -\frac{1}{2}\frac{\dot{W}}{W} = \frac{2\eta}{\rho R^2},$$

as it should be according to expression (3.52).

## 3.3 Convection and turbulence

### Problem 3.3.1

An ideal gas with molecular weight $\mu$ and adiabatic ratio $\gamma$ is in a static equilibrium in the gravity field $\vec{g} = -g\vec{e}_z$. Determine the criterion for convective instability in such a gas by using the energy principle.

The starting point is the basic equations of gas dynamics, which comprise the equation of motion:

$$\rho\frac{d\vec{v}}{dt} = -\vec{\nabla}p + \rho\vec{g},$$

the continuity equation

$$\frac{\partial\rho}{\partial t} + \vec{\nabla}\cdot(\rho\vec{v}) = 0,$$

and, in this case of an ideal gas without any dissipation process, the adiabatic law

$$\frac{d}{dt}\left(\frac{p}{\rho^\gamma}\right) = 0$$

For the initial static equilibrium of the gas the last two of them are satisfied identically, while the first one yields the hydrostatic law $dp/dz = -\rho g$. In order to analyze the stabilty of this equilibrium, consider a small deviation from it, which is specified by the pressure perturbation $\delta p$, the density perturbation $\delta\rho$, and the gas velocity $\vec{v}$. Then, linearization of the basic equations with respect to the above-introduced perturbations yields

$$\rho\frac{\partial \vec{v}}{\partial t} = -\vec{\nabla}\delta p + \delta\rho\vec{g}, \tag{3.53}$$

$$\frac{\partial \delta\rho}{\partial t} + \vec{v}\cdot\vec{\nabla}\rho + \rho(\vec{\nabla}\cdot\vec{v}) = 0, \tag{3.54}$$

$$\frac{1}{\rho^{\gamma}}\frac{\partial \delta p}{\partial t} - \gamma\frac{p}{\rho^{(\gamma+1)}}\frac{\partial \delta\rho}{\partial t} + \vec{v}\cdot\vec{\nabla}\left(\frac{p}{\rho^{\gamma}}\right) = 0 \tag{3.55}$$

The energy principle (see also Problem 4.3.4) is formulated in terms of vector $\vec{\xi}$, the displacement of a gas element from its initial equilibrium position. Then, since the velocity of the gas $\vec{v} = \dot{\vec{\xi}}$, equation (3.54) can be written as

$$\delta\rho = -\rho(\vec{\nabla}\cdot\vec{\xi}) - \xi_z\frac{d\rho}{dz}, \tag{3.56}$$

while the pressure perturbation, according to equation (3.55), takes the form

$$\delta p = \frac{\gamma p}{\rho}\delta\rho - \xi_z\left(\frac{dp}{dz} - \frac{\gamma p}{\rho}\frac{d\rho}{dz}\right) = \frac{\gamma p}{\rho}\left(-\rho(\vec{\nabla}\cdot\vec{\xi}) - \xi_z\frac{d\rho}{dz}\right) -$$
$$\xi_z\left(\frac{dp}{dz} - \frac{\gamma p}{\rho}\frac{d\rho}{dz}\right) = -\gamma p(\vec{\nabla}\cdot\vec{\xi}) - \xi_z\frac{dp}{dz} = -\gamma p(\vec{\nabla}\cdot\vec{\xi}) + \rho g\xi_z \tag{3.57}$$

Finally, by using the above expressions for $\delta\rho$ and $\delta p$, the equation of motion (3.53) can be written as

$$\rho\ddot{\vec{\xi}} = \vec{g}\left(-\rho(\vec{\nabla}\cdot\vec{\xi}) - \xi_z\frac{d\rho}{dz}\right) - \vec{\nabla}(-\gamma p(\vec{\nabla}\cdot\vec{\xi}) + \rho g\xi_z) \equiv \vec{F}(\vec{\xi})$$

Here $\vec{F}(\vec{\xi})$, which is a linear functional of $\vec{\xi}$, is the force exerted per unit volume of the gas due to these displacements. Then, one can introduce the potential energy associated with the continuous field of displacements as

$$\delta W = -\frac{1}{2}\int dV\vec{F}(\vec{\xi})\cdot\vec{\xi} = \frac{1}{2}\int dV\left[\vec{\xi}\cdot\vec{g}\left(\rho(\vec{\nabla}\cdot\vec{\xi}) + \xi_z\frac{d\rho}{dz}\right) +$$
$$\vec{\xi}\cdot\vec{\nabla}(-\gamma p(\vec{\nabla}\cdot\vec{\xi}) + \rho g\xi_z)\right] \tag{3.58}$$

Therefore, the system is stable if for all possible displacements $\delta W > 0$, and it is unstable if there is an allowed displacement field for which $\delta W < 0$.

In order to proceed further, it is helpful to transform the second term in the right-hand side of equation (3.58) as follows:

$$\int dV \vec{\xi} \cdot \vec{\nabla}(-\gamma p(\vec{\nabla} \cdot \vec{\xi}) + \rho g \xi_z) =$$

$$\int dV \vec{\nabla}[\vec{\xi} \cdot (-\gamma p(\vec{\nabla} \cdot \vec{\xi}) + \rho g \xi_z)] - \int dV (\vec{\nabla} \cdot \vec{\xi})(-\gamma p(\vec{\nabla} \cdot \vec{\xi}) + \rho g \xi_z) =$$

$$\int d\vec{S} \cdot \vec{\xi}(-\gamma p(\vec{\nabla} \cdot \vec{\xi}) + \rho g \xi_z) - \int dV (\vec{\nabla} \cdot \vec{\xi})(-\gamma p(\vec{\nabla} \cdot \vec{\xi}) + \rho g \xi_z)$$

The surface integral here is related to the boundaries surrounding the gas or, in their absence, to a remote surface at infinity. Since at a rigid boundary $\vec{\xi} \cdot d\vec{S} = 0$, there is no normal component of the displacement vector, while at infinity both $\rho$ and $p$ tend to zero, this surface integral makes no contribution to the potential energy. Thus, the remaining terms in equation (3.58) can be re-written as

$$\delta W = \frac{1}{2} \int dV \left[ \gamma p(\vec{\nabla} \cdot \vec{\xi})^2 - 2\rho g \xi_z(\vec{\nabla} \cdot \vec{\xi}) - g\xi_z^2 \frac{d\rho}{dz} \right] \qquad (3.59)$$

The next step is to reveal the most "dangerous" perturbation, which provides the minimum to the potential energy. As seen from (3.59), $\delta W$ is actually a functional of $\xi_z$ and $(\vec{\nabla} \cdot \vec{\xi})$. Thus, its extremum with respect to the latter yields

$$(\vec{\nabla} \cdot \vec{\xi}) = \frac{\rho g}{\gamma p} \xi_z \qquad (3.60)$$

By inserting this expression for $(\vec{\nabla} \cdot \vec{\xi})$ into equation (3.59), one gets the following form for the potential energy:

$$\delta W = \frac{1}{2} \int dV g\xi_z^2 \left( -\frac{d\rho}{dz} - \frac{g\rho^2}{\gamma p} \right) \qquad (3.61)$$

It is seen now that the necessary and sufficient condition for the linear stability of a static equilibrium of a gas in the gravitational field is positiveness of the expression inside the brackets in the integrand of equation (3.61). Indeed, if this expression is negative in a vicinity of some elevation $z$, it is possible to make the potential energy negative by choosing the vertical displacement $\xi_z$ that is strongly localized at this location. Thus, the onset of convection in a ideal gas requires that

$$\frac{d\rho}{dz} + \frac{g\rho^2}{\gamma p} > 0,$$

which, by using the hydrostatic law $dp/dz = -\rho g$ together with the equation of state $p = R\rho T/\mu$ ($R$ is the universal gas constant, and $T$ is the gas temperature), yields the following well-known condition:

$$\frac{dT}{dz} < -\frac{\mu g(\gamma - 1)}{\gamma R}$$

Finally, consider the physical meaning of the most convectively unstable perturbation, which satisfies relation (3.60). By inserting the latter into equation (3.57), one finds that in this case $\delta p = 0$, which indicates the quasistatic interchange of the two neighboring gas elements.

## Problem 3.3.2

Derive the criterion for convective instability in an incompressible liquid.

The term "incompressible" here means that density of a fluid, $\rho$, which, generally speaking, is a function of its pressure, $p$, and temperature, $T$, has in this case very weak dependence on the pressure, so that one can neglect it and assume that $\rho \equiv \rho(T)$. Furthermore, if the temperature varies within a not so wide interval about some $T = T_0$, the above dependence can be written as $\rho(T) \approx \rho_0[1 - \alpha(T - T_0)]$, where $\alpha > 0$ is the thermal expansion coefficient. Since the resulting density variation is also small, it can be neglected in the continuity equation, which then reduces to $\vec{\nabla} \cdot \vec{v} = 0$. In the equation of motion

$$\dot{\vec{v}} = -\frac{1}{\rho}\vec{\nabla}p + \vec{g} + \nu\vec{\nabla}^2\vec{v}$$

the total pressure $p$ can be represented as $p = \rho_0\vec{g}\cdot\vec{r} + p_1$, where the first term is due to the hydrostatic equilibrium with a fixed density equal to $\rho_0$, while the term $p_1$, which is small, accounts both for the weak thermal expansion as well as the perturbation associated with convection. Then

$$-\frac{\vec{\nabla}p}{\rho} = -\frac{\vec{\nabla}(\rho_0\vec{g}\cdot\vec{r} + p_1)}{\rho_0[1 - \alpha(T - T_0)]} \approx -\vec{g}[1 + \alpha(T - T_0)] - \frac{\vec{\nabla}p_1}{\rho_0},$$

and the equation of motion takes the form

$$\dot{\vec{v}} = -\frac{1}{\rho_0}\vec{\nabla}p_1 - \alpha\vec{g}(T - T_0) + \nu\vec{\nabla}^2\vec{v} \tag{3.62}$$

In order to make the system of equations complete, one should consider also the thermal conduction equation, which governs evolution of the fluid temperature. Thus, thermal flux in the fluid, $\vec{q}$, which is comprised of the diffusive (thermal conductivity) and the conductive (heat advection with a flow) terms, has the form $\vec{q} = -\kappa\vec{\nabla}T + \rho_0 CT\vec{v}$, where $\kappa$ is the thermal conductivity of the fluid, and $C$ is its heat capacity. Then, the heat balance equation

$$\frac{\partial(\rho_0 CT)}{\partial t} + \vec{\nabla}\cdot\vec{q} = 0$$

yields

$$\rho_0 C\frac{\partial T}{\partial t} = \vec{\nabla}(\kappa\vec{\nabla}T - \rho_0 CT\vec{v}) = \kappa\vec{\nabla}^2 T - \rho_0 C\vec{v}\cdot\vec{\nabla}T$$

(the coefficients $C$ and $\kappa$ are considered here as constants), and the temperature evolution equation takes the form

$$\frac{\partial T}{\partial t} + \vec{v} \cdot \vec{\nabla} T = \chi \vec{\nabla}^2 T, \tag{3.63}$$

where $\chi \equiv \kappa/\rho_0 C$ is the fluid temperature diffusivity. Thus, equations (3.62-3.63), together with the condition $\vec{\nabla} \cdot \vec{v} = 0$, make a complete system of equations, called the **Boussinesq equations**, which describe convection in a liquid.

Consider now the particular model of convection, when a liquid is immersed between the two horizontal planes, $z = 0$ and $z = H$, where the temperature is fixed at $T = T_0$ and $T = T_H$, respectively, with $\Delta T = (T_0 - T_H) > 0$ (the liquid is heated from below). For a static equilibrium the temperature profile $T(z)$, which, according to equation (3.63), satisfies the equation $d^2 T/d^2 z = 0$, takes the form $T(z) = T_0 - \Delta T z/H$. It follows then from equation (3.62) that the respective extra pressure $p_1$ is equal to $p_1 = \rho_0 \alpha g \Delta T z^2/2H$. A small deviation from this static equilibrium can be described by the perturbations of the temperature, $\delta T$, of the pressure, $\delta p_1$, and the velocity of the liquid, $\vec{v}$. Then, in the linear approximation, the Boussinesq equations yield:

$$\vec{\nabla} \cdot \vec{v} = 0,$$
$$\frac{\partial \vec{v}}{\partial t} = -\frac{1}{\rho_0} \vec{\nabla} \delta p_1 - \alpha \vec{g} \delta T + \nu \vec{\nabla}^2 \vec{v},$$
$$\frac{\partial \delta T}{\partial t} - v_z \frac{\Delta T}{H} = \chi \vec{\nabla}^2 \delta T$$

It is helpful to re-write this system of equations by introducing dimension-free variables as follows: length is measured in units of $H$, time in $H^2/\nu$, velocity in $\nu/H$, $\delta T$ in $\Delta T$, $\delta p_1$ in $\rho_0 \nu^2/H^2$. Thus, one arrives to the following equations:

$$\vec{\nabla} \cdot \vec{v} = 0,$$
$$\frac{\partial \vec{v}}{\partial t} = -\vec{\nabla} \delta p_1 - \mathcal{G} \delta T \hat{\vec{z}} + \vec{\nabla}^2 \vec{v}, \tag{3.64}$$
$$\frac{\partial \delta T}{\partial t} - v_z = \frac{1}{\mathcal{P}} \vec{\nabla}^2 \delta T,$$

where, for brevity, the same notations are used for the dimension-free variables. As seen from (3.64), such a convection is characterized by two dimensionless parameters: the **Grashof number**, $\mathcal{G}$, and the **Prandtl number**, $\mathcal{P}$, which are defined as

$$\mathcal{G} \equiv \frac{\alpha g H^3 \Delta T}{\nu^2}, \quad \mathcal{P} \equiv \frac{\nu}{\chi}$$

It means that this process possesses the so-called **similarity**: for any values of $\rho_0$, $\alpha$, $\nu$, etc. all properties of convection remain the same, provided the numbers $\mathcal{G}$ and $\mathcal{P}$ do not change, and all quantities are measured by the introduced above units.

Since the initial static equilibrium is uniform in respect to the coordinates $\vec{r}_\perp = (x, y)$, the perturbations can be considered in the form of $f(z)\exp(\lambda t + i\vec{k}_\perp \cdot \vec{r}_\perp)$. Therefore, the necessary and sufficient condition for the convective stability is that $\Re(\lambda)$ is negative for all values of $\vec{k}_\perp$. Furthermore, it is always possible to choose the coordinate axis in the $(x, y)$ plane in such a way that $\vec{k}_\perp = (k, 0)$ with the non-zero velocity components $v_x$ and $v_z$, so that the condition $\vec{\nabla} \cdot \vec{v} = ikv_x + dv_z/dz = 0$ yields $v_x = (i/k)dv_z/dz$. Then, by taking the curl of the equation of motion in (3.64) (in order to eliminate the pressure perturbation there), one gets

$$\lambda\left(\frac{dv_x}{dz} - ikv_z\right) = \left(-k^2 + \frac{d^2}{dz^2}\right)\left(\frac{dv_x}{dz} - ikv_z\right) + ik\mathcal{G}\delta T,$$

and, by expressing there $v_x$ in terms of $v_z$,

$$\lambda\left(\frac{d^2v_z}{d^2z} - k^2 v_z\right) = \left(-k^2 + \frac{d^2}{dz^2}\right)\left(\frac{d^2v_z}{d^2z} - k^2 v_z\right) + k^2\mathcal{G}\delta T \qquad (3.65)$$

In its turn, the thermal conduction equation in (3.64) takes the form

$$\lambda\delta T - v_z = \frac{1}{\mathcal{P}}\left(-k^2 + \frac{d^2}{d^2z}\right)\delta T \qquad (3.66)$$

These two equations for $v_z$ and $\delta T$ should be supplemented with the boundary conditions for these quantities at $z = 0, 1$. Since the boundaries are kept at fixed temperatures, $(\delta T)_{z=0,1} = 0$. As far as the boundary conditions for $v_z$ are concerned, the simplest case occurs when the boundary planes are perfectly slippery surfaces, i.e., there are no friction forces. Therefore, on top of $v_z(z = 0, 1) = 0$, one also requires that

$$\left(\frac{dv_x}{dz}\right)_{z=0,1} = \frac{i}{k}\left(\frac{d^2v_z}{d^2z}\right)_{z=0,1} = 0$$

It is easy to verify that the appropriate solutions of equations (3.65-3.66) are as follows:

$$\delta T = A\sin(n\pi z), \quad v_z = B\sin(n\pi z), \quad n = 1, 2, ...,$$

and their compatibility yields the following eigenvalues of $\lambda$:

$$\lambda = -\frac{(k^2 + n^2\pi^2)}{2}\left(1 + \frac{1}{\mathcal{P}}\right) \pm$$

$$\left[\frac{(k^2 + n^2\pi^2)}{4}\left(\left(1 + \frac{1}{\mathcal{P}}\right)^2 + \left(\mathcal{G}\frac{k^2}{(k^2 + n^2\pi^2)} - \frac{(k^2 + n^2\pi^2)^2}{\mathcal{P}}\right)\right)\right]^{1/2} \qquad (3.67)$$

It follows then from expression (3.67) that the perturbation is unstable, $\Re(\lambda) > 0$, if the number $\mathcal{R} = \mathcal{G}\mathcal{P} > (k^2 + n^2\pi^2)^3/k^2 \equiv F(k, n)$, where

$\mathcal{R} \equiv \alpha g H^3 \Delta T/\nu\chi$ is the **Rayleigh number**. Thus, perturbations with different values of $k$ and $n$ have different instability thresholds. At a given value of $n$ the "most dangerous" perturbation, i.e., the one with the lowest threshold, corresponds to $k = k_* = n\pi/\sqrt{2}$, and it becomes unstable when the Rayleigh number $\mathcal{R}$ exceeds $\mathcal{R}_*(n) = F(k_*, n) = 27n^4\pi^4/4$. Therefore, the overall lowest instability threshold is associated with the perturbation with $n = 1$ and $k_\perp = \pi/\sqrt{2}$, and the respective critical Rayleigh number $\mathcal{R}_{cr} = 27\pi^4/4 \approx 658$.

Thus, the convective instability threshold in a liquid is determined by a single non-dimensional parameter, the Rayleigh number, which can be explained by the following energy balance consideration. Consider the gain in gravitational energy, $\Delta W_g$, which can be achieved by interchanging a hotter (and, hence, a slightly lighter) portion of the liquid with a denser one located above. As the depth of the liquid is equal to $H$, one can interchange two cubes of sides of $H/2$ each. Then $\Delta W_g = \Delta m g H/2$, where

$$\Delta m = \Delta\rho \left(\frac{H}{2}\right)^3 \approx \alpha\rho_0 \frac{\Delta T}{2} \left(\frac{H}{2}\right)^3,$$

yielding $\Delta W_g \approx \alpha\rho_0 g \Delta T H^4/32$. However, in a viscous liquid some amount of energy, $\Delta W_v$, will be inevitably dissipated in the course of such interchange. Thus, the overall energy gain is positive and, therefore, the instability is able to proceed, if $\Delta W_g > \Delta W_v$. According to equation (3.44), in an incompressible fluid the volumetric viscous dissipation power is equal to

$$Q = \frac{\rho_0\nu}{2} \left(\frac{\partial v_i}{\partial x_k} + \frac{\partial v_k}{\partial x_i}\right)^2 \approx \rho_0\nu \frac{v^2}{H^2},$$

so the dissipated energy $\Delta W_v$ can be estimated as

$$\Delta W_v \approx Q H^3 \Delta t \approx \rho_0\nu v H^2,$$

where $\Delta t \approx H/v$ is duration of the interchange. Since $\Delta W_v$ is proportional to the velocity $v$, while the gravitational energy gain $\delta W_g$ does not depend on it, at first glance one might conclude that the system is always unstable; indeed, it looks like $\Delta W_v$ can be made arbitrarily small by simply reducing the interchange velocity. However, this is not so due to the thermal conduction. In deriving the energy gain $\Delta W_g$ it was tacitly assumed that in the course of the interchange the densities (and, hence, the temperatures) of the interchanged volumes remain unchanged and equal to their initial values. This is the case only when the process is quick enough, so that its duration $\Delta t \approx H/v$ is short compared to the characteristic temperature equilibration time $\tau_\chi \sim H^2/\chi$, which requires $v \geq \chi/H$. This puts the lower limit on the dissipated energy: $\Delta W_v \geq \rho_0\nu\chi H$, and convection becomes energetically favorable when $\Delta W_g \approx \alpha\rho_0 g \Delta T H^4/32 > \rho_0\nu\chi H$, or

$$\frac{\alpha g H^3 \Delta T}{\nu\chi} \equiv \mathcal{R} > 32$$

It goes without saying that the exact number on the right-hand side of the above inequality, i. e the exact value of the critical Rayleigh number $\mathcal{R}_{cr}$, cannot be obtained with such a general qualitative consideration. This number is sensitive to the particular boundary conditions; for example, in the case of the free upper surface kept at a fixed temperature, the onset of convection occurs at $\mathcal{R} > 11000$.

## Problem 3.3.3

Determine the possible types of ordered structures (convective cells) arising in the marginal regime of convective instability, when the Rayleigh number is just above its critical value: $(\mathcal{R} - \mathcal{R}_{cr}) \ll \mathcal{R}_{cr}$.

When the Rayleigh number $\mathcal{R}$ is just above the critical value $R_{cr}$, only the most unstable modes are excited (for the model of Problem 3.3.2, those are perturbations with $n = 1$ and $k_\perp = k_* = \pi/\sqrt{2}H$). In this case, the motion of a liquid can be viewed as a linear superposition of these modes, so that, for instance,

$$v_z(x, y, z) = \sin\left(\frac{\pi z}{H}\right) \Re \int dk_x dk_y A(k_x, k_y) \times$$
$$\exp[i(k_x x + k_y y)]\delta(k_x^2 + k_y^2 - k_*^2) \tag{3.68}$$

For an arbitrary set of the amplitudes $A(k_x, k_y)$ the resulting motion would be rather irregular. However, it is well known experimentally that the marginal convection under consideration usually possesses a steady ordered periodic structure in the $(x, y)$ plane. Thus, the entire space between the bounding horizontal surfaces is divided into adjacent identical convective cells, in each of which the liquid is moving along the closed paths without passing from one cell to another. The horizontal cross section of the cells form the two-dimensional periodic lattice, which could be made either of rectangles (that in extreme cases degenerate into squares and bands), or of regular triangles, or of regular hexagons. Thus, the next task is to determine the set of amplitudes $A(k_x, k_y)$ in equation (3.68) that correspond to each of these structures.

The rectangular cells are formed by superposition of the two basic vectors in the $(k_x, k_y)$ plane with equal amplitudes. Indeed, consider two vectors, $\vec{k}_1$ and $\vec{k}_2$, of the same length equal to $k_*$ and making the angle $2\alpha$ between them (see Figure 3.8). Thus, in the coordinate system shown there one gets

$$\vec{k}_1 = (k_* \cos\alpha, k_* \sin\alpha), \quad \vec{k}_2 = (k_* \cos\alpha, -k_* \sin\alpha)$$

Then, for $A_1 = A_2 = v_0/2$, it follows from expression (3.68) that the vertical

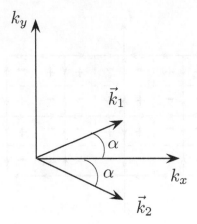

**FIGURE 3.8**

Basic wave vectors for the rectangular convective cells

velocity of the liquid is equal to

$$v_z = \frac{v_0}{2} \sin\left(\frac{\pi z}{H}\right) [\cos(k_{1x}x + k_{1y}y) + \cos(k_{2x}x + k_{2y}y)] =$$
$$v_0 \sin\left(\frac{\pi z}{H}\right) \cos(k_* x \cos\alpha) \cos(k_* y \sin\alpha)$$

Furthermore, according to the solution of Problem 3.3.2, its horizontal velocity in this case is

$$\vec{v}_\perp = \vec{v}_{\perp 1} + \vec{v}_{\perp 2} = -\frac{v_0}{\sqrt{2}} \cos\left(\frac{\pi z}{H}\right) \times$$
$$\left( \frac{\vec{k}_1}{k_*} \sin(k_{1x}x + k_{1y}y) + \frac{\vec{k}_2}{k_*} \sin(k_{2x}x + k_{2y}y) \right),$$

or, by components,

$$v_x = -v_0\sqrt{2} \cos\alpha \cos\left(\frac{\pi z}{H}\right) \sin(k_* x \cos\alpha) \cos(k_* y \sin\alpha),$$
$$v_y = -v_0\sqrt{2} \cos\left(\frac{\pi z}{H}\right) \cos(k_* x \cos\alpha) \sin(k_* y \sin\alpha)$$

The resulting flow pattern in the $(x, y)$ plane is shown on Figure 3.9. The dashed lines correspond to $v_z(x, y) = 0$, while the bold ones with arrows depict the horizontal streamlines at the boundaries of convective cells at $z = H$. It is assumed here that $v_0 > 0$, and the signs $\pm$ on Figure 3.9 correspond to the domains where the liquid flows up and down. Thus, the convective cell is a rectangular prism of a height equal to $H$, and with the sides $\Delta x = \sqrt{2}H/\cos\alpha$, $\Delta y = \sqrt{2}H/\sin\alpha$, which at $\alpha = 0$ transforms into convective rolls of width $\Delta x = \sqrt{2}H$. As shown, each convective cell is divided into four sub-domains: in two of them the liquid flows up, while in the other two down.

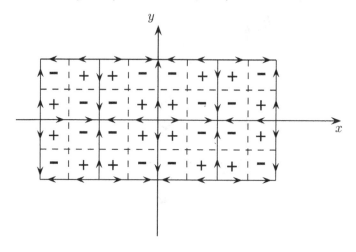

**FIGURE 3.9**
Structure of the rectangular convective lattice

In the case of the triangular lattice the solution is a superposition of the three basic vectors (see Figure 3.10): $\vec{k}_1 = k_*(0, 1)$, $\vec{k}_2 = k_*(-\sqrt{3}/2, -1/2)$, $\vec{k}_3 = k_*(\sqrt{3}/2, -1/2)$, with the amplitudes $A_1 = A_2 = A_3 = -iv_0$. It follows then from expression (3.68) that

$$v_z = v_0 \sin\left(\frac{\pi z}{H}\right)\left\{\sin(k_* y) + \sin\left[\frac{k_*}{2}(\sqrt{3}x - y)\right] - \sin\left[\frac{k_*}{2}(\sqrt{3}x + y)\right]\right\},$$

$$v_x = v_0\sqrt{\frac{3}{2}}\cos\left(\frac{\pi z}{H}\right)\left\{\cos\left[\frac{k_*}{2}(\sqrt{3}x - y)\right] - \cos\left[\frac{k_*}{2}(\sqrt{3}x + y)\right]\right\},$$

$$v_y = v_0\frac{1}{\sqrt{2}}\cos\left(\frac{\pi z}{H}\right)\left\{2\cos(k_* y) - \cos\left[\frac{k_*}{2}(\sqrt{3}x - y)\right] - \cos\left[\frac{k_*}{2}(\sqrt{3}x + y)\right]\right\}$$

The resulting flow pattern in the $(x, y)$ plane is shown in Figure 3.11. The lines of $v_z(x, y) = 0$ (the dashed lines) are straight lines dividing the $(x, y)$ plane into the regular triangles of a side equal to $4H/\sqrt{2/3}$, in each of which the fluid flows up $(+)$ or down $(-)$. The horizontal streamlines which correspond to the boundaries of convective cells (the bold lines with arrows) show that each cell represents a regular triangle prism with a side equal to $4\sqrt{2}H/3$, which is divided into two equal parts where the liquid flows up or down.

The hexagonal lattice (called Benard cells) is formed by the same basic vectors as the triangle one, but with the amplitudes $A_1 = A_2 = A_3 = v_0$,

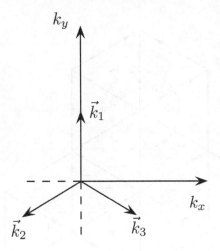

**FIGURE 3.10**
Basic wave vectors for the triangular and hexagonal convective cells

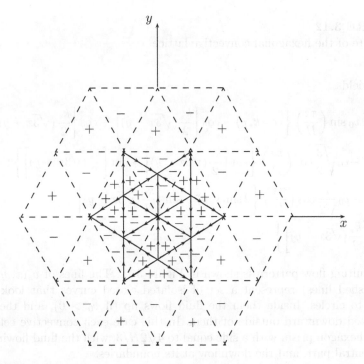

**FIGURE 3.11**
Structure of the triangular convective lattice

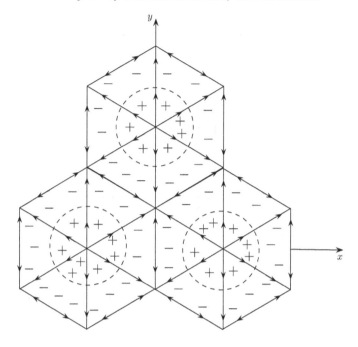

**FIGURE 3.12**
Structure of the hexagonal convective lattice

which yields

$$v_z = v_0 \sin\left(\frac{\pi z}{H}\right) \left\{ \cos(k_* y) + \cos\left[\frac{k_*}{2}(\sqrt{3}x - y)\right] + \cos\left[\frac{k_*}{2}(\sqrt{3}x + y)\right] \right\},$$

$$v_x = -v_0 \sqrt{\frac{3}{2}} \cos\left(\frac{\pi z}{H}\right) \left\{ \sin\left[\frac{k_*}{2}(\sqrt{3}x - y)\right] + \sin\left[\frac{k_*}{2}(\sqrt{3}x + y)\right] \right\},$$

$$v_y = -v_0 \frac{1}{\sqrt{2}} \cos\left(\frac{\pi z}{H}\right) \left\{ 2\sin(k_* y) - \sin\left[\frac{k_*}{2}(\sqrt{3}x - y)\right] + \right.$$
$$\left. \sin\left[\frac{k_*}{2}(\sqrt{3}x + y)\right] \right\}$$

The resulting flow pattern is shown in Figure 3.12. The lines of $v_z(x, y) = 0$ (the dashed lines) represent a set of isolated closed curves that look very similar to circles. Inside them the fluid flows up (if $v_0 > 0$), and the flow is directed downward outside of them. In this case each convective cell is a regular hexagon prism with a side equal to $4\sqrt{2}H/3$, with the fluid flowing up in its central part, and the downflow at its boundaries.

## Problem 3.3.4

Derive a relation between the net vertical convective heat flux and the total kinetic energy of a liquid in the case of square convective cells for the regime of the marginal convection, when $(\mathcal{R} - \mathcal{R}_{cr}) \ll \mathcal{R}_{cr}$.

The convective heat flux is equal to $\vec{q}^{(v)} = \rho_0 C T \vec{v}$, so its vertical component under consideration here reads $q_z^{(v)} = \rho_0 C T v_z$. Since the net vertical flux of the fluid in the convective cell is equal to zero, in deriving the net convective heat flux $Q = \int dx dy q_z^{(v)}$ one should take into account the temperature perturbation $\delta T$, hence $Q = \rho_0 C \Delta S \langle \delta T v_z \rangle$, where for the square cell the cross-section area $\Delta S = \Delta x \Delta y = 4H^2$, and the symbol $\langle \rangle$ means averaging over this cross section. The required perturbation $\delta T$ can be expressed in terms of the vertical fluid velocity $v_z$ with the help of the heat transport equation (3.63), where the time derivative $\partial \delta T / \partial t$ can be neglected for the marginal convection regime under consideration. Thus,

$$-v_z \frac{\Delta T}{H} = \chi \vec{\nabla}^2 \delta T = -\chi \left( k_*^2 + \frac{\pi^2}{H^2} \right) \delta T,$$

which yields

$$\delta T = v_z \frac{2H \Delta T}{3\pi^2 \chi}$$

Then, according to Problem 3.3.3, in the case of the square convective cells, the vertical fluid velocity takes the form

$$v_z = v_0 \sin \left( \frac{\pi z}{H} \right) \cos \left( \frac{\pi x}{2H} \right) \cos \left( \frac{\pi y}{2H} \right)$$

Therefore,

$$\langle \delta T v_z \rangle = \frac{\Delta T H v_0^2}{6\pi^2 \chi} \sin^2 \left( \frac{\pi z}{H} \right),$$

which yields the sought after net convective heat flux:

$$Q = \frac{2\rho_0 C H^3 \Delta T}{3\pi^2 \chi} \sin^2 \left( \frac{\pi z}{H} \right)$$

On the other hand, the total kinetic energy of the fluid in the square convective cell is equal to $E_{kin} = \rho_0 < v^2 > \Delta V / 2 = \rho_0 < v^2 > H \Delta S / 2 = 2\rho_0 < v^2 > H^3$. Since in this case the horizontal velocity components are equal (see Problem 3.3.3) to

$$v_x = -v_0 \cos \left( \frac{\pi z}{H} \right) \sin \left( \frac{\pi x}{2H} \right) \cos \left( \frac{\pi y}{2H} \right),$$

$$v_y = -v_0 \cos \left( \frac{\pi z}{H} \right) \cos \left( \frac{\pi x}{2H} \right) \sin \left( \frac{\pi y}{2H} \right),$$

the averaged $< v^2 >= 3v_0^2/8$, hence $E_{kin} = 3\rho_0 v_0^2 H^3/4$. Therefore, the maximum net convective heat flux, $Q_{max}$, which corresponds to the median plane $z = H/2$, relates to the total kinetic energy as

$$Q_{max} = \frac{8C\Delta T}{9\pi^2 \chi} E_{kin}$$

Note, that both $Q$ and $E_{kin}$ are proportional to the square of the fluid velocity amplitude $v_0$, which has to be derived from the non-linear equations of steady convection. It follows, however, from a rather general consideration (see, e. g., L. D. Landau and E. M. Lifshitz, *Fluid Mechanics*, §26, Pergamon Press, Oxford, 1987), that for the marginal regime, when $(\mathcal{R} - \mathcal{R}_{cr}) \ll \mathcal{R}_{cr}$, this amplitude scales as $\sqrt{\mathcal{R} - \mathcal{R}_{cr}}$; therefore in this case (the so-called "soft" regime of convection) the convective heat flux scales as $Q \propto (\mathcal{R} - \mathcal{R}_{cr})$.

## Problem 3.3.5

Estimate the net convective heat flux for turbulent convection, when the Rayleigh number $\mathcal{R}$ is well above the critical one: $\mathcal{R} \gg \mathcal{R}_{cr}$.

Under the regime of turbulent convection the transport of heat in the main body of a liquid is due to the convective heat flux $\vec{q}^{(v)} = \rho_0 C T \vec{v}$, while the diffusion of heat (thermal condictivity) plays a role only in the immediate vicinity of the bounding surfaces $z = 0, H$, where the vertical component of $\vec{q}^{(v)}$ is equal to zero. As a result, the mean, i.e., averaged over the turbulent pulsations, temperature profile $T(z)$ takes the form qualitatively depicted in Figure 3.13. Thus, the mean temperature is almost uniform in the main part of the volume, and the required temperature drop $\Delta T = T_0 - T_H$ occurs in the narrow boundary layers of width $h \ll H$, where the heat conductivity dominates. The magnitude of $h$ is established at such a value that the boundary layer remains at the state of the marginal convective stability, which means that the respective Rayleigh number $\mathcal{R}_h \sim \alpha g h^3 (\Delta T)/\nu\chi$ is of the order of the critical Rayleigh number $\mathcal{R}_{cr}$. Since the global Rayleigh number $\mathcal{R} = \alpha g H^3 (\Delta T)/\nu\chi$, this requirement yields $h \sim H(\mathcal{R}_{cr}/\mathcal{R})^{1/3}$. Then, the sought after heat flux $q_z$ can be estimated in the following way. Near the bounding surfaces the main contribution to $q$ comes from the heat conduction, hence

$$q = \kappa \left(\frac{dT}{dz}\right)_{z=0} \sim \kappa\frac{\Delta T}{h} \sim \kappa\frac{\Delta T}{H}\left(\frac{\mathcal{R}}{\mathcal{R}_{cr}}\right)^{1/3}$$

Since the respective heat flux without convection, i.e., in a stationary liquid, is equal to $q_0 = \kappa(\Delta T)/H$, it is seen now that turbulent convection amplifies the transport of heat between the boundary surfaces by a factor of $(\mathcal{R}/\mathcal{R}_{cr})^{1/3} \gg 1$.

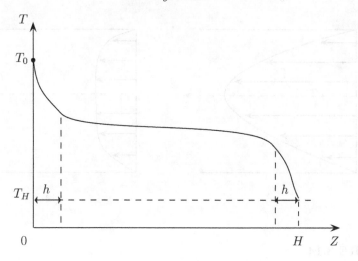

**FIGURE 3.13**
The mean temperature profile in a turbulent convection

## Problem 3.3.6

A steady flow of an incompressible viscous fluid of density $\rho$ and viscosity $\eta$ in a long pipe of radius $a$ and length $L$ is provided by the pressure drop $(\Delta p)$ between the pipe's ends. Determine the fluid discharge in the case of a small enough drop $(\Delta p)$, when the flow is laminar, and for a large pressure drop (in a fully developed turbulent regime).

In the case of the laminar flow the Navier-Stokes equation (3.42) yields the following equation for the fluid velocity component $v_z(r) \equiv v(r)$:

$$\frac{1}{r}\frac{d}{dr}\left(r\frac{dv}{dr}\right) = -\frac{(\Delta p)}{\eta L}$$

Its regular solution that satisfies the required boundary condition $v(r = a) = 0$ reads

$$v(r) = \frac{(\Delta p)}{4\eta L}(a^2 - r^2)$$

Thus, the velocity profile has a parabolic form as shown in Figure 3.14(a), and the fluid discharge is equal to

$$Q_l = 2\pi\rho\int\limits_0^a rdrv(r) = \frac{\pi(\Delta p)a^4}{8\nu L},$$

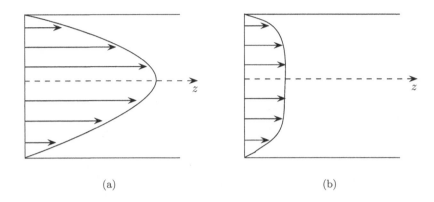

(a)                                              (b)

**FIGURE 3.14**
Fluid velocity profiles for the laminar (a) and the turbulent (b) flows in a pipe

which is known as the Poiseuille formula.

The fully developed turbulent flow in the pipe can be described in terms of the $z$-component of the momentum flux, $\sigma$, at $r = a$, which is equal to the viscous friction force per unit area of the pipe boundary surface. Therefore, it follows from the balance of forces that $\pi a^2(\Delta p) = \sigma 2\pi aL$, i.e., $\sigma = (\Delta p)a/2L$. Under a given $\sigma$ the mean velocity variation near the surface of the pipe follows the well-known logarithmic law (see, e.g., L. D. Landau and E. M. Lifshitz, *Fluid Mechanics*, §42, Pergamon Press, Oxford, 1987):

$$v(y) = \frac{1}{\kappa}\left(\frac{\sigma}{\rho}\right)^{1/2} \ln\left[\frac{y}{\nu}\left(\frac{\sigma}{\rho}\right)^{1/2}\right], \qquad (3.69)$$

where $y = (a - r)$ is the distance from the wall, and $\kappa \approx 0.4$ is a numerical constant. Such a logarithmic profile holds for $y \ll a$, where the momentum flux is approximately constant, but does not apply for a very narrow viscous layer at $r = a$, where the transition to $v(y = 0) = 0$ occurs. However, for a turbulent flow with a large Reynolds number $R \equiv av/\nu$ the expression (3.69) provides a correct magnitude (with logarithmic accuracy) of the flow velocity in the entire cross section of the pipe. The reason is that, in this case, even for $y \ll a$ the argument under the logarithm in (3.69) is already large, and, therefore, its variation with $y$ is very weak. Indeed, for $y \sim a$ this argument is of the order of $(a^3(\Delta p)/\rho\nu L)^{1/2}$, which is nothing else as the square root of the Reynolds number $R$ calculated with the laminar velocity $v \sim (\Delta p)a^2/\eta L$. Thus, under $R \gg 1$ the mean fluid velocity is approximately uniform in the main part of the pipe cross section (see Figure 3.14(b)), and is equal to

$$v \approx \frac{\ln R}{\kappa}\left(\frac{a(\Delta p)}{2L\rho}\right)^{1/2}$$

Then, one gets the fluid discharge

$$Q_t \approx \pi a^2 \rho v \approx \frac{\pi a^2 \rho \ln R}{\kappa} \left( \frac{a(\Delta p)}{2L\rho} \right)^{1/2}$$

Note, that while in the laminar regime the fluid discharge is proportional to the net pressure drop $\Delta p$, in the turbulent regime it scales as $(\Delta p)^{1/2}$, and, therefore, it becomes substantially reduced compared with the laminar one under the same $\Delta p$.

## Problem 3.3.7

Estimate the spatial scale of the viscous energy dissipation for the turbulent flow in a pipe (see Problem 3.3.6).

According to Problem 3.3.6, in the case of the turbulent flow the variation of the mean velocity is of the order of $\Delta v \sim (a\Delta p/\rho L)^{1/2}$. Therefore, the largest turbulent eddies have a spatial scale of the order of $a$ and a velocity amplitude $u \sim \Delta v \sim (a\Delta p/\rho L)^{1/2}$. Then, according to the **Kolmogorov-Obukhov's law** (see, e.g., L. D. Landau and E. M. Lifshitz, *Fluid Mechanics*, §33, Pergamon Press, Oxford, 1987), in the **inertial interval** eddies of size $\lambda$ have the velocity amplitude

$$v_\lambda \sim u \left( \frac{\lambda}{a} \right)^{1/3} \sim (a\Delta p/\rho L)^{1/2} \left( \frac{\lambda}{a} \right)^{1/3}$$

The lower boundary of the inertial interval, $\lambda \sim \lambda_*$, where the viscous dissipation takes place, is determined by the condition that the respective Reynolds number $R_{\lambda_*} \sim \lambda_* v_{\lambda_*}/\nu$ is of the order of unity. Hence, one gets

$$R_{\lambda_*} \sim (a\Delta p/\rho L)^{1/2}(\lambda_*/a)^{1/3}\lambda_*/\nu \sim 1,$$

which yields $\lambda_* \sim a(\nu^2 \rho L/a^3 \Delta p)^{3/8}$.

## Problem 3.3.8

Estimate the magnitude of the velocity variation, $(\Delta v)_\tau$, during a time interval $\tau$, for a fluid element, that moves about in a fully developed turbulent flow, when $\tau \ll L/u$. Here $L$ and $u$ are, respectively, the global spatial scale and the global velocity variation in the flow.

Under a fully developed turbulence a chaotic motion of the fluid can be viewed as a superposition of turbulent eddies of different size $\lambda$. In the inertial interval, for which $\lambda_* < \lambda < L$ (see Problem 3.3.7), the Kolmogorov-Obukhov's law states that the velocity amplitude $v_\lambda$ of eddies of size $\lambda$ scales as $v_\lambda \sim u(\lambda/L)^{1/3}$. Therefore, any fluid element is simultaneously involved in the motion of eddies with a wide range of spatial scales. Consider now what is the contribution of an eddy of some given size $\lambda$ to the velocity variation of the fluid element. The result essentially depends on the relations between the chosen time interval $\tau$ and the eddy characteristic turnover time $\tau_\lambda \sim \lambda/v_\lambda \sim (\lambda/L)^{2/3}L/u$. If $\tau \geq \tau_\lambda$, which holds for relatively small eddies, the velocity variation $(\Delta v)_\tau(\lambda)$ should be of the order of the velocity amplitude of the eddy, hence $(\Delta v)_\tau(\lambda) \sim v_\lambda \sim u(\lambda/L)^{1/3}$. On the other hand, for $\tau \leq \tau_\lambda$, i.e., when $\lambda$ is large enough, the velocity variation under discussion is equal only to a fraction of $v_\lambda$:

$$(\Delta v)_\tau(\lambda) \sim v_\lambda \frac{\tau}{\tau_\lambda} \sim u \frac{u\tau}{L} \left(\frac{\lambda}{L}\right)^{-1/3}$$

Thus, by using these estimates, one concludes that the main contribution to $(\Delta v)_\tau$ comes from the eddies with the turnover time $\tau_\lambda \sim \tau$, i.e., with $\lambda_\tau \sim L(\tau u/L)^{3/2}$, and hence

$$(\Delta v)_\tau \sim u\sqrt{\frac{u\tau}{L}} \qquad (3.70)$$

It follows from this consideration that for $\tau > L/u$ the velocity variation under discussion ceases to depend on the time interval $\tau$ and is equal to $u$. It is worth noting that relation (3.70) is also not valid for a very small $\tau$, when the respective $\lambda_\tau$ falls outside the inertial interval and becomes smaller than $\lambda_*$. In this case, i.e., for $\tau < (L/u)(\lambda_*/L)^{2/3}$, the motion of a fluid element is a regular one during such a time interval and, therefore, $(\Delta v)_\tau$ is proportional to $\tau$.

## Problem 3.3.9

Estmate temporal variation of the distance between the two adjacent fluid elements, moving about in a fully developed turbulent flow.

Consider two fluid elements whose initial separation is equal to $l(t = 0) = l_0$, with $\lambda_* \ll l_0 \ll L$. Their subsequent separation, $l(t)$, is determined by the difference in their velocities, $(\delta v)(l)$. The latter can be estimated by viewing the turbulent flow as a superposition of eddies with various spatial scales $\lambda$ and velocity amplitudes $v_\lambda \sim u(\lambda/L)^{1/3}$ (see Problem 3.3.8). If the size of some eddy is such that $\lambda \gg l$, its contribution to the velocity difference under question is $(\delta v)_\lambda \sim v_\lambda l/\lambda \sim u(l/L)(\lambda/L)^{-2/3}$, which, as seen, is increasing for smaller eddies. On the other hand, when the eddy is so small that $\lambda < l$, it ceases at all to affect $(\delta v)(l)$. Therefore, the major role in defining $(\delta v)(l)$ belongs to eddies with $\lambda \sim l$, and $(\delta v)(l) \sim u(l/L)^{1/3}$. It is worth noting here that the respective contribution of the mean flow, which by the order of magnitude is equal to $ul/L$, is small compared to $u(l/L)^{1/3}$ and, therefore, can be neglected. Thus, the separation of the closed fluid particles varies according to the following equation:

$$\frac{dl}{dt} \sim (\delta v)(l) \sim u \left( \frac{l}{L} \right)^{1/3}$$

This relation demonstrates that the turbulent flow pulls adjacent particles apart with a self-accelerating pace: after some time their separation becomes independent of its initial magnitude, and it grows with time according to the law $l(t) \sim (ut)^{3/2}/L^{1/2}$. Clearly, such a scaling holds until $l \leq L$.

---

## 3.4   Sound waves

In the linear (acoustic) approximation propagation of sound waves in a medium is described by the following wave equation for the velocity potential $\phi$ ($\vec{v} = -\vec{\nabla}\phi$):

$$\frac{\partial^2 \phi}{\partial^2 t} - c_s^2 \vec{\nabla}^2 \phi = 0,$$

where the speed of sound, $c_s$, is determined by the adiabatic compressibility of the medium: $c_s = \sqrt{(\partial p/\partial \rho)_s}$. For a plane sound wave propagating in the direction specified by a unit vector $\vec{n}$, the velocity potential $\phi$ has the following dependence on the coordinates and time:

$$\phi(\vec{r}, t) = f(\vec{n} \cdot \vec{r} - c_s t)$$

In this case the density perturbation, $\delta\rho$, and the perturbation of pressure, $\delta p$, relate to the velocity $\vec{v} = v\vec{n}$ as follows:

$$\delta\rho = \rho \frac{v}{c_s}, \ \delta p = \rho v c_s, \ v = -f'$$

The energy per unit volume, $E$, the flux of energy, $\vec{q}$, and the momentum flux tensor, $\Pi_{ik}$, associated with such a wave, are equal to:

$$E = \rho \langle v^2 \rangle, \ \vec{q} = c_s E \vec{n}, \ \Pi_{ik} = E n_i n_k \tag{3.71}$$

## Problem 3.4.1

Determine the subsequent motion induced in an ideal gas by a localized initial (at $t = 0$) perturbation comprising the velocity $v_x(x) = u(x)$ and the perturbed pressure $p(x) = p_0 + p_1(x)$. Assume the perturbation to be weak enough so that the acoustic approximation holds.

Clearly, the subsequent motion of the gas will be one-dimensional; hence the linearized continuity equation and the equation of motion read

$$\frac{\partial \delta\rho}{\partial t} + \rho\frac{\partial v}{\partial x} = 0, \quad \rho\frac{\partial v}{\partial t} = -\frac{\partial \delta p}{\partial x}, \tag{3.72}$$

where the density and pressure perturbations are related to each other by the adiabatic law:

$$\delta p = \delta\rho \left(\frac{\partial p}{\partial \rho}\right)_s = c_s^2 \delta\rho$$

These equations combine into a single wave equation for the gas velocity $v$:

$$\frac{\partial^2 v}{\partial^2 t} - c_s^2 \frac{\partial^2 v}{\partial^2 x} = 0$$

By introducing new variables, $\xi = x - c_s t$ and $\eta = x + c_s t$, it takes the form $\partial^2 v/\partial\xi\partial\eta = 0$. Its general solution is $v(\xi, \eta) = f(\xi) + g(\eta)$, where $f$ and $g$ are arbitrary functions; therefore, by reverting to the variables $(x, t)$, one gets

$$v(x, t) = f(x - c_s t) + g(x + c_s t)$$

The yet unknown functions $f$ and $g$ are to be determined from the initial conditions at $t = 0$, which are $v(x, 0) = u(x)$ and

$$\left(\frac{\partial v}{\partial t}\right)_{t=0} = -\frac{1}{\rho}\frac{\partial \delta p(x, 0)}{\partial x} = -\frac{1}{\rho}\frac{dp_1}{dx}$$

These yield

$$f(x) + g(x) = u(x), \quad c_s\left(\frac{dg}{dx} - \frac{df}{dx}\right) = -\frac{1}{\rho}\frac{dp_1}{dx}$$

Since all the perturbations vanish at infinity, a simple integration of the above equations finally yields

$$v(x, t) = \frac{1}{2}[u(x - c_s t) + u(x + c_s t)] + \frac{1}{2c_s\rho}[p_1(x - c_s t) - p_1(x + c_s t)]$$

## Problem 3.4.2

A plane compression impulse of length $l$ with a linear profile of the pressure enhancement that varies from $p_0$ to zero (see Figure 3.15(a)) propagates in a fluid toward its free boundary. When the impulse reaches the free boundary, a rarefaction impulse with a negative pressure starts to propagate into the fluid due to the boundary condition $p = 0$ at a free surface (see Figure 3.15(b)). When a negative pressure in the fluid exceeds some critical threshold equal to $p_c$, the fluid breaks off. Find out how many separate fluid layers will be formed in the course of this process, and what is the width of each of them, if $p_0 = Np_c$, when $N > 1$ is an integer number.

The kinematics of the waves becomes clear from Figure 3.15. Assume now that the impulse of compression reaches the boundary surface $x = 0$ at the instant $t = 0$. Then, since both impulses, the compressive one and the rarefactive one, propagate with the same velocity (equal to the speed of sound $c_s$) but in the opposite directions, at the instant $t = l/2Nc_s$ the net pressure at the fluid location $x = -l/2N$ becomes equal to $p_c = -p_0/N$. Therefore, the fluid there breaks off, releasing a separated layer of a width $\delta = l/2N$. After that the situation becomes similar to that at $t = 0$, except the amplitude of compression is reduced to $p_0' = (N - 1)p_0$. Thus, the total number of the identical separated fluid layers will be equal to $N$.

## Problem 3.4.3

Derive the energy and linear momentum (per unit volume) for a monochromatic sound wave propagating in a medium of density $\rho$ and with the speed of sound equal to $c_s$.

Consider the sound wave propagating along the $z$-axis with the respective fluid velocity

$$v_x = v_0 \cos(kz - \omega t),$$

where $\omega = kc_s$. Then, according to the first of equations (3.72), the density perturbation, $\delta\rho$, in such a wave is equal to

$$\delta\rho = \rho\frac{kv_0}{\omega} \cos(kz - \omega)$$

The averaged kinetic energy per unit volume of the fluid is

$$E_{kin} = \frac{\rho}{2}\langle v^2 \rangle = \frac{\rho v_0^2}{4},$$

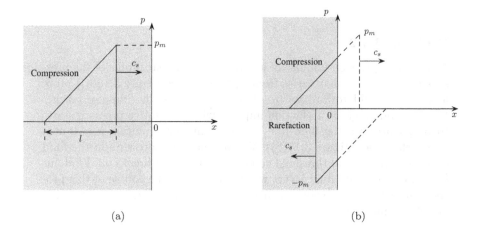

(a) (b)

**FIGURE 3.15**
Superposition of the compression and rarefaction pulses

and since for small amplitude oscillations and waves the potential energy is, on average, equal to the kinetic energy, the total energy of the wave $E = 2E_{kin} = \rho v_0^2/2$.

Furthermore, the only component of linear momentum present in this case, $P_z$, can be calculated as

$$P_x = \langle (\rho + \delta\rho)v_z \rangle = \langle \delta\rho v_z \rangle = \frac{k}{\omega} \frac{\rho v_0^2}{2},$$

which confirms the anticipated relation

$$\vec{P} = \frac{\vec{k}}{\omega} E$$

## Problem 3.4.4

A monochromatic sound wave propagating in a fluid of density $\rho_1$ and speed of sound $c_1$ is normally incident on the plane boundary with another fluid, whose density is $\rho_2$ and speed of sound $c_2$. Derive the wave reflection coefficient, and determine the resulting force exerted on the interface of the fluids.

This problem involves three waves: the incident wave (wave 1), the reflected wave (wave 2), and the transmitted wave (wave 3), with all of them having the same frequency $\omega$ as the incident wave. By denoting their velocity potential amplitudes as $A_1$, $A_2$, $A_3$, respectively, one can write them as follows:

$$\phi_1 = A_1 \exp\left[i\omega\left(\frac{-z}{c_1} - t\right)\right],$$

$$\phi_2 = A_2 \exp\left[i\omega\left(\frac{z}{c_1} - t\right)\right],$$

$$\phi_3 = A_3 \exp\left[i\omega\left(\frac{-z}{c_2} - t\right)\right]$$

The above coefficients are related to each other by the following boundary conditions, which must be satisfied at the interface $z = 0$: continuity of the normal velocity component, $v_z$, and of the fluid pressure perturbation, $\delta p$. Since, according to the linearized equation of motion, $\rho \partial \vec{v}/\partial t = -\vec{\nabla}\delta p$, the pressure perturbation $\delta p = \rho \partial \phi/\partial t = -i\omega\rho\varphi$, and the above boundary conditions take the following form:

$$\rho_1(\phi_1 + \phi_2)_{z=0} = \rho_2(\phi_3)_{z=0},$$

$$\left(\frac{\partial\phi_1}{\partial z} + \frac{\partial\phi_2}{\partial z}\right)_{z=0} = \left(\frac{\partial\phi_3}{\partial z}\right)_{z=0}$$

Therefore,

$$A_2 = A_1 \frac{(c_2\rho_2 - c_1\rho_1)}{(c_2\rho_2 + c_1\rho_1)}, \quad A_3 = A_1 \frac{2c_2\rho_1}{(c_2\rho_2 + c_1\rho_1)} \qquad (3.73)$$

The sought after reflection coefficient, $R$, is equal to the ratio of the fluxes of energy, $\vec{q}_{1,2} = q_{1,2}\vec{e}_z$, carried by waves (2) and (1). Since $q_{1,2} = (\rho_1\omega^2/2c_1)A_{1,2}^2$, it follows from expressions (3.73) that

$$R = \left(\frac{A_2}{A_1}\right)^2 = \left[\frac{(c_2\rho_2 - c_1\rho_1)}{(c_2\rho_2 + c_1\rho_1)}\right]^2$$

For example, if the fluid (2) is a liquid with a very small compressibility, the respective speed of sound $c_2$ is very large, and such a fluid acts as an "acoustic mirror" with $R \approx 1$.

It is quite obvious that in the absence of any dissipation the incident energy flux $q_1$ is divided between the reflected and transmitted waves so that $q_1 = q_2 + q_3$, which can be verified directly by using expressions (3.73) and the relation $q_3 = (\rho_2\omega^2/2c_2)A_3^2$. However, the respective fluxes of the linear momentum, $\Pi_{zz} = q/c_s$, do not balance each other, which results in the force $F$ exerted on the boundary surface:

$$F = \Pi_{zz}^{(1)} + \Pi_{zz}^{(2)} - \Pi_{zz}^{(3)} = \frac{1}{c_1}(q_1 + q_2) - \frac{1}{c_2}q_3$$

By using relations (3.73), one finds that

$$F = \frac{2q_1}{c_1} \frac{c_1^2\rho_1^2 + c_2^2\rho_2^2 - 2\rho_1\rho_2c_1^2}{(c_1\rho_1 + c_2\rho_2)^2},$$

which in the case of an "acoustic mirror" yields the expected result that $F = 2q_1/c_1$.

## Problem 3.4.5

An ideal gas of temperature $T$, molecular weight $\mu$, and adiabatic index $\gamma$ is confined in a vertical tube of length $L$, both ends of which are plugged. Determine eigenfrequencies of the longitudinal (vertical) oscillations of the gas in the presence of the gravity field $\vec{g}$.

In the state of unperturbed equilibrium the pressure of a stationary gas, $p(z)$, varies with height according to the equation $dp/dz = -\rho g$. Together with the equation of state, $p = RT\rho/\mu$, it yields the barometric formula for the density and pressure distribution along the tube:

$$\rho(z) = \rho_0 \exp\left(-\frac{z}{h}\right), \quad p(z) = \frac{\rho_0 RT}{\mu} \exp\left(-\frac{z}{h}\right), \quad h = \frac{RT}{\mu g}$$

The vertical oscillation results in the density perturbation, $\delta\rho(z, t)$, the pressure perturbation, $\delta p(z, t)$, and the gas vertical velocity, $v_z(z, t) \equiv v(z, t)$. By assuming that all these perturbations vary in time proportional to $\exp(-i\omega t)$, one arrives to the following linearized equations of continuity, motion, and the entropy conservation:

$$-i\omega\delta\rho + \frac{\partial}{\partial z}(v\rho_0 \exp(-z/h)) = 0,$$

$$-i\omega\rho_0 v \exp(-z/h) = -\frac{\partial\delta p}{\partial z} - g\delta\rho, \qquad (3.74)$$

$$\frac{ds}{dt} = \frac{\partial\delta s}{\partial t} + v\frac{ds_0}{dz} = 0$$

Since for an ideal gas the entropy $s \propto p/\rho^\gamma$, the last of equations (3.74) takes the form

$$-i\omega\left(\frac{\delta\rho}{p^\gamma} - \frac{\gamma p\delta\rho}{\rho^{\gamma+1}}\right) + v\frac{d}{dz}\left(\frac{p}{\rho^\gamma}\right) = 0$$

By expressing now $\delta\rho$ and $\delta p$ in terms of $v$, and substituting the results into the equation of motion in (3.74), one arrives to the following single equation for the quantity $u(z) \equiv v(z)\exp(-z/h)$:

$$-\omega^2 u = c_s^2\frac{d^2 u}{d^2 z} + \gamma g\frac{du}{dz},$$

where $c_s = \sqrt{\gamma RT/\mu}$ is the speed of sound in the gas. Since the coefficients of this equation do not depend on the variable $z$, one can seek its solution in the form $u(z) \propto \exp(ikz)$, which yields

$$k = i\frac{\gamma g}{2c_s^2} \pm \left(\frac{\omega^2}{c_s^2} - \frac{\gamma^2 g^2}{4c_s^4}\right)^{1/2}$$

Therefore, a general solution of this equation reads

$$u(z) = A\exp\left(-\frac{\gamma g z}{2c_s^2}\right)\exp\left[iz\left(\frac{\omega^2}{c_s^2} - \frac{\gamma^2 g^2}{4c_s^4}\right)^{1/2}\right] +$$

$$B\exp\left(-\frac{\gamma g z}{2c_s^2}\right)\exp\left[-iz\left(\frac{\omega^2}{c_s^2} - \frac{\gamma^2 g^2}{4c_s^4}\right)^{1/2}\right]$$

Then, the first boundary condition, $u(0) = 0$, requires $B = -A$, while the second one, $u(L) = 0$, determines the eigenfrequencies $\omega$:

$$\left(\frac{\omega^2}{c_s^2} - \frac{\gamma^2 g^2}{4c_s^4}\right)^{1/2} = n\frac{\pi}{L}, \ n = 1, 2, 3...$$

This relation, which can be re-written as $\omega_n = \sqrt{\omega_g^2 + n^2\omega_p^2}$, where $\omega_g = \gamma g/2c_s$ and $\omega_p = \pi c_s/L$, indicates two different mechanisms behind such oscillation of the gas. The first frequency, $\omega_g$, is due to the gravity and, hence, is related to the so-called g-modes. On the other hand, the frequency $\omega_p$ corresponds to p-modes, which are supported by the gas pressure. The relative role of these two effects depends on the parameter $\alpha \equiv h/L = RT/\mu g L \sim \omega_p/\omega_g$. If $\alpha \gg 1$, the effect of gravity is negligible, $\omega_n \approx n\omega_p$, which signifies standing sound waves in a uniform gas in the tube of a finite length. In the opposite limit, when $\alpha \ll 1$, the gas pressure plays a role only for the high-number overtones with $n \geq \alpha^{-1} \gg 1$. Otherwise, the oscillation is the g-mode with the frequency $\omega \approx \omega_g$ and the vertical velocity

$$v(z) = v_0\exp\left(\frac{z}{2h}\right)\sin\left(\frac{n\pi z}{L}\right),$$

which is sharply increasing toward the upper end of the tube, where the gas density is small.

## Problem 3.4.6

Estimate the rate of the "radiative" damping of a small amplitude oscillation of a gas bubble immersed into an inviscid fluid (see Problem 3.2.6), when the latter has a finite compressibility characterized by the speed of sound equal to $c_s$.

Damping of the radial oscillation of a gas bubble in an inviscid liquid occurs due to emission of sound waves, which takes place in a liquid with a large but finite speed of sound $c_s$. The respective damping rate can be estimated by the following consideration. In an incompressible liquid the variation of the bubble radius as $r_b(t) = R + a \exp(-i\omega t)$ induces a liquid flow with the velocity

$$v_r(r,\, t) = -i\omega a \exp(-i\omega t)\frac{R^2}{r^2},$$

which instantly adjusts to the variation of $r_b(t)$ in the entire body of the liquid because in this case $c_s \to \infty$. If a finite value of $c_s$ is taken into account, such an adjustment has enough time to establish itself only at distances $r \le r_* \sim c_s/\omega$, i.e., in a so-called "quasistationary" zone. At larger distances, where $r > r_*$, in a "wave" zone, the motion of the liquid represents the spherically expanding sound waves. Therefore, the amplitude of sound waves at the boundary of these two domains, $v_s(r_*) = v_s^*$, can be estimated as

$$v_s^* \sim v(r_*) \sim \omega a R^2/r_*^2$$

Then, by knowing this amplitude, one can calculate the total flux of energy, $S$, which is carried away by sound waves. Indeed, according to equations (3.71), the density of this energy flux, $q$, is equal to $q \sim c_s\rho(v_s^*)^2$, hence

$$S \sim qr_*^2 \sim \rho c_s \omega^2 a^2 \frac{R^4}{r_*^2} \sim \rho\omega^4 a^2 \frac{R^4}{c_s}$$

The sought after damping rate is $\Gamma \sim S/W$, where $W$ is the oscillation energy, which, according to Problem 3.2.6, is of the order of $W \sim \rho\omega^2 a^2 R^3$. Therefore, $\Gamma \sim \omega^2 R/c_s$. Note that this radiative damping is weak in the sense that $\Gamma/\omega \ll 1$. Indeed, since, according to Problem 3.2.6, $\omega = \sqrt{3\gamma p/\rho R^2}$, one gets that

$$\frac{\Gamma}{\omega} \sim \frac{\omega R}{c_s} \sim \frac{\sqrt{p/\rho}}{c_s},$$

which is small because in a weakly compressible liquid the speed of sound $c_s$ is large compared with $\sqrt{p/\rho}$ (see also Problem 3.2.6).

## Problem 3.4.7

Determine the rate of damping for a sound wave in a gas due to thermal conductivity.

Consider a plane monochromatic sound wave propagating along the $z$-axis with the wave vector $k$ and the frequency $\omega$. Then, the linearized equations for perturbations of the density, $\delta\rho$, of the pressure, $\delta p$, of the entropy, $\delta s$, and the gas velocity $v_z \equiv v$ read:

$$-i\omega\delta\rho + ikv\rho = 0,$$
$$i\omega\rho v = ik\delta p, \qquad (3.75)$$
$$-i\omega\rho T\delta s = \delta Q$$

Here $\delta s$ is perturbation of the entropy per unit mass of the gas, and $\delta Q = \vec{\nabla} \cdot (\kappa\vec{\nabla}\delta T) = -\kappa k^2\delta T$ is the heat deposition per unit volume due to the thermal conductivity. Since the entropy per unit mass of an ideal gas with the adiabatic index $\gamma$ is equal to (see, e.g., L. D. Landau and E. M. Lifshitz, *Statistical Physics*, Part 1, §43, Pergamon Press, Oxford, 1985):

$$s = \frac{R}{(\gamma-1)\mu}\ln\left(\frac{p}{\rho^\gamma}\right),$$

equations (3.75) yield

$$\delta p = \frac{\omega^2}{k^2}\delta\rho, \quad -i\omega\rho T\frac{R}{(\gamma-1)\mu}\left(\frac{\delta p}{p} - \gamma\frac{\delta\rho}{\rho}\right) = -\kappa k^2\delta T$$

Together with the equation of state for an ideal gas, $p = R\rho T/\mu$, these result in the following dispersion relation, $\omega(k)$:

$$\frac{\omega^2}{k^2 c_s^2} = 1 - \frac{\kappa k^2(\gamma-1)/\gamma}{[\kappa k^2 - i\omega\rho R/(\gamma-1)\mu]}, \quad c_s^2 = \gamma\frac{p}{\rho} \qquad (3.76)$$

Note, that in order to derive the sought after wave damping rate $\Gamma$, there is, actually, no need to solve the third order equation (3.76) for $\omega$. The reason is that within the domain of applicability of the macroscopic equations of gas dynamics, which requires $\lambda \gg l$, where $l$ is the mean free path of molecules of the gas, the damping is weak. Therefore, the frequency $\omega(k)$ can be written as $\omega \approx kc_s - i\Gamma$, with the decrement $\Gamma \ll kc_s$, and the terms in equation (3.76), which are proportional to the heat conductivity $\kappa$, can be considered as small. Then, a simple derivation yields $\Gamma \approx (\gamma-1)^2\kappa\mu k^2/2\gamma\rho R$.

Consider now the estimates confirming that the above damping is weak indeed. Firstly, the heat conductivity coefficient in a gas, $\kappa$, can be estimated as $\kappa \sim R\rho v_t l/\mu$, where $v_t$ is the thermal velocity of molecules. Thus, the ratio $\Gamma/kc_s \sim v_t kl/c_s \sim v_t l/c_s\lambda$, and since the speed of sound in the gas is of the order of $v_t$, one gets that $\Gamma/kc_s \sim l/\lambda$, which must be small by the very meaning of the macroscopic description of the gas.

## Problem 3.4.8

Determine the damping rate for a sound wave in a liquid, which is due to its resonant interaction with a small fraction of gas bubbles present in the liquid (D. D. Ryutov, JETP Letters, v.22(9), 215, 1975).

Consider a liquid with gas bubbles that are randomly but, on average, uniformly distributed in space. They can be described by the distribution function $f(r)$, so that $f(r)dr$ is the number of bubbles per unit volume whose radius falls into the interval $(r, r+dr)$. It is assumed that gas bubbles occupy only a small fraction of the total volume, which in quantitative terms means that their mean radius $R = \int_0^\infty f(r)rdr / \int_0^\infty f(r)dr$ is small compared to the mean distance $l$ between the bubbles $(l = (\int f(r)dr)^{-1/3})$. If the wavelength $\lambda$ of a sound wave propagating in such a liquid is much longer than $l$, the medium can be considered as a continuous one and, therefore, described by the macroscopic equations for quantities that are averaged over the distance large compared to $l$ but much smaller than $\lambda$. Thus, the linearized equations of continuity and motion take the following form:

$$\frac{\partial}{\partial t}\langle\delta\rho\rangle + \langle\rho\rangle\frac{\partial v}{\partial x} = 0, \ \langle\rho\rangle\frac{\partial v}{\partial t} = -\frac{\partial}{\partial x}\langle\delta p\rangle,$$

or

$$\frac{\partial^2}{\partial^2 t}\langle\delta\rho\rangle = \frac{\partial^2}{\partial^2 x}\langle\delta p\rangle, \tag{3.77}$$

where $\langle\delta\rho\rangle$ and $\langle\delta p\rangle$ are the averaged density and pressure perturbations, while $\langle\rho\rangle$ is the mean density of this medium: $\langle\rho\rangle = \rho(1-\alpha)$. Here $\rho$ is the density of liquid, and

$$\alpha = \frac{4\pi}{3}\int_0^\infty f(r)r^3 dr \sim \left(\frac{R}{l}\right)^3 \ll 1$$

is the fraction of the total volume occupied by bubbles (density of the gas is negligibly small). Then, in order to find the dispersion relation $\omega(k)$ and, hence, the sought after damping rate of the wave, one needs to obtain the relation between $\langle\delta\rho\rangle$ and $\langle\delta p\rangle$. In a pure liquid it simply reads $\delta p = c_s^2\delta\rho$, where $c_s$ is the speed of sound in the liquid. However, in the present case a perturbation of pressure induces forced oscillations of the bubbles and, therefore, variation of the radius of a bubble, $\xi_r$, also contributes to the perturbation of density (since under a given $\delta p$ this variation, $\xi$, depends on the bubble radius $r$, a subscript $r$ is added to it). Thus, for a small value of parameter $\alpha$ one gets that

$$\langle\delta\rho\rangle = \frac{\langle\delta p\rangle}{c_s^2} - 4\pi\rho\int_0^\infty f(r)r^2\xi_r dr,$$

where the value of $\xi_r$ is determined by the forced oscillation equation that follows from Problem 3.2.6:

$$\frac{d^2\xi_r}{d^2t} + \omega_0^2(r)\xi_r = -\frac{\langle\delta p\rangle}{r\rho},$$

where $\omega_0(r) = \sqrt{3\gamma p/\rho r^2}$ is the eigenfrequency of the radial oscillation of a bubble of radius $r$. Thus,

$$\xi_r = -\frac{\langle\delta p\rangle}{r\rho[\omega_0^2(r) - \omega^2]},$$

$$\langle\delta\rho\rangle = \langle\delta p\rangle\left(\frac{1}{c_s^2} + 4\pi\int\limits_0^\infty \frac{f(r)rdr}{[\omega_0^2(r) - \omega^2]}\right),$$

which being inserted into equation (3.77) yields the following result:

$$\omega^2 = k^2c_s^2\left(1 + 4\pi c_s^2 \int_0^\infty \frac{f(r)rdr}{[\omega_0^2(r) - \omega^2]}\right)^{-1} \tag{3.78}$$

In the absence of bubbles $\omega = kc_s$, so for a small value of $\alpha$ one can try the perturbation approach by putting $\omega = kc_s$ in the denominator of the integrand in equation (3.78), which yields the dispersion law

$$\omega \approx kc_s\left(1 - 2\pi c_s^2 \int_0^\infty \frac{f(r)rdr}{[\omega_0^2(r) - k^2c_s^2]}\right) \tag{3.79}$$

Now one can estimate the resulting correction to the frequency of the sound wave, $\Delta\omega$, caused by the presence of gas bubbles. Thus, assume at first that there is no singularity in the denominator of equation (3.79), in which case

$$\int_0^\infty \frac{f(r)rdr}{[\omega_0^2(r) - k^2c_s^2]} \sim \frac{\alpha}{R^2\omega_0^2(R)} \sim \frac{\alpha\rho}{p},$$

hence

$$\frac{\Delta\omega}{\omega} \sim \frac{\alpha}{\epsilon},$$

where $\epsilon \equiv p/\rho c_s^2 \ll 1$. Therefore, even a small fraction of gas bubbles, when $\alpha \ll 1$, can significantly affect the properties of sound waves in such a medium. Consequently, the suggested perturbation approach does hold only when $\alpha \ll \epsilon \ll 1$.

However, even under this condition the presence of gas bubbles in a liquid can bring about a qualitatively new effect: the resonant damping of sound waves, which is associated with the bubbles whose oscillation eigenfrequency $\omega_0(r)$ is equal to the frequency of the sound wave. Therefore, such a damping, which occurs in the absence of any dissipation process, is completely analogous to the Landau damping of Langmuir waves in a plasma (see Problem

2.1.17), and to the resonant wind-surface wave interaction considered in Problem 3.1.15. Its formal derivation can be carried out as follows. The singularity in the integrand in equation (3.79) is due to the divergence of $\xi_r$, which tends to infinity for the resonant bubbles. Therefore, in order to avoid this singularity one can add an infinitesimally weak damping to the bubble eigenmode (for example, due to viscosity of the liquid as in Problem 3.2.6). Then, the eigenfrequency $\omega_0(r)$ acquires a small imaginary part: $\omega_0(r) = \omega_0 - i\nu$, $\nu > 0$, and the integral in (3.79) takes the form

$$\int_0^\infty \frac{f(r)r\,dr}{(\omega_0 + kc_s)(\omega_0 - kc_s - i\nu)} = \int_0^\infty \frac{f(r)r\,dr[(\omega - kc_s) + i\nu]}{(\omega_0 + kc_s)[(\omega_0 - kc_s)^2 + \nu^2]}$$

Since

$$\lim_{\nu \to 0} \frac{\nu}{(\omega_0 - kc_s)^2 + \nu^2} = \pi\delta(\omega_0 - kc_s),$$

this integral acquires the imaginary part

$$\Im \int_0^\infty \frac{f(r)r\,dr}{(\omega_0 + kc_s)(\omega_0 - kc_s - i\nu)} \approx \pi \int_0^\infty \frac{f(r)r\,dr}{2\omega_0(r)} \delta[\omega_0(r) - kc_s]$$

Therefore, equation (3.79) yields the following expression for the resonant damping rate $\Gamma$:

$$\Gamma = \pi^2 c_s^2 \int_0^\infty f(r)r\,dr\,\delta[\omega_0(r) - kc_s] = \frac{\pi^2 c_s^2 f(r_*)r_*}{|d\omega_0(r)/dr|_{r_*}} =$$

$$\frac{\pi^2 c_s}{k} f(r_*)r_*^2, \quad \omega_0(r_*) = kc_s$$

If the distribution function of the bubbles, $f(r)$, is a smooth function with a scale of variation of the order of $r_*$, the radius of the resonant bubbles, a simple estimate shows that $\Gamma \sim \alpha c_s^2/r_*^2\omega_0(r_*)$ and, hence, the ratio $\Gamma/\omega \sim \alpha c_s^2/r_*^2\omega^2 \sim \alpha/\epsilon$. Therefore, all these results, which are obtained with the help of the perturbation approach, are valid only when $\alpha \ll \epsilon \ll 1$. Otherwise, the presence of gas bubbles in a liquid results in a significant alteration of the properties of sound waves in such a medium.

## 3.5 Shock waves. Solitons

### Problem 3.5.1

In a uniform ideal gas of pressure $p_0$, density $\rho_0$, and adiabatic index $\gamma$, the following initial static density perturbation, $\delta\rho$, is formed at the moment of time $t = 0$:

$$\delta\rho = \begin{cases} \rho_1 \sin\left(\frac{\pi x}{L}\right), & |x| \leq L \\ 0, & |x| \geq L \end{cases}$$

Determine the location and the moment of time when a discontinuity (shock wave) appears in such a gas, assuming that the initial perturbation is weak, i.e., $\rho_1 \ll \rho_0$.

In the linear (acoustic) approximation such a perturbation during the time interval $\tau_0 \sim L/c_0$ splits into two acoustic impulses of constant shape that are separated in space and propagate with the speed equal to $c_0 = (\gamma p_0/\rho_0)^{1/2}$ in the opposite directions along the $x$-axis. According to Problem 3.4.1, the velocity profiles in these pulses are as follows:

$$v_+(x,\,t) = \frac{c_0\rho_1}{2\rho_0} \sin\left[\frac{\pi}{L}(x - c_0 t)\right],$$

$$|x \pm c_0 t| \leq L, \tag{3.80}$$

$$v_-(x,\,t) = -\frac{c_0\rho_1}{2\rho_0} \sin\left[\frac{\pi}{L}(x + c_0 t)\right]$$

Here $v_+$ and $v_-$ define the impulses propagating, respectively, to the right and to the left. Steepening of the profile of these impulses and, eventually, formation of the discontinuity is a non-linear process, which under the condition $\rho_1 \ll \rho_0$ proceeds on the time scale $\tau_1 \gg \tau_0$. Therefore, the non-linear distortion of the two impulses in equation (3.80) can be considered independently of each other. The actual steepening mechanism is that the propagation velocity is not the same for different points of the impulse: it depends on the local fluid velocity and, hence, differs, though only slightly in the present case, from the sound speed $c_0$ in the unperturbed gas. The exact non-linear solution of the gas dynamics equations, the so-called Riemann wave (see, e.g., L.D. Landau and E. M. Lifshitz, *Fluid Mechanics*, §104, Pergamon Press, Oxford, 1987), shows that this propagation velocity $u$ is equal to

$$u(v) = v \pm c_s(v), \tag{3.81}$$

where $c_s(v)$ is the local sound speed, and two signs correspond to the impulses propagating, respectively, to the right and to the left. In the case of a weak non-linearity, when $\rho_1 \ll \rho_0$, the value of $c_s(v)$ can be obtained by the perturbation method. Thus, in the linear approximation the perturbations of density and pressure in the impulses (3.80), which follow from the continuity equation, are equal to

$$\delta\rho_\pm = \pm\frac{\rho_0}{c_0}v_\pm = \frac{\rho_1}{2}\sin\left[\frac{\pi}{L}(x \pm c_0 t)\right],$$

$$\delta p_\pm = c_0^2 \delta\rho_\pm$$

Therefore, the local speed of sound, $c_s(v)$, can be expanded as

$$c_s(v) = \left(\gamma\frac{p}{\rho}\right)^{1/2} \approx \left(\gamma\frac{p+\delta p}{\rho+\delta\rho}\right)^{1/2} \approx$$

$$c_0\left(1 + \frac{\delta p}{2p_0} - \frac{\delta\rho}{2\rho_0}\right) \approx c_0 \pm \frac{(\gamma-1)}{2}v,$$

and the propagation velocity (3.81) takes the form

$$u(v) = \pm c_0 + \frac{(\gamma+1)}{2}v \tag{3.82}$$

(note that, in the case of an ideal gas with a constant adiabatic index $\gamma$, expression (3.82), obtained above for a weak non-linearity, actually reproduces the exact result). Then, the sought after location and instant of the discontinuity formation in the gas can be derived by the method already used in Problem 3.1.3 for the dust cloud. The only difference is that unlike the dust, where each fluid particle continues to move with its initial velocity $v$, in a compressible gas a point with a given fluid velocity $v$ propagates in space with the speed given by expression (3.82). Therefore, if the initial coordinate of such a point is equal to $x_0$, its subsequent position is

$$x(x_0, t) = x_0 + u(v)t = x_0 \pm c_0 t + \frac{1}{2}(\gamma+1)v_\pm t =$$

$$x_0 \pm c_0 t \pm \frac{(\gamma+1)}{2}\frac{c_0 t}{2}\frac{\rho_1}{\rho_0}\sin\left(\frac{\pi x_0}{L}\right)$$

A discontinuity is formed when the derivative $(\partial x/\partial x_0)_t$ becomes equal to zero. Thus, in the present case

$$\left(\frac{\partial x}{\partial x_0}\right)_t = 1 \pm \frac{(\gamma+1)}{2}\frac{c_0 t}{2}\frac{\rho_1}{\rho_0}\frac{\pi}{L}\cos\left(\frac{\pi x_0}{L}\right)$$

Therefore, in the impulse moving to the right (the sign plus in the above expression) the discontinuity forms first at the impulse ends, $x_0 = \pm L$, at the moment of time $t = \tau_1$, with

$$\tau_1 = \frac{4L}{\pi c_0(\gamma+1)}\frac{\rho_0}{\rho_1} \gg \tau_0$$

In the impulse moving to the left the discontinuity occurs at the same time but in the middle of the impulse, $x_0 = 0$.

## Problem 3.5.2

A shock wave with the Mach number $\mathcal{M}$ propagates in a gas of pressure $p_1$, density $\rho_1$. Determine the pressure, $p_2$, and the density, $\rho_2$, of the gas behind the shock wave, assuming that the adiabatic index of the gas, $\gamma$, remains constant.

The states of a gas in front of the shock wave and behind it are related by continuity of the fluxes of mass, linear momentum, and energy at both sides of the shock front. Thus, in the rest frame of the shock one gets the following conditions:

$$\rho_1 v_1 = \rho_2 v_2,$$
$$p_1 + \rho_1 v_1^2 = p_2 + \rho_2 v_2^2, \qquad (3.83)$$
$$w_1 + \frac{v_1^2}{2} = w_2 + \frac{v_2^2}{2}$$

Here $w$ is the heat function, which for an ideal gas with a constant adiabatic index $\gamma$ is equal to $w = \gamma p / (\gamma - 1)\rho$. Then, by taking into account a definition of the Mach number: $\mathcal{M} = v_1/c_1$, where $c_1 = \sqrt{\gamma p_1/\rho_1}$ is the speed of sound in the gas in front of the shock, one gets from equations (3.83) that

$$1 + \gamma \mathcal{M}^2 = y + \frac{\gamma \mathcal{M}^2}{x},$$
$$\frac{1}{(\gamma - 1)} + \frac{1}{2}\mathcal{M}^2 = \frac{y}{(\gamma - 1)x} + \frac{1}{2x^2}\mathcal{M}^2,$$

where $x \equiv \rho_2/\rho_1$ and $y \equiv p_2/p_1$. After eliminating a trivial solution $x = y = 1$, one finally arrives to

$$\frac{\rho_2}{\rho_1} = \frac{(\gamma + 1)\mathcal{M}^2}{[(\gamma - 1)\mathcal{M}^2 + 2]}, \quad \frac{p_2}{p_1} = \frac{[2\gamma \mathcal{M}^2 - (\gamma - 1)]}{(\gamma + 1)} \qquad (3.84)$$

It follows from these relations that for a weak shock, when $(\mathcal{M} - 1) \ll 1$,

$$\frac{\rho_2}{\rho_1} \approx 1 + \frac{4(\mathcal{M} - 1)}{(\gamma + 1)}, \quad \frac{p_2}{p_1} \approx 1 + \frac{4\gamma(\mathcal{M} - 1)}{(\gamma + 1)} \qquad (3.85)$$

As seen, in this case $\delta p/p \approx \gamma \delta \rho/\rho$, which is in accordance with the adiabatic law. Therefore, in a weak shock wave the change (increase) in the gas entropy is of a higher order of the small parameter $(\mathcal{M} - 1)$. In the opposite limit of a strong shock wave, for which the Mach number $\mathcal{M} \gg 1$, the expressions (3.84) yield

$$\frac{\rho_2}{\rho_1} \approx \frac{(\gamma + 1)}{(\gamma - 1)}, \quad \frac{p_2}{p_1} \approx \frac{2\gamma \mathcal{M}}{(\gamma + 1)} \qquad (3.86)$$

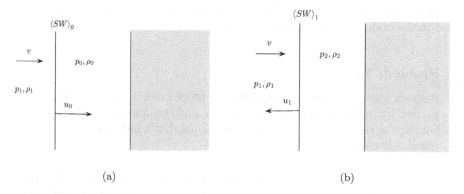

**FIGURE 3.16**
Reflection of the shock wave

Thus, behind a strong shock there is only a finite degree of compression, and, therefore, the prescribed increase in the pressure is mainly due to heating of the gas.

### Problem 3.5.3

A strong shock wave with the Mach number $\mathcal{M}_0 \gg 1$ is reflected backward by a plane rigid wall as shown in Figure 3.16. Determine the Mach number $\mathcal{M}_1$ of the reflected shock wave.

In the rest frame of the wall, where the gas next to it is also stationary, the incident shock propagates toward the wall with the velocity $u_0 = \mathcal{M}_0 c_0$, while the gas behind it has a velocity $v$ as shown in Figure 3.16(a). As this shock is strong, the relations (3.84) hold, and the first of them yields

$$\frac{\rho_1}{\rho_0} = \frac{u_0}{(u_0 - v)} \approx \frac{(\gamma + 1)}{(\gamma - 1)},$$

so that

$$v \approx \frac{2}{(\gamma + 1)} \mathcal{M}_0 c_0 \tag{3.87}$$

Then, the reflected shock, shown in Figure 3.16(b), propagates relative to the gas in front of it with the velocity equal to $(u_1 + v) = \mathcal{M}_1 c_1$, where $c_1$ is the speed of sound in this gas. It follows from expressions (3.86) that

$$c_1 = \left(\frac{\gamma p_1}{\rho_1}\right)^{1/2} \approx c_0 \frac{\mathcal{M}_0}{(\gamma + 1)} \sqrt{2\gamma(\gamma - 1)} \tag{3.88}$$

On the other hand, by using the first relations in (3.84), the degree of compression in the reflected shock, $\rho_2/\rho_1$, is equal to

$$\frac{\rho_2}{\rho_1} = \frac{(u_1 + v)}{u_1} = \frac{\mathcal{M}_1 c_1}{(\mathcal{M}_1 c_1 - v)} = \frac{(\gamma + 1)\mathcal{M}_1^2}{[(\gamma - 1)\mathcal{M}_1^2 + 2]}$$

By substituting here expressions (3.87) and (3.88) for $v$ and $c_1$, one arrives to the following equation for the Mach number $\mathcal{M}_1$:

$$\mathcal{M}_1^2 - \frac{(\gamma + 1)}{\sqrt{2\gamma(\gamma - 1)}} - 1 = 0,$$

with the appropriate solution (the second root of the above equation is negative) $\mathcal{M}_1 = \sqrt{2\gamma/(\gamma - 1)}$. As seen, $\mathcal{M}_1 \sim 1$, i.e., the reflected shock wave is significantly attenuated.

## Problem 3.5.4

Determine the internal structure of a weak shock wave (Mach number $(\mathcal{M} - 1) \ll 1$) in a viscous gas.

As follows from the results obtained in Problem 3.4.1, in an ideal gas a non-linear impulse, propagating, say, to the right along the $x$-axis, is described by the following equation:

$$\frac{\partial v}{\partial t} + \left(c_0 + \frac{(\gamma + 1)}{2}v\right)\frac{\partial v}{\partial x} = 0, \tag{3.89}$$

and in the course of its evolution the steepening of the impulse profile eventually results in its breaking up. However, even a weak dissipation (in a form of viscosity or thermal conductivity), which is always present in a real gas, prevents it, leading instead to formation of a regular structure of finite width. In order to account for this effect, one can add the viscous term to equation (3.89), so that

$$\frac{\partial v}{\partial t} + \left(c_0 + \frac{(\gamma + 1)}{2}v\right)\frac{\partial v}{\partial x} = \nu\frac{\partial^2 v}{\partial^2 x},$$

where $\nu$ is kinematic viscosity of the gas. By making transition to the reference frame moving with the sound speed $c_0$, and by introducing the new function $u = (\gamma + 1)v/2$, the above-written equation takes the form

$$\frac{\partial u}{\partial t} + u\frac{\partial u}{\partial x} = \nu\frac{\partial^2 u}{\partial^2 x}, \tag{3.90}$$

which is known as **Burgers' equation**. Since the interest here is in a steady shock wave propagating with a constant velocity, one can seek a solution of equation (3.90) in the form $u(x, t) = f(x - v_0 t)$, where $v_0$ is the speed of the shock in the chosen reference frame. By substituting it into (3.90), one gets the following equation for $f(z)$:

$$-v_0 \frac{df}{dz} + f \frac{df}{dz} = \nu \frac{d^2 f}{d^2 z},$$

which after being integrated once yields

$$\nu \frac{df}{dz} = \frac{1}{2}(f - v_0)^2 - \frac{w_0^2}{2}$$

(note that such a choice of the constant of integration, $w_0^2/2 > 0$, ensures that the solution $f(z)$ tends to some constants at $z \to \pm\infty$ while $df/dz$ tends to zero). Then, for the function $\phi(z) \equiv (f - v_0)/w_0$ one gets the equation

$$\phi^2 - 1 = \frac{2\nu}{w_0} \frac{d\phi}{dz},$$

whose solution is

$$\phi(z) = -\tanh\left(\frac{w_0}{2\nu} z\right) \tag{3.91}$$

In the weak shock the discontinuity of the gas velocity, $v_1 - v_2$, is equal, according to relations (3.85), to

$$\Delta v = (v_1 - v_2) \approx c_0 \frac{4(\mathcal{M} - 1)}{(\gamma + 1)}$$

On the other hand, the solution (3.91) yields $\Delta v = 2w_0 \Delta\phi/(\gamma+1) = 4w_0/(\gamma+1)$. Therefore, the parameter $w_0 = c_0(\mathcal{M} - 1)$, and, according to expression (3.91), the width of a weak shock wave, $\Delta x$, is equal to

$$\Delta x = \frac{2\nu}{w_0} = \frac{2\nu}{c_0(\mathcal{M} - 1)} \tag{3.92}$$

It is worth emphasizing that the above-given derivation does not hold for a sufficiently strong shock with a Mach number $(\mathcal{M} - 1) \geq 1$. The reason is that the width of such a shock is of the order of the mean free path $l$ in a gas, and, therefore, its internal structure cannot be described by the macroscopic fluid equations. Indeed, in a gas the kinematic viscosity $\nu$ is of the order of $\nu \sim v_t l$, where $v_t$ is the thermal velocity of molecules. Then, since the speed of sound in a gas $c_0 \sim v_t$, equation (3.92) yields $\Delta x \sim l/(\mathcal{M} - 1)$, which is formally less than $l$ for a strong shock.

## Problem 3.5.5

Derive the dispersion equation for a gravity wave propagating on the free surface of an ideal incompressible fluid of a finite depth equal to $H$.

Unlike the case of an infinitely deep fluid considered in Problem 3.1.8, now in the general solution of the Laplace equation for the velocity potential $\phi$ one should retain both $z$-components, hence

$$\phi(x,\, z,\, t) = [A\exp(kz) + B\exp(-kz)]\exp[i(kx - \omega t)]$$

The constant $B$ is determined by the boundary condition at the bottom surface $z = -H$, where the normal (vertical) component of the fluid velocity should vanish:

$$v_z(z = -H) = -\left(\frac{\partial\phi}{\partial z}\right)_{z=-H} = 0,$$

which yields

$$\phi = A\cosh[k(z + H)]\exp[i(kx - \omega t)] \tag{3.93}$$

Then, the linearized equation of motion (3.14) yields the following distribution of pressure in the fluid:

$$p = -\rho g z - i\omega\rho\phi = -\rho g z - i\omega\rho A\cosh[k(z + H)]\exp[i(kx - \omega t)] \tag{3.94}$$

If vertical displacement of the free surface of the fluid is equal to $\xi(x,\, t) = \xi_0\exp[i(kx - \omega t)]$, the constants $A$ and $\xi_0$ are related to each other by the condition

$$\frac{\partial\xi}{\partial t} = -\left(\frac{\partial\phi}{\partial z}\right)_{z=0},$$

which yields $A = i\omega\xi_0/k\sinh(kH)$. Finally, the sought after dispersion law $\omega(k)$ follows from the boundary condition at the free surface of the fluid: $p_{z=\xi} = 0$, which, by using equation (3.94) with the required linear accuracy, yields

$$\omega(k) = \sqrt{kg\tanh(kH)} \tag{3.95}$$

As seen from this dispersion law, a finite depth of the fluid affects only long wavelength surface wave, for which $\lambda = 2\pi/k \geq H$. For short wavelength motions, when $\lambda \leq H$, the dispersion relation (3.95) reduces to $\omega = \sqrt{gk}$ — deep water gravity surface waves. In the limit of a very long wavelength, $\lambda \gg H$, (3.95) yields

$$\omega \approx \pm u_0 k, \quad u_0 = \sqrt{gH}, \tag{3.96}$$

which is a shallow water gravity wave that propagates with the speed equal to $u_0$.

## Problem 3.5.6

Derive an evolution equation for a small amplitude shallow-water gravity wave, $(\lambda \gg H)$, by taking into account weak effects of the non-linearity and dispersion.

In this problem it is helpful to explore from the very beginning all simplifications arising in the limit of a very long wavelength of the fluid motion. Thus, it follows from equations (3.93) and (3.94) that in this case $v_x \gg v_z$, and that both significant perturbations, $v_x$ and $\delta p$, have a very weak dependence on the $z$-coordinate. Therefore, one can consider the problem to be as the one-dimensional, with $v_x \equiv v(x)$ and $\delta p(x)$. Furthermore, since under the vertical displacement of the surface, $\xi(x, t)$, the total depth of the fluid becomes equal to $h(x, t) = H + \xi(x, t)$, the mass conservation law, which substitutes here the continuity equation, takes the form

$$\frac{\partial h}{\partial t} + \frac{\partial}{\partial x}(hv) = 0 \tag{3.97}$$

In the equation of motion,

$$\frac{\partial v}{\partial t} + v\frac{\partial v}{\partial x} = -\frac{1}{\rho}\frac{\partial \delta p}{\partial x},$$

the pressure perturbation $\delta p$ can be expressed in terms of the surface vertical displacement $\xi$ by using equation (3.94) and the boundary condition $p_{z=\xi} = 0$, which yield $\delta p = \rho g \xi$, hence

$$\frac{\partial v}{\partial t} + v\frac{\partial v}{\partial x} = -g\frac{\partial \xi}{\partial x} = -g\frac{\partial h}{\partial x} \tag{3.98}$$

In the linear approximation equations (3.97) and (3.98) reduce to

$$\frac{\partial h}{\partial t} + H\frac{\partial v}{\partial x} = 0, \quad \frac{\partial v}{\partial t} + g\frac{\partial h}{\partial x} = 0,$$

which combine into the wave equation

$$\frac{\partial^2 h}{\partial^2 t} - gH\frac{\partial^2 h}{\partial^2 x} = 0,$$

with the dispersion law (3.96).

The next task is to use equations (3.97-3.98) for derivation of the nonlinear equation describing the impulse that is analogous to the Riemann wave in the gas dynamics (see Problem 3.5.1). In this case there is a one-to-one correspondence between the functions $h(x, t)$ and $v(x, t)$, so that $h(x, t) \equiv h(v)$, and it follows from (3.97-3.98) that

$$\frac{dh}{dv}\left(\frac{\partial v}{\partial t} + v\frac{\partial v}{\partial x}\right) = -h\frac{\partial v}{\partial x},$$

$$\frac{\partial v}{\partial t} + v\frac{\partial v}{\partial x} = -g\frac{dh}{dv}\frac{\partial v}{\partial x} \tag{3.99}$$

By dividing the left-hand and right-hand sides of these equations by each other one gets

$$\left(\frac{dh}{dv}\right)^2 = \frac{h}{g}$$

Thus, $(dh/dv) = \pm\sqrt{h/g}$, and equation (3.99) takes the form

$$\frac{\partial v}{\partial t} + [v \pm \sqrt{gh}]\frac{\partial v}{\partial x} = 0 \qquad (3.100)$$

This means that a point with a given value of the fluid velocity $v$ (and, hence, $h$) propagates along the $x$-axis with the speed equal to

$$u = v \pm \sqrt{gh} \qquad (3.101)$$

As seen from expression (3.101), the quantity $\sqrt{gh}$ plays here the role of the sound speed in gas dynamics (see equation (3.81)). In the linear approximation $u = \pm\sqrt{gH}$, and in the case of a weak non-linearity one gets, by expanding (3.101), that

$$u \approx v \pm \sqrt{gH} \pm \frac{1}{2}\left(\frac{g}{H}\right)^{1/2}(h - H),$$

and since

$$h = H + \left(\frac{dh}{dv}\right)_{h=H} \qquad v = H \pm \left(\frac{H}{g}\right)^{1/2}v,$$

finally

$$u \approx \frac{3}{2}v \pm \sqrt{gH} \qquad (3.102)$$

By comparing this expression with that of (3.82), one concludes that weakly non-linear shallow-water surface waves are analogous to non-linear impulses in an ideal gas with the adiabatic index $\gamma = 2$. In particular, they are subjected to the same steepening of the impulse profile as it happens in a gas (see Problem 3.5.1). However, the ultimate outcome of this process in a gas and in a shallow water is completely different. In a gas, the steepening proceeds until some dissipation (viscosity or thermal conductivity) comes into play, and a shock wave is formed (see Problem 3.5.4). On the contrary, in a shallow water the steepening is terminated well before that due to the effect of dispersion. It has been mentioned above that equations (3.97-3.98) are actually the first-order approximation with respect to the small parameter $H/\lambda$. Since steepening of the wave results in the effective reduction of $\lambda$ (the short wavelength harmonics are formed), the effects of higher-order in $H/\lambda$ come into play. Therefore, the propagation velocity of a perturbation becomes dependent on the harmonic wavelength, which brings about spreading of the impulse (see Problem 2.3.4). This process impedes the non-linear steepening, and a competition between the two processes results in the formation of the non-linear structure that propagates on the shallow water surface without changing its shape (see Problem 3.5.7).

In the case of the weak non-linearity and dispersion these two effects can be considered separately, and the latter can be accounted for by using the linear dispersion equation (3.95), where the term of the next order in the small parameter $kH$ is retained. Thus, instead of relation (3.95), one gets

$$\omega(k) \approx \pm ku_0\left(1 - \frac{k^2H^2}{6}\right) \qquad (3.103)$$

Then, in order to incorporate this effect into the evolution equation (3.100), one should modify the latter in such a way that in the linear approximation it reproduces the dispersion law (3.103). Hence, the required form is as follows:

$$\frac{\partial v}{\partial t} + u_0 \frac{\partial v}{\partial x} + \frac{u_0 H^2}{6} \frac{\partial^3 v}{\partial^3 x} = 0,$$

which after accounting also for the weak non-linearity reads

$$\frac{\partial v}{\partial t} + u_0 \frac{\partial v}{\partial x} + \frac{u_0 H^2}{6} \frac{\partial^3 v}{\partial^3 x} + \frac{3}{2} v \frac{\partial v}{\partial x} = 0$$

The second term of this equation describes propagation of the impulse as a whole with the speed $u_0$ and, therefore, has no effect on the impulse shape. Thus, after transition to the reference frame moving with the velocity $u_0$, and introduction of the new function $V = 3v/2$, one arrives to

$$\frac{\partial V}{\partial t} + V \frac{\partial V}{\partial x} + \beta \frac{\partial^3 V}{\partial^3 x} = 0, \tag{3.104}$$

with $\beta = u_0^2 H^2 / 6$, which is known as the **Korteweg-deVries equation**, or simply KdV. It has been intensively investigated during the last half a century because of its relavance to numerous problems in the physics of non-linear waves. The remarkable new mathematical method of solving the non-linear evolution equations, the inverse scattering transform, was first discovered (C. S. Gardner et al., Phys. Rev. Letters, v.19, 1095, 1967) by studing this equation (see also, e.g., S. P. Novikov et al., *Theory of Solitons*, Springer, Berlin, 1984).

### Problem 3.5.7

Determine a steady profile of the non-linear dispersive solitary impulse (soliton), which is governed by the Korteweg-de Vries equation.

Consider the solution of the KdV equation in the form of an impulse of steady shape that propagates with a constant speed: $V = f(x - ut)$. Its substitution into equation (3.104) yields $-uf' + ff' + \beta f''' = 0$, which after one integration results in

$$f'' = \frac{u}{\beta} f - \frac{f^2}{2\beta} + C \tag{3.105}$$

Here $C$ is a constant of integration, which can be made equal to zero by an appropriate choice of the reference frame. This equation can be analyzed with the help of the mechanical analogy by writing (3.105) (with $C = 0$) in the form

$$\frac{\partial^2 f}{\partial^2 x} = \frac{u}{\beta} f - \frac{f^2}{2\beta} \equiv F(f) \tag{3.106}$$

By interpreting $f$ as the coordinate and $x$ as time, it becomes the equation of motion for a point of unit mass under a force $F(f)$. Hence, the respective potential energy, $W(f)$, which is defined as $F = -dW/df$, is equal to

$$W(f) = -\frac{u}{2\beta}f^2 + \frac{1}{6\beta}f^3 \qquad (3.107)$$

It is plotted in Figure 3.17, where $u > 0$ is assumed. The interest here is in the "bounded" trajectories, which correspond to the energy interval $-2u^3/3\beta < E \le 0$. For a low energy the "particle" oscillates at the bottom $f = 2u$ of the potential well, where the shape of $W(f)$ is close to parabolic. This yields

$$V(x, t) \approx 2u + V_0 \exp\left[i\left(\frac{u}{\beta}\right)^{1/2}(x - ut)\right], \qquad (3.108)$$

which represents a linear wave with the dispersion law that follows from the "linearized" KdV equation

$$\frac{\partial V}{\partial t} + 2u\frac{\partial V}{\partial x} + \beta\frac{\partial^3 V}{\partial^3 x} = 0$$

(note that in equation (3.108) $k = \sqrt{u/\beta}$). A role of the non-linearity increases with the increasing energy $E$, when a non-symmetric shape of the potential energy $W(f)$ (steeper at $f > 2u$, and flatter at $f < 2u$) becomes important. Therefore, in the non-linear periodic wave the maxima of $V$ become relatively narrow (a "particle" passes through respective interval of $f$ "quickly"), while the minima are wider (a "particle" moves there "slowly"). In the limit of $E \to 0$ the left-hand side "reflection" point $f = 0$ is reached asymptotically at $t \to \infty$, as the period of oscillation diverges at this energy. In terms of $f(x)$ it means that while $x$ varies from $-\infty$ to $+\infty$, the solution $f(x)$ varies from $f(-\infty) = 0$ to the maximum $f(0) = 3u$, and then back to $f = 0$ at $x \to +\infty$. This particular solution is called the **"soliton"** (from a solitary wave). Thus, on the surface of the fluid, which is at rest at infinity, such a soliton propagates with the "supersonic" speed equal to $u_0 + u$, where $u = f_{max}/3$.

Consider now the shape of the soliton. The first integral of equation (3.106) reads

$$\frac{1}{2}\left(\frac{df}{dx}\right)^2 + W(f) = const = E = 0 \qquad (3.109)$$

It follows then from equations (3.109) and (3.107) that

$$\frac{df}{dx} = \pm\left(\frac{u}{\beta}\right)^{1/2}f\left(1 - \frac{f}{3u}\right)^{1/2} \qquad (3.110)$$

If the soliton apex, where $f = f_{max} = 3u$, is located at $x = 0$, in the region $x > 0$ the derivative $df/dx < 0$, and one gets from equation (3.110) that

$$\int_{f/f_{max}}^{1} \frac{dz}{z\sqrt{1 - z}} = \left(\frac{u}{\beta}\right)^{1/2}x,$$

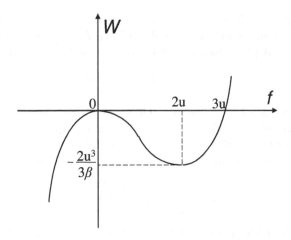

**FIGURE 3.17**
The effective potential energy for waves in the KdV equation

which after a standard integration yields

$$\frac{1 + \sqrt{1 - f/f_{max}}}{\sqrt{f/f_{max}}} = \exp\left[\left(\frac{u}{4\beta}\right)^{1/2} x\right]$$

Thus, the shape of the soliton is discribed by the following expression

$$f(x) = \frac{3u}{\cosh^2(x/l)}, \quad l = \sqrt{4\beta/u} \tag{3.111}$$

As seen from expression (3.111), the width of the soliton, $l$, decreases with the increasing amplitude, $u$, as $l \propto 1/\sqrt{u}$. Such a scaling can be explained by the following consideration. A steady shape of the soliton is established by competition of two processes: the non-linearity, that leads to the steepening of the impulse, and the dispersion, that causes its spreading. Thus, for the soliton, where these two effects balance each other, the non-linear and the dispersive terms in the KdV equation (3.104) should be of the same order of magnitude. Hence,

$$V\frac{\partial V}{\partial x} \sim \frac{u^2}{l} \approx \beta\frac{\partial^3 V}{\partial^3 x} \sim \beta\frac{u}{l^3},$$

which yields $l \sim \sqrt{\beta/u}$.

# 4

# Magnetohydrodynamics

## 4.1 Ideal MHD. Magnetostatic equilibria. Excess magnetic energy.

Basic MHD equations are as follows.

The equation of motion for a conducting fluid is

$$\rho \frac{d\vec{v}}{dt} = -\vec{\nabla}p + \frac{1}{c}(\vec{j} \times \vec{B}) = -\vec{\nabla}(p + \frac{B^2}{8\pi}) + \frac{1}{4\pi}(\vec{B} \cdot \vec{\nabla}) \cdot \vec{B} \qquad (4.1)$$

The magnetic induction equation is given by

$$\frac{\partial \vec{B}}{\partial t} = \vec{\nabla} \times (\vec{v} \times \vec{B}) + \eta \nabla^2 \vec{B}, \qquad (4.2)$$

and the continuity equation is

$$\frac{\partial \rho}{\delta t} + \vec{\nabla}(\rho \vec{v}) = 0$$

In the case of negligible dissipation, the evolution equation for thermal pressure reduces to the adiabatic law, namely

$$\frac{d}{dt}s(p, \rho) = 0$$

In the equation of motion (4.1) the relative role of the thermal pressure $p$ is characterized by the non-dimensional parameter $\beta = 8\pi p/B^2$. In the case of small $\beta$ (i.e., when $\beta \ll 1$), the magnetic force dominates. In this regime the dynamic timescale is the Alfven time, $\tau_A = L/v_A$, where $L$ is the spatial scale of the system, and $v_A = B/(4\pi\rho)^{1/2}$ is the **Alfven velocity**.

In the magnetic induction equation (4.2) the relative role of magnetic diffusion due to finite fluid resistivity $\eta$ is given by the **magnetic Reynolds number** $\mathcal{R}_m = Lv/\eta$. It is also common to use another non-dimensional parameter called the **Lundquist number**, $\mathcal{S}$. This is the ratio of $\tau_A$ and the resistive diffusion timescale, $\tau_\eta = L^2/\eta$, so that $\mathcal{S} = \tau_\eta/\tau_A$. If these parameters are large, which is the case for a wide class of laboratory and astrophysical applications, the induction equation (4.2) reduces to

$$\frac{\partial \vec{B}}{\partial t} = \vec{\nabla} \times (\vec{v} \times \vec{B}) \qquad (4.3)$$

This is the limit of **ideal magnetohydrodynamics (MHD)**, when magnetic field is "frozen" to the fluid flow. It implies conservation of the magnetic field lines connectivity and topology, as well as fixing the magnetic flux through any surface moving with the fluid. Formally, the "frozen-in" condition can be formulated as follows: if two infinitely close fluid particles are on the same magnetic line of force at some time $t_0$, they will remain on the same line of force in the course of fluid motion. The separation between these fluid elements, $\delta \vec{l}(t)$, is related to the magnetic field, $\vec{B}(t)$, as

$$B_i = \frac{\rho}{\rho_0} B_k^{(0)} \frac{\delta l_i}{\delta l_k^{(0)}}, \qquad (4.4)$$

where $\vec{B}^{(0)}$, $\delta \vec{l}^{(0)}$, and $\rho^{(0)}$ correspond to $t = t_0$. As can be seen from equation (4.4), in a perfectly conducting fluid the magnetic field can be amplified by fluid compression and/or stretching of magnetic field lines.

## Problem 4.1.1

A straight cylindrical magnetic flux tube with uniform axial magnetic field, $B_z = B_0$, is embedded into a perfectly conducting fluid.

  a) The tube is twisted around its axis by a twist angle $\phi(z)$ as shown in Figure 4.1. What is the deformed magnetic field?

  b) After such, the tube undergoes uniform radial expansion from $r \to 2r$. Derive the resulting magnetic field.

a) Consider two neighboring fluid particles, $A$ and $B$, that are initially separated by a distance $dz$ along the field (see Figure 4.1). By using cylindrical coordinates $(r, \theta, z)$, their separation before and after the twisting deformation is

$$\delta \vec{l}^{(0)} = (0, 0, dz); \quad \delta \vec{l}^{(a)} = \left(0, r_0 \frac{d\phi}{dz} dz, dz\right)$$

Since such a deformation is incompressible, it follows from Equation (4.4) that

$$B_z^{(a)} = B_0 \frac{\delta l_z^{(a)}}{\delta l_z^{(0)}} = B_0; \; B_\theta^{(a)} = B_0 \frac{\delta l_\theta^{(a)}}{\delta l_z^{(0)}} = B_0 r_0 \frac{d\phi}{dz}$$

b) Radial expansion from $r \to 2r$ leads to the separation vector $\delta \vec{l}^{(b)} =$

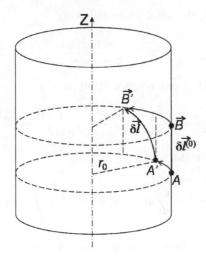

**FIGURE 4.1**
Twisting of the magnetic flux tube

$(0, 2r_0\frac{d\phi}{dz}dz, dz)$ and a density reduction $\rho = \rho_0/4$. Therefore, by applying the transformation rule (4.4) once again, one gets:

$$B_z^{(b)} = \frac{B_0}{4}; \quad B_\theta^{(b)} = \frac{B_\theta^{(a)}}{4}\frac{\delta l_\theta^{(b)}}{\delta l_\theta^{(a)}} = \frac{B_0}{2}r_0\frac{d\phi}{dz} = \frac{B_0}{4}r\frac{d\phi}{dz}$$

## Problem 4.1.2

A Z-pinch device contains a cylindrical plasma discharge in which the axial electric current, $j_z(r)$, is concentrated at the radial center and falls to zero at the external boundary, $r = R$. The plasma thermal pressure, $p(r)$, also peaks at $r = 0$ and gradually decreases to zero at $r = R$. Show, that for magnetostatic equilibrium, when the radial plasma thermal pressure gradient is balanced by the magnetic force, the net discharge current, given by $I = \int j_z dS$, and the plasma diamagnetic response, $P = \int p dS$, are related to each other as $P = I^2/2c^2$ (the so-called Bennett relation).

The radial component of the magnetostatic equilibrium condition reads

$$-\frac{dp}{dr} + \frac{1}{c}(\vec{j} \times \vec{B})_r = -\frac{dp}{dr} - \frac{1}{c}j_z B_\theta = 0 \qquad (4.5)$$

Since $\vec{j} = \frac{c}{4\pi}(\vec{\nabla} \times \vec{B})$, the axial current density

$$j_z = \frac{c}{4\pi r} \frac{d(rB_\theta)}{dr}, \tag{4.6}$$

and, thus, equation (4.5) can be re-written as

$$\frac{dp}{dr} = -\frac{B_\theta}{4\pi r} \frac{d(rB_\theta)}{dr} = -\frac{1}{8\pi r^2} \frac{d(r^2 B_\theta^2)}{dr}$$

Itegration of this equation yields

$$\frac{R^2 B_\theta^2(R)}{8} = -\pi \int_0^R r^2 \frac{dp}{dr} dr = -(\pi r^2 P)\big|_0^R + \int_0^R 2\pi r p(r) dr = P \tag{4.7}$$

On the other hand, it follows from equation (4.6) that

$$RB_\theta(R) = \frac{2}{c} \int_0^R j_z 2\pi r dr = \frac{2I}{c},$$

which, together with equation (4.7), results in the Bennett relation.

## Problem 4.1.3

Prove that a potential magnetic field is the minimum magnetic energy state that can be achieved by magnetic relaxation inside some fixed spatial domain.

The amount of magnetic energy contained within some spatial domain $V$ is given by the volumetric integral

$$W_M = \int_V \frac{B^2}{8\pi} dV \tag{4.8}$$

At first glance, one may conclude that the minimum energy is achieved simply by $\vec{B} = 0$ everywhere in the volume. However, this cannot be reached by the process of magnetic relaxation within the volume if the magnetic field at the exterior remains unchanged. Indeed, since $\vec{\nabla} \cdot \vec{B} = 0$, the normal component of $\vec{B}$ must be continuous at the boundary surface. Therefore, only magnetic fields with a prescribed boundary condition

$$B_n|_S = B_n^{(ext)} \tag{4.9}$$

are allowed to compete in minimizing the magnetic energy in equation (4.8). Consider now a potential magnetic field, $\vec{B}_p$, inside this volume, which satisfies the required boundary condition (4.9). By presenting this field as $\vec{B}_p = \vec{\nabla}\phi$, and applying the condition $\vec{\nabla} \cdot \vec{B} = 0$, one finds that the magnetic potential $\phi$ satisfies the Laplace equation. Thus, one must solve $\nabla^2\phi = 0$ subject to the boundary condition $\frac{\partial\phi}{\partial n}\big|_S = B_n^{(ext)}$. This is a well-known mathematical problem (the so-called Neumann problem), which always has a unique solution (see, e.g., R. Courant and D. Hilbert, *Methods of Mathematical Physics*, v.2, Ch. 4, Interscience Publishers, 1989). Now, any admissible magnetic field can be represented as

$$\vec{B} = \vec{B}_p + \vec{b},$$

with the boundary condition $b_n|_S = 0$. Inserting this expression for $\vec{B}$ into equation (4.8) results in

$$W_M = \frac{1}{8\pi} \int_V (B_p^2 + b^2 + 2\vec{B}_p \cdot \vec{b})dV$$

However, since $\vec{\nabla} \cdot \vec{b} = 0$, the last term here vanishes: $\int_V (\vec{B}_p \cdot \vec{b})dV = \int_V (\vec{\nabla}\phi \cdot \vec{b})dV = \int_V [\vec{\nabla} \cdot (\phi\vec{b})])dV = \int_S \phi b_n dS = 0$. Thus, the potential magnetic field corresponds to the state of the minimum magnetic energy. Any deviation from this potential field, which by definition, has associated electric currents, results in an excess magnetic energy

$$\Delta W_M = \int_V \frac{b^2}{8\pi}dV$$

## Problem 4.1.4

If the plasma thermal pressure is negligibly small compared to the pressure of the magnetic field (i.e., for $\beta \to 0$), and the plasma is in magnetostatic equilibrium, then $(\vec{\nabla} \times \vec{B}) \times \vec{B} = 0$; the magnetic force is absent. Hence, such a field is called a **force-free magnetic field**. Derive the general form for a force-free field that does not vary along one of the cooordinates (say, $z$).

The most general form for an arbitrary magnetic field, $\vec{B}(x, y)$, that satisfies identically a necessary condition $\vec{\nabla} \cdot \vec{B} = 0$ is

$$\vec{B}(x, y) = [\vec{\nabla}\Psi(x, y) \times \vec{e}_z] + B_z(x, y)\vec{e}_z, \tag{4.10}$$

where $\psi(x, y)$ is called the poloidal flux function, and $B_z$ the toroidal magnetic field. Now, $B_x = \frac{\partial\Psi}{\partial y}$, $B_y = -\frac{\partial\Psi}{\partial x}$, so that the equation $\Psi(x, y) = const$ defines projections of magnetic field lines on the $(x, y)$ plane. According to equation (4.10),

$$(\vec{\nabla} \times \vec{B}) = [\vec{\nabla}B_z(x, y) \times \vec{e}_z] - \vec{\nabla}^2\Psi(x, y)\vec{e}_z, \tag{4.11}$$

and since, for a force-free field, vectors (4.10) and (4.11) must be parallel to each other, this should be also the case for vectors $\vec{\nabla}\Psi(x,y)$ and $\vec{\nabla}B_z(x,y)$. This implies that $B_z(x,y) = F(\Psi)$. Furthermore, the ratio of the poloidal components of (4.11) and (4.10), which is $dF/d\Psi$, should be equal to the ratio of their toroidal components, which is $-\vec{\nabla}^2\Psi/F$. Hence, we have

$$\vec{\nabla}^2\Psi = -F\frac{dF}{d\Psi} \tag{4.12}$$

This is the sought after **Grad-Shafranov equation** for the force-free magnetic field (4.10). In general, this is a non-linear equation that is hardly tractable analytically. It becomes greatly simplified for the special case of the so-called linear force-free fields, for which $F(\Psi) = \alpha\Psi$, where $\alpha = const.$ With this choice the equation (4.12) reduces to

$$\vec{\nabla}^2\Psi = -\alpha^2\Psi \tag{4.13}$$

## Problem 4.1.5

Consider a linear force-free magnetic field with the poloidal flux function

$$\Psi(x,y) = \frac{B_0}{k}cos(kx)\exp(-\kappa y), \kappa^2 = k^2 - \alpha^2 > 0, \tag{4.14}$$

and with $B_z = \alpha\Psi$, inside the domain given by $y > 0, -\pi/2k < x < +\pi/2k$. It represents a magnetic "arcade" as shown in Figures 4.2 and 4.3.

a) Derive the excess magnetic energy of this configuration as a function of the force-free parameter $\alpha$.

b) Prove by a direct calculation, that, if $\alpha$ is changing with time, the variation of the excess magnetic energy is equal to the Poynting flux injected into the base of the arcade at $y = 0$.

a) From equation (4.14), the magnetic field components of this configuration are equal to:

$$B_x = \frac{\partial\Psi}{\partial y} = -B_0\frac{\kappa}{k}\cos(kx)\exp(-\kappa y),$$

$$B_y = -\frac{\partial\Psi}{\partial x} = -B_0\sin(kx)\exp(-\kappa y), \tag{4.15}$$

$$B_z = \alpha\Psi = B_0\frac{\alpha}{k}\cos(kx)\exp(-\kappa y)$$

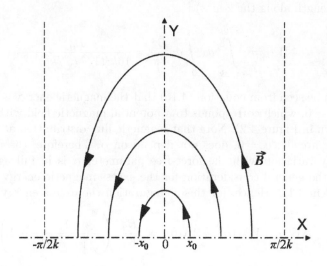

**FIGURE 4.2**
Magnetic arcade with the potential magnetic field

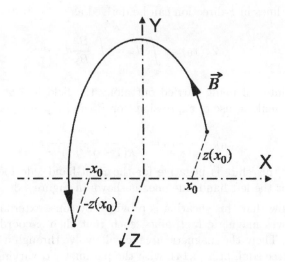

**FIGURE 4.3**
Shearing of magnetic field lines

A straightforward calculation yields the following magnetic energy (per unit length along the z-axis),

$$W_M = \int_{-\pi/2k}^{\pi/2k} dx \int_0^\infty dy \frac{B^2}{8\pi} = \frac{B_0^2}{16k^2(1 - \alpha^2/k^2)^{1/2}} \qquad (4.16)$$

It can be seen from equation (4.16) that the magnetic energy is minimal for $\alpha = 0$, which corresponds to a potential magnetic field with $B_z = 0$ (shown in Figure 4.2). Note that magnetic flux distribution at the base of the arcade, $y = 0$, does not depend on $\alpha$. Therefore, the magnetic energy variation with the force-free parameter $\alpha$ is in full agreement with the general consideration for the excess magnetic energy given in Problem 4.1.3. Hence, for this configuration the excess energy is equal to

$$\Delta W_M = W_M(\alpha) - W_M(0) = \frac{B_0^2}{16k^2} \left( \frac{1}{\sqrt{1 - \alpha^2/k^2}} - 1 \right)$$

b) Each magnetic field line of the arcade, defined by the equation $\Psi(x, y) = const$, can be more conveniently labeled by $x_0$: the x-coordinate of its footpoints at $y = 0$ (see Figure 4.3) (they relate to each other by $\Psi = \frac{B_0}{k} \cos(kx_0)$). A non-potential field with a non-zero $\alpha$ has these footpoints being shifted in z-direction on $\pm z(x_0)$. The non-potential arcade as a whole bulges upwards since the characteristic height, $\kappa^{-1} = k^{-1}(1-\alpha^2/k^2)^{-1/2}$, increases with $\alpha$. Quantatively, the shearing of magnetic field lines in $z$-direction can be derived as

$$2z(x_0) = \int_{-x_0}^{+x_0} dz = \int_{-x_0}^{+x_0} \frac{B_z}{B_x} dx$$

with the integral being carried out along the field line $\Psi = const = \Psi(x_0)$. By making use of expressions for $B_z$ and $B_x$ in equation (4.15), one gets

$$z(x_0) = -\frac{\alpha}{\kappa} x_0 = -\frac{\alpha}{k(1 - \alpha^2/k^2)^{1/2}} x_0 \qquad (4.17)$$

Note that this shift is negative for the right-hand side footpoint and positive for the left-hand side one, as shown in Figure 4.3.

Assume now that this shearing is provided by some external force that slowly moves magnetic footpoints, such that their $z$-coordinates vary with time. Then, the magnetic arcade will evolve through a sequence of the force-free equilibria (4.14), with the parameter $\alpha$ varying with time in accordance with the relation (4.17). Thus, we have

$$v_z(x_0) = \frac{dz(x_0)}{dt} = -\frac{x_0}{k} \frac{d}{dt} \left[ \frac{\alpha}{(1 - \alpha^2/k^2)^{1/2}} \right] = -\frac{d\alpha/dt}{k(1 - \alpha^2/k^2)^{3/2}} x_0 \qquad (4.18)$$

These flows at the base of the arcade lead to Poynting flux of energy $\vec{P}$ into the arcade, with the relevant component $P_y$ equal to $P_y|_{y=0} = \frac{c}{4\pi}(E_z B_x - E_x B_z)_{y=0}$. In a perfectly conducting fluid $\vec{E} = -\frac{1}{c}(\vec{v} \times \vec{B})$. Thus, for the z-velocity of equation (4.18), $E_z|_{y=0} = 0, E_x|_{y=0} = \frac{1}{c}(v_z B_y)|_{y=0}$, and hence

$$P_y|_{y=0} = -\frac{v_z}{4\pi}(B_y B_z)|_{y=0} = \frac{B_0^2}{4\pi k^2}\frac{\alpha d\alpha/dt}{(1 - \alpha^2/k^2)^{3/2}}x_0 \sin(kx_0) \cos(kx_0)$$

This yields the total power supply into the arcade

$$\frac{dW}{dt} = \int\limits_{-\pi/2k}^{\pi/2k} P_y|_{y=0}dx_0 = \frac{B_0^2}{16k^4}\frac{\alpha d\alpha/dt}{(1 - \alpha^2/k^2)^{3/2}},$$

which, according to equation (4.16), is equal to the rate of change of its magnetic energy.

---

## 4.2    MHD waves. Alfven resonance.

### Problem 4.2.1

Consider small-amplitude ideal MHD waves supported by a zero-$\beta$ plasma of density $\rho$ in a uniform magnetic field $\vec{B}$.

For a small amplitude wave, the linearized equation of motion and magnetic induction equation for the magnetic field perturbation $\vec{b}$ and the plasma velocity $\vec{v}$ take the following form:

$$\rho\frac{\partial\vec{v}}{\partial t} = \frac{1}{4\pi}[-\vec{\nabla}(\vec{B}\cdot\vec{b}) + (\vec{B}\cdot\vec{\nabla})\vec{b}], \tag{4.19}$$

$$\frac{\partial\vec{b}}{\partial t} = \vec{\nabla} \times (\vec{v} \times \vec{B}) \tag{4.20}$$

The two terms on the right-hand side of (4.19) represent magnetic pressure and magnetic tension forces, respectively. Assuming that $\vec{v}$ and $\vec{b}$ vary in space and time as $\exp[i(\vec{k}\cdot\vec{r} - \omega t)]$, that the initial magnetic field $\vec{B}$ is directed along the z-axis, and that the wave-vector, $\vec{k}$, is lying in the $(x, z)$ plane, eqs. (4.19-4.20) reduce to

$$4\pi\omega\rho\vec{v} = \vec{k}Bb_z - k_z B\vec{b}, \tag{4.21}$$

$$-\omega \vec{b} = \vec{k} \times (\vec{v} \times \vec{B})$$

Writing these in component form gives

$$v_z = 0, \; v_y = -\frac{k_z B}{4\pi \rho \omega} b_y, \; v_x = \frac{B}{4\pi \rho \omega}(k_x b_z - k_z b_x) \tag{4.22}$$

$$b_x = -\frac{k_z B}{\omega} v_x, \; b_y = -\frac{k_z B}{\omega} v_y, \; b_z = \frac{k_x B}{\omega} v_x \tag{4.23}$$

As seen from equations (4.22-4.23), these perturbations make two separate groups, $(v_y, b_y)$ and $(v_x, b_x, b_z)$, indicating that this system supports two types of waves. The first type, called the **shear Alfven wave**, involves $(v_y, b_y)$ and has dispersion relation

$$\omega = \frac{k_z B}{4\pi \rho} = \vec{k} \cdot \vec{V}_A, \; \vec{V}_A = \vec{B}/4\pi\rho \tag{4.24}$$

Hence, its group velocity $\vec{v}_g = \partial \omega / \partial \vec{k} = \vec{V}_A$, i.e., such a wave propagates along the initial magnetic field with the Alfven speed. As seen from equation (4.21), the restoring force in this case is entirely due to magnetic tension.

For the second type of wave, which involves $(v_x, b_x, b_z)$, the dispersion relation reads:

$$\omega = k V_A \tag{4.25}$$

The group velocity, $\vec{v}_g = V_A \vec{k}/k$, is directed along the wave-vector $\vec{k}$. In such a wave, called the **compressional Alfven wave**, both magnetic pressure and magnetic tension contribute to the restoring force (see equation (4.21)). In the limit $k_x = 0$ this wave becomes identical to the shear Alfven wave, while for $k_z = 0$ its propagation is entirely due to magnetic pressure.

## Problem 4.2.2

The magnetic field perturbation due to a shear Alfven wave, propagating in a uniform magnetic field $\vec{B}_0 = B_0 \vec{e}_z$, is equal to $b_y = \epsilon B_0 \cos(kz - \omega t)$. Consider the magnetic field line which in the absence of the wave is defined by $(x = 0, y = 0)$. Derive the coordinates of this field line when the wave is present, and verify that the magnetic field is frozen into the fluid, i.e., that the fluid particle displacements indeed follow the magnetic field line.

A magnetic field line is defined by the following equations:

$$dx/B_x = dy/B_y = dz/B_z,$$

or

$$dx/dz = B_x/B_z, dy/dz = B_y/B_z.$$

In our case $B_x = 0$, $B_y = \epsilon B_0 \cos(kz - \omega t)$, $B_z = B_0$. Hence,

$$dx/dz = 0, \; dy/dz = \epsilon \cos(kz - \omega t)$$

Integrating gives the sought after coordinates,

$$x = 0, \; y = \frac{\epsilon}{k} \sin(kz - \omega t) = \lambda \frac{\epsilon}{2\pi} \sin(kz - \omega t),$$

where $\lambda = 2\pi/k$ is the perturbation wavelength. Also, it follows from equation (4.22) that the fluid velocity is equal to

$$v_y = -\frac{\epsilon k B_0^2}{4\pi \rho \omega} \cos(kz - \omega t).$$

The latter results in the fluid particle displacement

$$\xi_y = \frac{\epsilon k B_0^2}{4\pi \rho \omega^2} \sin(kz - \omega t),$$

which is equal to the magnetic field line y-coordinate since $\omega^2/k^2 = B_0^2/4\pi\rho$.

## Problem 4.2.3

a) Calculate the energy per unit volume for shear and compressional Alfven waves.

b) Derive the flux of energy for each of the two types of waves, and verify that the respective energy propagation velocities are equal to the respective group velocities obtained in Problem 4.2.1.

Consider first a shear Alfven wave, for which the magnetic field and fluid velocity perturbations can be written as (see Problem 4.2.1):

$$b_y = b_1 \cos(\vec{k} \cdot \vec{r} - \omega t), \; v_y = -\frac{k_z B}{4\pi \rho \omega} b_1 \cos(\vec{k} \cdot \vec{r} - \omega t)$$

Such a wave possesses the additional kinetic and magnetic energies per unit volume

$$\Delta W_k = \frac{\rho < v_y^2 >}{2} = \frac{\rho k_z^2 B^2}{4(4\pi\rho)^2 \omega^2} b_1^2 = \frac{b_1^2}{16\pi},$$

$$\Delta W_M = \frac{< (\vec{B} + \vec{b})^2 >}{8\pi} - \frac{B^2}{8\pi} = \frac{b_1^2}{16\pi},$$

where the symbol $< >$ means an average over spatial coordinates, and the dispersion law $\omega^2 = k_z^2 V_A^2$ has been used. Hence, the energy per unit volume of this wave is

$$\Delta W = \Delta W_k + \Delta W_M = \frac{b_1^2}{8\pi},$$

which is evenly divided between the kinetic and magnetic contributions. This is not a coincidence, but comes from the general rule, that for a small amplitude oscillation the associated potential energy (in this case the magnetic one) and kinetic energy are, on average, equal to each other. Thus, it is also the case for a compressional Alfven wave (see below).

The flux of energy in the wave is given by the averaged Poynting vector $< \vec{P} >= \frac{c}{4\pi} < (\vec{E} \times \vec{B}) >$. Since in the linear approximation $\vec{E} = -\frac{1}{c}(\vec{v} \times \vec{B})$, in this case its only non-zero component is $E_x = -Bv_y/c$. Hence

$$< P_z >= \frac{c}{4\pi} < E_x b_y >= \frac{k_z B^2}{32\pi^2 \rho\omega} b_1^2 = \frac{b_1^2}{8\pi} V_A,$$

which, when represented as $< \vec{P} >= (\Delta W)\vec{V_A}$, confirms the group velocity of this wave derived in Problem 4.2.1.

In the case of a compressional Alfven wave it is convenient to express the wave energy and its flux in terms of the velocity amplitude, so that

$$
\begin{aligned}
v_x &= v_1 \cos(\vec{k} \cdot \vec{r} - \omega t),\ b_x = -\frac{k_z B}{\omega} v_1 \cos(\vec{k} \cdot \vec{r} - \omega t), \\
b_z &= \frac{k_x B}{\omega} v_1 \cos(\vec{k} \cdot \vec{r} - \omega t)
\end{aligned}
$$

Then, the energies are given by $\Delta W_k = \rho < v_x^2 > /2 = \rho v_1^2/4$, $\Delta W_M = (< b_x^2 > + < b_z^2 >)/8\pi = \rho v_1^2/4 = \Delta W_k$ (the dispersion equation $\omega^2 = k^2 V_A^2$ is used here), so that the total energy

$$\Delta W = \Delta W_k + \Delta W_M = \frac{1}{2}\rho v_1^2$$

The only non-zero component of the electric field is $E_y = Bv_x/c$, therefore

$$< P_x >= \frac{B}{4\pi} < v_x b_z >= \frac{B^2 k_x}{8\pi\omega} v_1^2, \ < P_z >= -\frac{B}{4\pi} < v_x b_x >= \frac{B^2 k_z}{8\pi\omega} v_1^2$$

This yields

$$< \vec{P} >= \frac{\rho v_1^2}{2} V_A \frac{\vec{k}}{k} = (\Delta W)V_A \frac{\vec{k}}{k},$$

again in agreement with the group velocity derived from the dispersion relation in Problem 4.2.1.

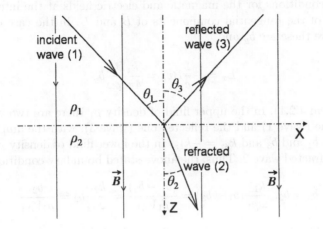

**FIGURE** 4.4

Reflection and refraction of MHD waves

### Problem 4.2.4

Derive the refraction law and reflection coefficients at the interface between two perfecly conducting media of different density, for both shear and compressional Alfven waves, as shown in Figure 4.4.

The refraction law is determined by the two following requirements: the frequency of the refracted wave (wave 2) is equal to that of the incident wave (wave 1), and the $x$-components of the wave-vectors, $\vec{k_1}$ and $\vec{k_2}$, are also equal to each other. Thus, in the case of a shear Alfven wave, one gets: $\omega_1 = k_{1z}V_{A1} = \omega_2 = k_{2z}V_{A2}$; $k_{1x} = k_1 \sin\theta_1 = k_{2x} = k_2 \sin\theta_2$. Since $V_A \propto \rho^{-1/2}$, it follows from the first of these equations that $k_2 \cos\theta_2 = k_1 \cos\theta_1 (\rho_2/\rho_1)^{1/2}$, which together with the second equation yield

$$(\rho_2)^{1/2} \tan\theta_2 = (\rho_1)^{1/2} \tan\theta_1$$

For a compressional Alfven wave with the dispersion relation $\omega = kV_A$, the above requirements result in

$$(\rho_2)^{1/2} \sin\theta_2 = (\rho_1)^{1/2} \sin\theta_1,$$

which looks like the standard Snell's law, with the refractive index of a medium being proportional to a square root of its density. Hence, there is no refracted wave if $\rho_2 < \rho_1$ and the angle of incidence exceeds $\theta_r = \sin^{-1}(\rho_2/\rho_1)^{1/2}$, the angle of total reflection.

In order to obtain reflection coefficients of waves one should consider the boundary conditions for the magnetic and electric fields at the interface: the continuity of the tangential components of $\vec{B}$ and $\vec{E}$. In the case of a shear Alfven wave these are $b_y$ and

$$E_x = -\frac{1}{c}Bv_y = \frac{k_z B^2}{4\pi c\rho\omega}b_y$$

(see Problem 4.2.1). In the upper fluid of density $\rho_1$ there are two waves, the incident one (wave 1) and the reflected one (wave 3), with the amplitudes of $b_y$ equal to $b_1$ and $b_3$ and $k_{3z} = -k_{1z}$. In the lower fluid of density $\rho_2$ there is only the refracted wave 2. Thus, the above-stated boundary conditions imply:

$$b_1 + b_3 = b_2, \quad \frac{k_{1z}}{\omega\rho_1}(b_1 - b_3) = \frac{(b_1 - b_3)}{\rho_1 V_{A1}} = \frac{k_{2z}}{\omega\rho_2}b_2 = \frac{b_2}{\rho_2 V_{A2}} \qquad (4.26)$$

Then, by taking into account that $V_{A2}/V_{A1} = (\rho_1/\rho_2)^{1/2}$, it follows from (4.26) that the reflection coefficient

$$R = \left(\frac{b_3}{b_1}\right)^2 = \left[\frac{1 - (\rho_2/\rho_1)^{1/2}}{1 + (\rho_2/\rho_1)^{1/2}}\right]^2$$

Note that this reflection coefficient does not depend on the angle of incidence.

Similarly, in the case of a compressional Alfven wave, one should require the continuity of $b_x$, $b_z$ and $E_y = -Bv_x/c$. Thus, if $v_{1,2,3}$ are velocity amplitudes of each of the three waves shown in Figure 4.4, the above requirements yield (with the help of equations (4.22-4.23)):

$$v_1 + v_3 = v_2, \quad k_{1z}(v_1 - v_3) = k_{2z}v_2$$

(note that the continuity of $b_z$ is ensured by the first of the above relations, because all three waves share the same $\omega$ and $k_x$). By using the dispersion law (4.25), one gets then the following reflection coefficient:

$$R = \left(\frac{v_3}{v_1}\right)^2 = \left(\frac{1 - \delta}{1 + \delta}\right)^2, \quad \delta = \frac{(\rho_1/\rho_2)^{1/2}\cos\theta_1}{[1 - (\rho_1/\rho_2)\sin^2\theta_1]^{1/2}}, \quad \theta_1 < \theta_r.$$

## Problem 4.2.5

A sheared force-free magnetic field

$$\vec{B}^{(0)} = (0, B_0\sin(\theta(x)), B_0\cos(\theta(x))) \qquad (4.27)$$

is embedded in a slab, $|x| < l$, of zero-$\beta$ perfectly conducting fluid of density $\rho$. Assume now that the right-hand side boundary surface of

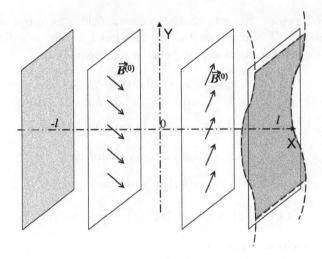

**FIGURE 4.5**
Boundary deformation of the sheared force-free magnetic configuration

the system, $x_b^{(+)} = l$, is subjected to a continuous deformation as (see Figure 4.5):

$$x_b^{(+)} = l + a\cos(ky)\exp(-i\omega t), \ a << l \qquad (4.28)$$

Derive the energy absorption rate associated with the **Alfven resonances**, occurring at $\omega = \pm\vec{k}\cdot\vec{V}_A(x)$, where $\vec{V}_A = \vec{B}^{(0)}/4\pi\rho$.

The external perturbation results in fluid motion, which can be described in terms of the displacement vector $\vec{\xi}(\vec{r}, t) = \vec{\xi}(x,y)\exp(-i\omega t)$. For the weak perturbation under discussion, ($a << l$), the linearized ideal MHD equations apply, so that equation of motion takes the form

$$\rho\frac{\partial^2\vec{\xi}}{\partial t^2} = -\omega^2\rho\vec{\xi} = \frac{1}{c}(\vec{j}^{(0)}\times\vec{b}) + \frac{1}{c}(\vec{j}^{(1)}\times\vec{B}^{(0)}), \qquad (4.29)$$

where $\vec{b}$ is the magnetic field perturbation with the associated current $\vec{j}^{(1)}$, and

$$\vec{j}^{(0)} = \frac{c}{4\pi}\frac{d\theta}{dx}\vec{B}^{(0)}$$

is the initial electric current of the sheared magnetic field (4.27). The magnetic induction equation yields

$$\vec{b} = \vec{\nabla}\times(\vec{\xi}\times\vec{B}^{(0)}), \qquad (4.30)$$

which, after being inserted into (4.29), results in a single equation for the displacement vector $\vec{\xi}$. Since the initial magnetic field is force-free and, hence, $\vec{j}^{(0)}$ is parallel to $\vec{B}^{(0)}$, it follows from (4.29) that $\vec{\xi}$ is perpendicular to $\vec{B}^{(0)}$. Therefore, the displacement vector $\vec{\xi}$ can be represented as

$$\xi_x(x, y) = \xi_x(x) \cos(ky) \exp(-i\omega t),$$
$$\xi_\perp(x, y) = \xi_y \cos\theta - \xi_z \sin\theta = \xi_\perp(x) \sin(ky) \exp(-i\omega t)$$

Then, the magnetic field perturbation of (4.30) takes the form:

$$b_x = -kB_0\xi_x(x) \sin(ky) \sin\theta \exp(-i\omega t),$$
$$b_y = -B_0 \cos(ky) \frac{d}{dx}[\sin\theta\xi_x(x)] \exp(-i\omega t), \quad (4.31)$$
$$b_z = -B_0 \cos(ky) \exp(-i\omega t)\{k\xi_\perp(x) + \frac{d}{dx}[\cos\theta\xi_x(x)]\}$$

Finally, the equation of motion (4.29) yields

$$\xi_\perp(x) = \frac{k\cos\theta}{(\omega^2/V_A^2 - k^2)} \frac{d\xi_x}{dx},$$

where $\xi_x(x)$ satisfies the following equation:

$$\frac{d}{dx}\left[\delta(x)\frac{d\xi_x}{dx}\right] + \left(\frac{\omega^2}{V_A^2} - k^2\right)\delta(x)\xi_x = 0, \quad (4.32)$$

where $\delta(x) \equiv \omega^2 - k^2 V_A^2 \sin^2\theta(x)$. This equation acquires a singularity, if at some location $x = x_r$ $\delta(x_r) = 0$, i.e., $\omega^2 = k^2 V_A^2 \sin^2\theta(x_r) = (\vec{k}\cdot\vec{V}_A)^2$. The latter equality is nothing but the dispersion relation for shear Alfven waves (see Problem 4.2.1). Thus, in this case the external perturbation resonates at $x = x_r$ with the shear Alfven wave, causing the magnetic energy dissipation even in an apparently ideal fluid.

The resulting dissipation power can be derived in the following way (L. Chen and A. Hasegawa, Physics of Fluids, vol.17, 1399, 1974). In the vicinity of the resonant point the frequency shift $\delta(x)$ can be approximated as

$$\delta(x) \approx \left(\frac{d\delta}{dx}\right)_{x_r}(x - x_r) = \delta'(x_r)(x - x_r),$$

and, therefore, the displacement $\xi_x(x)$ behaves at this point, according to equation (4.32), as

$$\xi_x(x) \approx \xi_r \log(x - x_r), \quad (4.33)$$

where the amplitude $\xi_r$ is determined by a global solution of (4.32) with the boundary condition specified by (4.28). Consider now the energy balance inside a thin layer enclosing the resonant point, i.e., $x_r - \epsilon < x < x_r + \epsilon$, and per unit area in the $(y, z)$ plane:

$$\frac{dW_A}{dt} = < P_x(x_r - \epsilon) > - < P_x(x_r + \epsilon) >, \quad (4.34)$$

where $< P_x >$ is an $x$-component of the Poynting vector $\vec{P} = c(\vec{E} \times \vec{B})/4\pi$, averaged over its variation with time and $y$-coordinate. Since in ideal MHD $\vec{E} = -(\vec{v} \times \vec{B}/c) \approx i\omega(\vec{\xi} \times \vec{B}^{(0)})/c$, a non-vanishing averaged contribution to $< P_x >$ is due to the magnetic field perturbation $\vec{b}$. Hence, the respective part of $P_x$ is $P_x = c[\Re(E_y)\Re(b_z) - \Re(E_z)\Re(b_y)]/4\pi$, which, by using (4.31), can be written as

$$P_x = \frac{\omega B_0^2}{4\pi} \cos^2(ky)\Re \left( i\xi_x \exp(-i\omega t) \frac{\delta(x)}{(\omega^2 - k^2 V_A^2)} \right) \Re \left( \frac{d\xi_x}{dx} \exp(-i\omega t) \right)$$
(4.35)

Therefore, in the vicinity of the resonant point $x_r$, where $\delta(x) \approx \delta'(x_r)(x - x_r)$ and $d\xi_x/dx = \xi_r(x - x_r)^{-1}$, Equation (4.35) reduces to

$$P_x = \frac{\omega B_0^2}{4\pi} \cos^2(ky)Re \left( i\xi_x \exp(-i\omega t) \frac{\delta'(x_r)\xi_r \cos(\omega t)}{(\omega^2 - k^2 V_A^2)} \right)$$

As seen from this expression, the power (4.34) supplied into the resonant area, if non-zero, should be due to a discontinuity of $\xi_x(x)$ at $x = x_r$, i.e., $\Delta\xi_x = \xi_x(x_r - \epsilon) - \xi_x(x_r + \epsilon)$, so that

$$\frac{dW_A}{dt} = \frac{\omega B_0^2}{8\pi} \frac{\delta'(x_r)\xi_r}{(\omega^2 - k^2 V_A^2)} < \Re[i\Delta\xi_x \exp(-i\omega t)) \cos(\omega t)] >$$

In order to derive $\Delta\xi_x$, it is convenient to resolve the singularity in equation (4.32) by introducing an infinitesimally small positive imaginary part to the driving frequency, i.e., by replacing $\omega$ with $\omega + i\gamma$, $\gamma \to 0$, which in physical terms means that the external perturbation is adiabatically "switched-on" at $t \to -\infty$. It follows then from the expression (4.32) for $\delta(x)$ that such a procedure results in the resonant point being now displaced into the complex plane of variable $x$ by $\Delta x_r = -2i\gamma\omega/\delta'(x_r)$. Thus, the resonant point becomes located below the real axis if $\delta'(x_r) > 0$, and vice versa. In other words, introduction of $\gamma > 0$ provides one with a rule of passing the resonant singularity (see also Problem 3.4.8): it should be passed above if $\delta'(x_r) > 0$, and below if $\delta'(x_r) < 0$. Thus, while passing from $x = x_r - \epsilon$ to $x = x_r + \epsilon$ the variable $(x - x_r)$ acquires additional phase factor $\exp(i\Delta\phi)$ with $\Delta\phi = -\pi sign\delta'(x_r)$, which, according to expression(4.33), yields $\Delta\xi_x = i\pi\xi_r sign[\delta'(x_r)]$. Therefore, the Alfven resonance power supply is equal to

$$\frac{dW_A}{dt} = \frac{\omega B_0^2}{16(k^2 V_A^2 - \omega^2)} |\delta'(x_r)|\xi_r^2$$

which, by using expression (4.32) for $\delta(x)$ and the resonant condition $\delta(x_r) = 0$, takes the form

$$\frac{dW_A}{dt} = \frac{\omega^2 B_0^2}{8(k^2 V_A^2 - \omega^2)^{1/2}} |\frac{d\theta}{dx}|\xi_r^2$$
(4.36)

Then, always present finite dissipation effects such as fluid resistivity or viscosity result in this power being absorbed inside a narrow layer at the resonant surface $x = x_r$.

As an example, consider a linear force-free magnetic field for which $\theta(x) = \alpha x$, assuming, for simplicity, that the perturbation length-scale $k^{-1}$ and the shear length $\alpha^{-1}$ are comparable to the size of the system, i.e., $kl \sim \alpha l \sim 1$ (as shown in Problem 4.3.5, such a field becomes MHD unstable when $\alpha l > \pi/2$). In this case there are two Alfven resonances located at $x_r = \pm l \sin^{-1}(\omega/kV_A)$, which allow a broad interval of the driving frequency $\omega$, ranging from $\omega_{max} \sim kV_A \sim V_A/l \sim \tau_A^{-1}$, the inverse Alfven transit time, down to $\omega = 0$. A general dimensional consideration yields that in this case the resonance amplitude $\xi_r \sim a$. Thus, for the low driving frequencies, when $\omega \tau_A \ll 1$, the dissipated power can be estimated from equation (4.36) as

$$\frac{dW_A}{dt} \sim V_A \frac{B_0^2}{8\pi} (\omega \tau_A)^2 (a/l)^2. \tag{4.37}$$

## 4.3    Magnetic reconnection

### Problem 4.3.1

A linear force-free magnetic field

$$\vec{B}^{(0)} = [0, B_0 \sin(\alpha x), B_0 \cos(\alpha x)] \tag{4.38}$$

is embedded in a zero-$\beta$ plasma and bounded by two perfectly conducting surfaces $x_b^{(\pm)} = \pm l$. Find a deformed magnetostatic equilibrium configuration, which is established due to a small bending of the right-hand side boundary as

$$x_b^{(+)} = l + a \cos(ky), \quad a \ll l, \tag{4.39}$$

as shown in Figure 4.5.

Since the system remains invariant with respect to $z$-coordinate, the deformed magnetic field can be represented in the form (4.10), with the initial field (4.38) corresponding to

$$\psi_0(x) = \frac{B_0}{\alpha} \cos(\alpha x), \quad B_z^{(0)} = F_0(\psi) = \alpha \psi_0$$

By using the perturbation approach, one can write for the deformed field:

$$\psi = \psi_0(x) + \delta\psi(x, y), \quad F(\psi) = F_0(\psi) + \delta F(\psi). \tag{4.40}$$

Since the deformed boundary surface (4.39) should remain a magnetic surface,

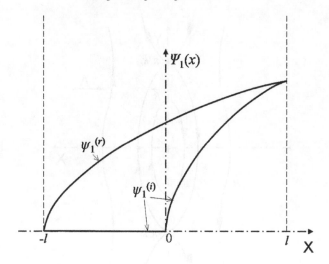

**FIGURE 4.6**
Poloidal flux function perturbations for the two magnetic equilibria

the poloidal flux function $\psi$ is constant there; thus its perturbation must be of the form $\delta\psi(x, y) = \psi_1(x) \cos ky$. Furthermore, the new equilibrium is force-free and, hence, by applying the Grad-Shafranov equation (4.12), one finds that in the linear approximation the functional dependence $F(\psi)$ remains unchanged, i.e., $\delta F(\psi) = 0$ and $F(\psi) = F_0(\psi) = \alpha\psi$, so that $\psi_1(x)$ satisfies the following equation:

$$\psi_1'' + (\alpha^2 - k^2)\psi_1 = 0 \tag{4.41}$$

As already mentioned, the boundary conditions for the proper solution of equation (4.41) come from the requirement that $\psi(x, y)$ in (4.40) is constant at $x = x_b^{(-)} = -l$ and $x = x_b^{(+)} = l + a \cos ky$. The former condition implies that $\psi_1(-l) = 0$, while the latter yields $\psi_0'(l)a \cos ky + \psi_1(l) \cos ky = 0$, i.e., $\psi_1(l) = aB_0 \sin(\alpha l)$. The respective regular solution of equation (4.41) is

$$\psi_1^{(r)}(x) = \frac{aB_0 \sin(\alpha l)}{\sin(2\kappa l)} \sin[\kappa(x + l)], \tag{4.42}$$

where $\kappa^2 = (\alpha^2 - k^2)$ is assumed to be positive, and $\alpha l < \pi/2$ (the latter ensures MHD stability of the initial magnetic configuration (see Problem 4.3.5)).

At first glance, there is nothing special about this solution shown in Figure 4.6. However, it turns out that since $\psi_1^{(r)}(0) \neq 0$, the respective deformed magnetic field acquires a topology which is different from that of the initial configuration of (4.38). Indeed, in the vicinity of $x = 0$ the poloidal flux func-

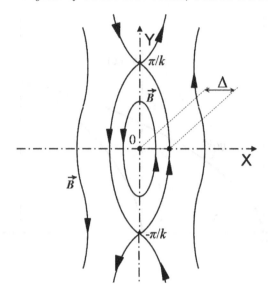

**FIGURE 4.7**
Formation of magnetic islands

tion $\psi(x, y)$ takes the form

$$\psi(x, y) \approx \frac{B_0}{\alpha}(1 - \alpha^2 x^2/2) + \psi_1^{(r)}(0) \cos(ky),$$

which indicates formation there of a periodic chain of identical magnetic islands as shown in Figure 4.7. The $O$-point with the coordinates $x = 0$, $y = 0$ at the center of the island corresponds to $\psi = \psi^{(0)} = (B_0/\alpha) + \psi_1^{(r)}(0)$, while the two $X$-points located at $x = 0$, $y = \pm\pi/k$ have $\psi = \psi_S = (B_0/\alpha) - \psi_1^{(r)}(0)$, the separatrix flux function. Therefore, the total amount of magnetic flux confined inside each island is equal to $\Delta\psi = \psi^{(0)} - \psi_S = 2\psi_1^{(r)}(0)$. The width of islands is $\Delta = 2[\psi_1^{(r)}(0)/B_0\alpha]^{1/2}$.

Clearly, such a cofiguration cannot be formed in a perfectly conducting fluid with a frozen-in magnetic field. Therefore, another, ideal MHD solution, $\psi_1^{(i)}$, which preserves the magnetic field topology, should exist. It then follows from the consideration given above, that this solution must satisfy the additional requirement, namely that $\psi_1^{(i)}(0) = 0$, which makes it equal to zero at $-l \le x \le 0$, and

$$\psi_1^{(i)}(x) = \frac{aB_0 \sin(\alpha l)}{\sin(\kappa l)} \sin(\kappa x), \qquad (4.43)$$

at $0 \le x \le l$ (see Figure 4.6). As seen from (4.43), this solution is singular: a current sheet with a discontinuity of $B_y = -\partial\psi/\partial x$ is formed at $x = 0$. The latter is the so-called resonance surface, location of which, in a general case of

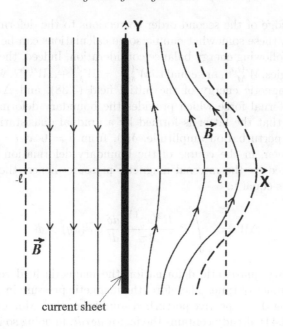

**FIGURE 4.8**
Ideal MHD equilibrium with the current sheet

the perturbation with $\vec{k} = (0, k_y, k_z)$, is defined by the condition $\vec{k} \cdot \vec{B}^{(0)} = 0$. The poloidal magnetic field pattern of the ideal MHD solution is depicted in Figure 4.8.

It is shown below (see next Problem) that in a highly conducting medium magnetic reconnection acts as a mechanism of magnetic relaxation from the state of higher magnetic energy associated with $\psi_1^{(i)}$, to the lower-energy configuration of $\psi_1^{(r)}$.

## Problem 4.3.2

Derive magnetic energy of the two magnetostatic equilibria obtained in Problem 4.3.1.

The magnetic energy of the system under consideration is equal (per unit area in the $(y, z)$ plane) to $W_M = (8\pi)^{-1} \int_{-l}^{l} < B^2 > dx$, where the symbols $< >$ mean averaging over field variation along the $y$-coordinate. Since, for a weak perturbation, the change of magnetic energy of each of the two equilibria, $\psi_1^{(i)}$ and $\psi_1^{(r)}$, is of the order of $(a/l)^2$; their straightforward derivations would require knowledge of the second order corrections to the deformed magnetic field. However, these somewhat cumbersome calculations can be bypassed by exploring the following energy balance consideration. Indeed, the sought after magnetic energies, $W_M^{(i,r)}$, are equal to $W_M^{(i,r)} = W_M^{(0)} + \Delta W^{(i,r)}$, where $W_M^{(0)} = B_0^2 2l/8\pi$ is magnetic energy of the initial field (4.38), and $\Delta W^{(i,r)}$ is the work of an external force which provides the boundary deformation (4.39). Thus, assume that the latter is formed by a gradual quasistatic increase of the boundary perturbation amplitude, $\delta(t)$, from $\delta = 0$ at $t = 0$ to $\delta = a$ at $t \to \infty$. Then, in the course of the boundary deformation the external force is balanced by the internal magnetic pressure; hence, the work under consideration is equal to

$$\Delta W = -\int\limits_0^\infty < \frac{B^2(x_b^{(+)})}{8\pi} \frac{d\delta}{dt} \cos(ky) > dt \qquad (4.44)$$

Under the above quasistatic deformation the magnetic field remains force-free at any instance of time. Therefore, the magnetic pressure in (4.44) can be derived by using the respective perturbed equilibria $\psi^{(i,r)}$. Moreover, since the integrand in (4.44) already contains the factor $d\delta/dt$, in doing so it is sufficient to use the flux perturbation $\psi_1^{(i,r)}$ obtained in the linear approximation.

Thus, the procedure is as follows. The deformed magnetic field is $\vec{B} = \vec{B}^{(0)} + \vec{b}$, where the field perturbation $\vec{b}$ is related to the flux function perturbation $\psi_1$ as

$$b_x = \frac{\partial \delta\psi}{\partial y} = -k\psi_1(x)\sin(ky),$$

$$b_y = -\frac{\partial \delta\psi}{\partial x} = -\psi_1'(x)\cos(ky),$$

$$b_z = \alpha\delta\psi = \alpha\psi_1(x)\cos(ky)$$

Then, the required magnetic pressure at the deformed boundary takes the form

$$\frac{B^2(x_b^{(+)})}{8\pi} = \frac{B_0^2}{8\pi} + \frac{B_0}{4\pi}[\alpha\psi_1(l)\cos(\alpha l) - \psi_1'(l)\sin(\alpha l)]\cos(ky) \qquad (4.45)$$

Obviously, only the last term on the right-hand side of (4.45) makes a non-zero contribution in (4.44). By using expressions (4.42) and (4.43) for $\psi_1^{(r,i)}$ (with $a$ now replaced by $\delta(t)$), equations (4.44) and (4.45) yield for $\Delta W^{(i,r)}$:

$$\Delta W^{(i)} = \frac{B_0^2 \sin^2(\alpha l)}{16\pi l} a^2 [(\kappa l)\cot(\kappa l) - (\alpha l)\cot(\alpha l)]$$

$$\Delta W^{(r)} = \frac{B_0^2 \sin^2(\alpha l)}{16\pi l} a^2 [(\kappa l)\cot(2\kappa l) - (\alpha l)\cot(\alpha l)] \qquad (4.46)$$

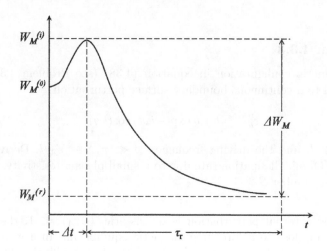

**FIGURE 4.9**
Reconnective magnetic relaxation of the system

It follows from the above expressions that the reconnected magnetic configuration always has a lower magnetic energy than its ideal MHD counterpart:

$$\Delta W_M = \Delta W^{(i)} - \Delta W^{(r)} = \frac{B_0^2 \sin^2(\alpha l)}{16\pi l} a^2(\kappa l)[\cot(\kappa l) - \cot(2\kappa l)] > 0, \quad (4.47)$$

which is, therefore, released in the process of magnetic reconnection. Moreover, the released energy, $\Delta W_M$, can greatly exceed the amount of energy, $\Delta W^{(i)}$, that needs to be supplied to the system by an external force in order to form the ideal MHD equilibrium with the current sheet (note that, according to (4.46), the energy difference $\Delta W^{(i)}$ is always positive). Indeed, as seen from (4.47), the released energy formally tends to infinity when the parameter $(\kappa l)$ approaches $\pi/2$, and the system becomes MHD unstable in respect to reconnective tearing mode (see Problem 4.3.5). This means that the released energy $\Delta W_M$ is not supplied externally but is tapped from the excess magnetic energy stored in the initial magnetic configuration (4.38). Therefore, even a weak external perturbation can trigger a substantial internal reconnective magnetic relaxation in a highly conducting fluid. An overall scenario of this process, called **forced magnetic reconnection**, occurs as follows (Figure 4.9). A system undergoes external perturbation, during which ($t \sim \Delta t$) a small plasma resistivity still plays no role. Therefore, an ideal MHD equilibrium with a current sheet and slightly increased magnetic energy $W_0 + \Delta W^{(i)}$ is formed. Then, on a much longer time scale $\tau_r$ (see next Problem), resistive effects intervene, destroying the current sheet and causing transition to an equilibrium with different magnetic topology and lower magnetic energy.

## Problem 4.3.3

The magnetic configuration in equation (4.38) (see Problem 4.3.1) is subjected to a continuous boundary surface perturbation

$$x_b^{(+)} = l + a\exp(-i\omega t)\cos(ky),$$

with $a \ll l$, and the driving frequency $\omega \ll \tau_A^{-1} \equiv V_A/l$. Derive the resulting Ohmic dissipation rate due to a small plasma resistivity, when Lundquist number $S \gg 1$.

Since the external perturbation is quasistatic, $(\omega\tau_A) \ll 1$, the system should remain close to a force-free magnetic equilibrium. In a general case it can be represented as a superposition of the two equilibria obtained in Problem 4.3.1: the ideal MHD one, (i), and the reconnected one, (r). Thus, in the linear approximation, $(a \ll l)$, the flux function perturbation $\psi_1(x,y,t)$ takes the form:

$$\psi_1 = [A\psi_1^{(i)} + (1-A)\psi_1^{(r)}]\exp(-i\omega t), \qquad (4.48)$$

where $\psi_1^{(i,r)}$ are given by equations (4.42) and (4.43) of Problem 4.3.1. Here $A$ is the yet unknown amplitude, the magnitude of which is determined by the pace of magnetic reconnection. The latter can be characterized by the reconnection time, $\tau_r$, derived below. Thus, if $\tau_r$ is long compared with the period of external driving, i.e., $\omega\tau_r \gg 1$, one may expect the system to be close to the ideal MHD solution, hence, $A \to 1$. In the opposite limit of a relatively fast reconnection, $\omega\tau_r \ll 1$, the reconnected state should be established, i.e., $A \to 0$. The reconnection time $\tau_r$ depends on the plasma dynamics inside the current sheet at $x = 0$, which is present because of the singularity associated with the ideal MHD solution $\psi_1^{(i)}$. However, in the present case this current sheet acquires a finite width due to a non-zero plasma resistivity.

To proceed further, consider first plasma motion generated by the oscillating boundary surface. In the linear approximation, both plasma acceleration and velocity vectors are directed perpendicular to the initial magnetic field (4.38), so the velocity components can be written as

$$
\begin{aligned}
v_x &= v(x)\cos(ky)\exp(-i\omega t), \\
v_y &= u(x)\cos(\alpha x)\sin(ky)\exp(-i\omega t), \\
v_z &= -u(x)\sin(\alpha x)\sin(ky)\exp(-i\omega t)
\end{aligned}
$$

The amplitudes $v(x)$ and $u(x)$ have to be derived from the magnetic induction equation (4.2), which in the linear approximation yields:

$$-i\omega\psi_1(x) = B_0 v(x)\sin(\alpha x) + \eta(\psi_1'' - k^2\psi_1) \qquad (4.49)$$

$$-i\omega B_{z1} = \alpha B_0 v(x)\sin(\alpha x) - B_0 v'(x)\cos(\alpha x) - B_0 k u(x) + \eta(B''_{z1} - k^2 B_{z1})$$
$$(4.50)$$

Outside the current sheet (in the so-called external region), $B_{z1} = \alpha\psi_1$ and the plasma resistivity is negligible. Hence, it follows from the above equations that there

$$v(x) = -i\omega[A\psi_1^{(i)}(x) + (1-A)\psi_1^{(r)}(x)]/B_0\sin(\alpha x),$$
$$v'(x)\cos(\alpha x) + ku(x) = 0 \qquad (4.51)$$

As seen from (4.51), both $v(x)$ and $u(x)$ formally diverge at the current sheet (at $x \to 0$). There, they must be matched with the solution inside the current sheet (the internal solution), for which the plasma resistivity and inertia play a role. This internal solution can be simplified by noting that the second of equations (4.51) at $x \to 0$ becomes equivalent to $\vec{\nabla} \cdot \vec{v} = 0$, indicating in this way that plasma flow inside the current sheet is almost incompressible. Thus, the latter can be described in terms of the stream function $\phi(x, y, t) = \phi(x)\sin(ky)\exp(-i\omega t)$ as $\vec{v} = (\vec{\nabla}\phi \times \vec{z})$, so that $v(x) = k\phi(x)$, $u(x) = -\phi'(x)$. Then, the linearized equation of motion (4.1), given by

$$-i\omega\rho v = \frac{B_0}{4\pi}\cos(\alpha x)(\alpha\psi'_1 - B'_{z1}) + \frac{B_0}{4\pi}\sin(\alpha x)[\psi''_1 - k^2\psi_1 + \alpha B_{z1}],$$
$$-i\omega\rho u = -\frac{kB_0}{4\pi}(\alpha\psi_1 - B_{z1}),$$

can be reduced into a single equation for the stream function $\phi$:

$$-i\omega\phi'' + \frac{\alpha B_0}{4\pi}kx\psi''_1 = 0 \qquad (4.52)$$

(where the inequalities $\alpha x \ll 1$; $\psi''_1 \gg k^2\psi_1, \alpha B'_{z1}$; $\phi'' \gg k^2\phi$ have been used). Similarly, the reduced equation (4.49) reads:

$$-i\omega\psi_1(x) = \alpha B_0 kx\phi(x) + \eta\psi''_1, \qquad (4.53)$$

which, together with (4.52), defines the internal solution. By using non-dimensional variables and functions with $x$ scaled by $\alpha^{-1}$ (and $k$ by $\alpha$), $\psi_1$ by $B_0\alpha^{-1}$ and $\phi$ by $-i\omega\alpha^{-2}$, one gets from (4.52) and (4.53):

$$-i\psi_1(x) = -ikx\phi + \psi''_1(\omega\tau_\eta)^{-1} \qquad (4.54)$$

$$-(\omega\tau_A)^2\phi'' + kx\psi''_1 = 0 \qquad (4.55)$$

Here $\tau_A = [\alpha B_0/(4\pi\rho)^{1/2}]^{-1}$ is the Alfven time, and $\tau_\eta = (\alpha^2\eta)^{-1} = \tau_A\mathcal{S}$, the global resistive time. Since the governing equations (4.54) and (4.55) contain two small parameters: $(\omega\tau_A) \ll 1$, $(\omega\tau_\eta)^{-1} \ll 1$, it is convenient to rescale the variables in such a way that the resulting equations become parameter-free. Thus, after the substitutions:

$$x = (\omega\tau_A/\mathcal{S}k^2)^{1/4}\xi, \quad \phi = \psi_1(0)[\mathcal{S}/(\omega\tau_A)k^2]^{1/4}\chi(\xi), \qquad (4.56)$$

one gets from (4.54-4.55) a "standard" equation for the function $\chi(\xi)$:

$$\chi'' - i\xi^2\chi + i\xi = 0 \qquad (4.57)$$

The above rescaling indicates a small width, $(\Delta x)$, of the internal current sheet. Indeed, it follows from (4.56 and (4.57) that $(\Delta\xi) \sim 1$ corresponds to $(\Delta x) \sim (\omega\tau_A/\mathcal{S})^{1/4} \ll 1$ (it is assumed here that $k \sim 1$). This justifies the so-called "constant-$\psi$" approximation, which was tacitly used in obtaining equation (4.57): because of a very narrow current sheet the flux function variation there is small, i.e., $\psi(x) \approx \psi(0) = const$ (see Problem 4.3.5 for more details). Moreover, such a transformation allows one to estimate the reconnection time $\tau_r$ without actually solving equation (4.57). Indeed, since the latter is parameter-free, its appropriate solution $\chi(\xi) \sim 1$, and, hence, according to (4.56),

$$\phi \sim \psi_1(0)\mathcal{S}^{1/4}(\omega\tau_A)^{-1/4}$$

It follows then from (4.55) that

$$\psi_1'' \sim \frac{(\omega\tau_A)^2\phi}{(\Delta x)^3} \sim \psi_1(0)(\omega\tau_A)\mathcal{S}$$

Thus, the internal solution determines the magnitude of electric current density supported in the current sheet (recall that for the magnetic field (4.10) $j_z^{(1)} = -c\nabla^2\psi_1/4\pi \sim -c\psi_1''$). Therefore, the total current in the current sheet, which is $i_z \sim j_z^{(1)}(\Delta x)$, should match the discontinuity of $B_y^{(1)}$, i.e., the jump of $\psi_1'$ at $x = 0$, which is imposed by the external solution (4.48). It follows then from the latter and eqs. (4.42) and (4.43) that

$$\Delta\psi_1' = \psi_1'(0 + \epsilon) - \psi_1'(0 - \epsilon) = A\left[\left(\frac{d\psi_1^i}{dx}\right)_{0+\epsilon} - \left(\frac{d\psi_1^{(i)}}{dx}\right)_{0-\epsilon}\right]$$

$$= \frac{2\kappa A}{(1 - A)}\cot(\kappa l)\psi_1(0) \sim \frac{\alpha A}{(1 - A)}\psi_1(0)$$

(it is assumed here, for simplicity, that $k \sim \alpha \sim l^{-1}$). Therefore, the required matching condition reads $\psi_1''(\Delta x) \sim \frac{A}{(1-A)}\psi_1(0)$, i.e.,

$$\frac{A}{(1 - A)} \sim (\omega\tau_A)\mathcal{S}[(\omega\tau_A)/\mathcal{S}]^{1/4},$$

which yields

$$A \approx \frac{1}{1 + (\omega\tau_r)^{-5/4}}, \quad \tau_r = \tau_A\mathcal{S}^{3/5}, \qquad (4.58)$$

where $\tau_r$ is the effective reconnection time. As it is expected beforehand, $A \to 0$ when $\omega\tau_r \to 0$, and $A \to 1$ for $\omega\tau_r \to \infty$.

A rigorous solution of (4.57) (G. Vekstein and R. Jain, Physics of Plasmas, vol.6, 2897, 1999) yields

$$A = A_1 + iA_2 = \frac{(\omega\tau_r)^{5/4}(a^2 + b^2)\{[a + (\omega\tau_r)^{5/4}(a^2 + b^2)] + ib\}}{[a + (\omega\tau_r)^{5/4}(a^2 + b^2)]^2 + b^2}, \qquad (4.59)$$

with

$$a + ib = \frac{2\pi\Gamma(3/4)}{\Gamma(1/4)}\exp(i3\pi/8)$$

As seen from equation (4.59), in the general case both $\psi_1^{(i)}$ and $\psi_1^{(r)}$ contribute to the external equilibrium (4.48), and the amplitude $A$ is a complex number. The latter means a phase shift (a temporal lag) between the external driver and the internal response, and it is this lag that determines the energy dissipation rate in the system.

Since all parameters of the system vary periodically with time, the energy dissipation rate can be derived as the mean power of an external force which provides the continuous boundary deformation. Thus, similarly to (4.44), the sought after dissipation power is equal to

$$\frac{dW_r}{dt} = - << \Re[\frac{B^2(x_b^{(+)})}{8\pi}]\Re[\frac{dx_b^{(+)}}{dt}] >>,$$

where the symbols $<<>>$ mean averaging over variations along the $y$-coordinate and over the period of oscillations. A straightforward calculation yields the following result:

$$\frac{dW_r}{dt} = \frac{\Delta W_M}{2\tau_r}\mathcal{F}(\omega\tau_r), \mathcal{F}(\omega\tau_r) = -(\omega\tau_r)A_2(\omega\tau_r), \qquad (4.60)$$

where $\Delta W_M$ is the excess magnetic enegy given in (4.47), and $A_2$ is the imaginary part of the amplitude $A$ given in (4.59). The relaxation function $\mathcal{F}(\omega\tau_r)$ is plotted in Figure 4.10. In general, it exhibits features typical for a relaxation process, with the dissipation being most effective when the time-scale of the external driving is equal to the characteristic time of the internal relaxation. In this sense, the plasma heating via forced magnetic reconnection is analogous to a well-known effect of the enchanced attenuation of sound waves in a gas or liquid due to the "second" viscosity (see, e.g., L. D. Landau and E. M. Lifshitz, *Fluid Mechanics*, §81, Pergamon Press, 1987). Therefore, faster magnetic reconnection, i.e., shorter reconnection time $\tau_r$, does not necessarily result in increased plasma heating rate. Indeed, as seen from the above-given derivations, $dW_r/dt$ is actually decreasing with a shorter $\tau_r$ when $(\omega\tau_r) < 1$. The reason is that too fast reconnection does not allow build-up of the excess magnetic energy by bringing magnetic configuration too close to the relaxed state $\psi_1^{(r)}$.

Finally, it is worth to note an interesting link between this process and the energy dissipation by a low-frequency, $(\omega\tau_A) \ll 1$, Alfven resonance discussed

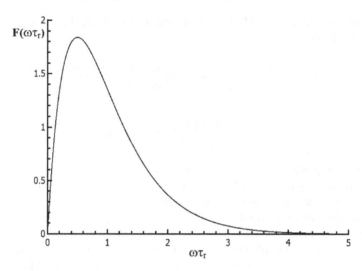

**FIGURE 4.10**
Reconnective relaxation function $F(\omega\tau_r)$

in Problem 4.2.5. It can be understood in the following way. In applying above the quasistatic external equilibrium solution (4.48), it was tacitly assumed that there are no Alfven resonances outside of the reconnecting current sheet. This is true if the driving frequency $\omega$ is low enough, so that the respective Alfven resonances, formally located at $x_r \sim \pm l(\omega\tau_A)$ (see Problem 4.2.5), fall inside the current sheet and are, therefore, destroyed there by a finite plasma resistivity. This is the case if $l(\omega\tau_A) < (\Delta x) \sim l(\omega\tau_A)^{1/4}S^{-1/4}$ (see Equation (4.56)). Therefore, Alfven resonances survive if

$$\omega > \omega_* \sim \tau_A^{-1}S^{-1/3},$$

while in the opposite limit forced magnetic reconnection takes over (G. Vekstein, Physics of Plasmas, vol.7, 3808, 2000). It is, therefore, not surprising that the two dissipation rates, the resonant one of (4.37) and the reconnective one given by (4.60), match each other at $\omega \sim \omega_*$. Indeed, since $(\omega_*\tau_r) \sim S^{4/5} \gg 1$, the relaxation function $\mathcal{F}$ in (4.60) at $\omega \sim \omega_*$ is of the order of $S^{-1/5}$ ($\mathcal{F}(\omega\tau_r)$ behaves as $(\omega\tau_r)^{-1/4}$ at $(\omega\tau_r) \gg 1$), which yields the reconnective dissipation power

$$\frac{dW_r}{dt} \sim \frac{\Delta W_M}{\tau_r}S^{-1/15} \sim \frac{B_0^2 a^2}{l}\frac{S^{-1/15}}{\tau_A S^{3/5}} \sim \frac{B_0^2 a^2}{\tau_A l}S^{-2/3}$$

(expression (4.47) for $\Delta W_M$ is used here). On the other hand, the Alfven resonance counterpart (4.37) yields at $\omega \sim \omega_*$ the same result:

$$\frac{dW_A}{dt} \sim V_A B_0^2(\omega_*\tau_A)^2\frac{a^2}{l^2} \sim \frac{B_0^2 a^2}{\tau_A l}S^{-2/3}$$

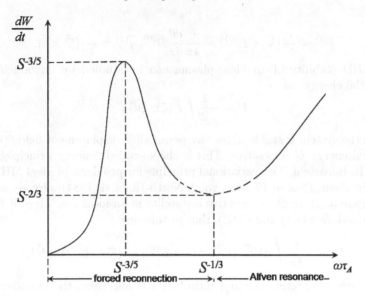

**FIGURE 4.11**
Combined diagram for the resonant and reconnective magnetic energy dissipation power

A combined diagram, which incorporates both resonant and reconnective dissipations, is shown in Figure 4.11.

### Problem 4.3.4

Investigate linear MHD stability of a planar force-free magnetic field

$$\vec{B}^{(0)} = [0, B_0 \sin \theta(x), B_0 \cos \theta(x)], \qquad (4.61)$$

(here $\theta(x)$ is an arbitrary regular function) embedded into a zero-$\beta$ plasma and bounded with two perfectly conducting surfaces $x_b^{(\pm)} = \pm l$, by using the energy principle.

In terms of the fluid displacement vector $\vec{\xi}$, in the ideal MHD the perturbed magnetic field in the linear approximation takes the form:

$$\vec{B} = \vec{B}^{(0)} + \vec{b}, \ \vec{b} = \vec{\nabla} \times (\vec{\xi} \times \vec{B}^{(0)}) \qquad (4.62)$$

Thus, the respective evolution equation for $\vec{\xi}$, the equation of motion, is

$$\rho \frac{\partial^2 \vec{\xi}}{\partial t^2} = \frac{1}{c}[(\vec{j}^{(0)} \times \vec{b}) + (\vec{j}^{(1)} \times \vec{B}^{(0)}] \equiv \vec{F}(\xi),$$

where

$$\vec{j}^{(0)} = \frac{c}{4\pi}(\vec{\nabla} \times \vec{B}^{(0)}) = \frac{c}{4\pi}\frac{d\theta}{dx}\vec{B}^{(0)}, \quad \vec{j}^{(1)} = \frac{c}{4\pi}(\vec{\nabla} \times \vec{b}).$$

The MHD stability of an ideal plasma can be studied by introducing the "potential energy" as

$$U = -\frac{1}{2}\int \vec{F}(\xi) \cdot \vec{\xi} dV,$$

so that the system is stable if, for any permissible displacement field $\vec{\xi}(\vec{r})$, the potential energy $U$ is positive. This is the so-called "energy principle" (see, e.g., I. B. Bernstein, The variational principle for problems of ideal MHD stability, in *Basic Plasma Physics*, vol.2, North-Holland, 1984), already applied in Problem 3.3.1 for the convective instability in an ideal gas. Thus, it follows from equations (4.61) and (4.62), that in this case

$$U = -\frac{1}{8\pi}\int dV \vec{\xi} \cdot \{[(\vec{\nabla} \times \vec{b}) \times \vec{B}^{(0)}] + [(\vec{\nabla} \times \vec{B}^{(0)}) \times \vec{b}]\},$$

which, after integrating the first term by parts and using the boundary condition $\vec{\xi} \cdot \vec{dS} = 0$, reduces to

$$U = \frac{1}{8\pi}\int dV\{[\vec{\nabla} \times (\vec{\xi} \times \vec{B}^{(0)})]^2 - [\vec{\xi} \times (\vec{\nabla} \times \vec{B}^{(0)})] \cdot [\vec{\nabla} \times (\vec{\xi} \times \vec{B}^{(0)})]\} \quad (4.63)$$

Since the first term on the right-hand side of equation (4.63) is always positive (or equal to zero), the MHD instability, if any, is caused by a non-zero electric current in the initial magnetic field, $(\vec{\nabla} \times \vec{B}^{(0)}) \neq 0$. This is another demonstration that such a current provides excess magnetic energy that can make the system unstable (see Problem 4.1.3).

In the particular case of the initial magnetic field (4.61), the perturbation of the vector potential $\vec{A} = (\vec{\xi} \times \vec{B}^{(0)})$ is equal to

$$\begin{aligned} A_x &= B_0(\xi_y \cos\theta - \xi_z \sin\theta) \equiv B_0\xi_\perp, \quad A_y = -\xi_x B_0 \cos\theta, \\ A_z &= \xi_x B_0 \sin\theta \end{aligned} \quad (4.64)$$

Without any loss of generality, one can represent the displacement components in the following way:

$$\xi_x = \xi_1(x)\cos(ky), \quad \xi_\perp = \xi_2(x)\cos(ky) + \xi_3(x)\sin(ky),$$

which makes the magnetic field perturbation $\vec{b} = (\vec{\nabla} \times \vec{A})$ equal to

$$\begin{aligned} b_x &= -k\xi_1 B_0 \sin\theta \sin(ky), \\ b_y &= -B_0\left(\sin\theta\frac{d\xi_1}{dx} + \cos\theta\frac{d\theta}{dx}\xi_1\right)\cos(ky), \\ b_z &= B_0\left(\sin\theta\frac{d\theta}{dx}\xi_1 - \cos\theta\frac{d\xi_1}{dx}\right)\cos(ky) + \\ &\quad kB_0(\xi_2 \sin(ky) - \xi_3 \cos(ky)) \end{aligned} \quad (4.65)$$

Then, the potential energy (4.63), which is now derived per unit area in the $(y, z)$ plane, takes the form:

$$U = \frac{1}{8\pi} \int\limits_{-l}^{l} dx \left[ < b^2 > - \frac{d\theta}{dx} < \vec{A} \cdot \vec{b} > \right],$$

where the symbols $<>$ mean averaging over variations along the $y$-coordinate. By making here use of expressions (4.64)-(4.65), a straightforward calculation results in

$$U = \frac{B_0^2}{16\pi} \int\limits_{-l}^{l} dx \left[ k^2 \sin^2 \theta \xi_1^2 + k^2 \xi_2^2 + k^2 \xi_3^2 + \left(\frac{d\xi_1}{dx}\right)^2 - 2k \cos\theta \xi_3 \frac{d\xi_1}{dx} \right]$$

Its minimization with respect to $\xi_2$ and $\xi_3$ yields

$$\xi_2 = 0, \xi_3 = -\frac{\cos\theta}{k}\frac{d\xi_1}{dx},$$

so finally one gets

$$U = \frac{B_0^2}{8\pi} \int\limits_{-l}^{l} dx \sin^2 \theta \left[ \left(\frac{d\xi_1}{dx}\right)^2 + k^2 \xi_1^2 \right] \geq 0 \qquad (4.66)$$

Therefore, this magnetic configuration is stable in the framework of the ideal MHD whatever the shear distribution $\theta(x)$.

However, it may become unstable in respect to the wider class of perturbations that are not allowed in the ideal MHD with the frozen-in magnetic field. This can be demonstrated in the following way (J. Goedblaed and R. Dagazian, Physical Review A, vol.4, 1554, 1971). Consider the potential energy (4.66) re-written in terms of the poloidal flux function perturbation $\psi_1$ introduced in Problem 4.3.1. Since such a flux function is equivalent to the $z$-component of the vector potential $\vec{A}$, it follows from expressions (4.64) that $\xi_1(x) = \psi_1(x)/B_0 \sin\theta(x)$, so a simple integration by parts in (4.66) results in

$$U = \frac{1}{8\pi} \int\limits_{-l}^{l} dx[(\psi_1')^2 - (\alpha^2 - k^2)\psi_1^2] \qquad (4.67)$$

(here and in what follows it is assumed, for simplicity, that the initial force-free magnetic configuration (4.61) is a linear one, i.e., $\theta(x) = \alpha x$). Considering now a minimum of the functional $U(\psi_1)$ defined by equation (4.67), one arrives to the Euler-Lagrange equation for $\psi_1(x)$:

$$\psi_1'' + (\alpha^2 - k^2)\psi_1 = 0, \qquad (4.68)$$

which, not surprisingly, is the equation of magnetostatic equilibrium obtained in Problem 4.11. Its regular solution with boundary conditions $\psi_1(x = \pm l) = 0$ is $\psi_1(x) = 0$ and, hence, does not lead to instability, which requires $U < 0$. However, a negative $U$ can be achieved if $\psi_1(x)$ has a discontinuous first derivative $\psi_1'$. Since the latter is equal to $b_y$, the $y$-component of the magnetic field perturbation $\vec{b}$, such a singularity implies a current sheet with a surface current $i_z$. Therefore, this current sheet can be located only where its current is directed along the initial magnetic field $\vec{B}^{(0)}$, i.e., where $B_y^{(0)} = 0$ (otherwise an infinite volumetric magnetic force will be exerted on a plasma), which in our case is $x = 0$. Bearing this in mind, one can transform expression (4.67) for the potential energy by integrating it by parts in the intervals $(-l, 0 - \epsilon)$ and $(0 + \epsilon, l)$ with $\epsilon \to 0$. Thus,

$$
\begin{aligned}
U &= \frac{1}{8\pi}\Bigg\{ \int_{-l}^{0-\epsilon} dx \left[ \left(\frac{d\psi_1}{dx}\right)^2 + (k^2 - \alpha^2)\psi_1^2 \right] \\
&\quad + \int_{0+\epsilon}^{l} dx \left[ \left(\frac{d\psi_1}{dx}\right)^2 + (k^2 - \alpha^2)\psi_1^2 \right] \Bigg\} = \\
&= \frac{1}{8\pi}\Bigg\{ -\int_{-l}^{0-\epsilon} dx\,\psi_1[\psi_1'' + (\alpha^2 - k^2)\psi_1] + \left(\psi_1\frac{d\psi_1}{dx}\right)_{-l}^{0-\epsilon} \\
&\quad + \left(\psi_1\frac{d\psi_1}{dx}\right)_{0+\epsilon}^{l} - \int_{0+\epsilon}^{l} dx\,\psi_1[\psi_1'' + (\alpha^2 - k^2)\psi_1] \Bigg\}
\end{aligned}
$$
(4.69)

If flux function $\psi_1(x)$ satisfies the equilibrium equation (4.68) and boundary conditions $\psi_1(-l) = \psi_1(l) = 0$, it follows from (4.69) that

$$
U = -\frac{1}{8\pi}\psi_1(0)\left[\left(\frac{d\psi_1}{dx}\right)_{0+\epsilon} - \left(\frac{d\psi_1}{dx}\right)_{0-\epsilon}\right] = -\frac{1}{8\pi}\psi_1(0)^2\Delta',
$$
(4.70)

where

$$
\Delta' = \frac{[(d\psi_1/dx)_{0+\epsilon} - (d\psi_1/dx)_{0-\epsilon}]}{\psi_1(0)}
$$
(4.71)

Therefore, if $\Delta' > 0$, such a perturbation reduces magnetic energy of the system and, hence, can lead to its MHD instability. The essential point here is that the respective energy reduction (4.70) requires $\psi_1(0) \neq 0$. However, as it has been demonstrated in Problem 4.3.1, the latter condition is not compatible with the ideal MHD, because it results in the change of the magnetic field topology due to formation of magnetic islands. Therefore, this instability, which is called the **tearing mode**, cannot proceed without a finite albeit small plasma resistivity, which breaks the frozen-in constraint of ideal MHD evolution. Therefore, the tearing mode growth rate is determined by the plasma resistivity (see next Problem), although the latter has no effect on the instability threshold.

## Problem 4.3.5

Derive the tearing instability threshold and growth rate for a linear force-free magnetic configuration (4.38) of Problem 4.3.1.

According to the recipe obtained in Problem 4.3.4, tearing stability/instability of a magnetic configuration is determined by the sign of the parameter $\Delta'$ defined by expression (4.71). Thus, the first step is derivation of a non-trivial solution of the equilibrium equation (4.68), which satisfies the boundary conditions $\psi_1(x = \pm l) = 0$ and, hence, has a discontinuous first derivative at $x = 0$. (It should be remembered here that this location of the discontinuity presumes that the perturbation wave-vector in the $(y, z)$-plane is $\vec{k} = (k, 0)$. In a general case, tearing current sheet is located at $x = x_s$ where $\vec{k} \cdot \vec{B}^{(0)}(x_s) = 0$). The result is as follows:

$$\psi_1(x) = \begin{cases} \psi_1(0) \sin[\kappa(x + l)]/\sin(\kappa l), & -l \le x \le 0 \\ \psi_1(0) \sin[\kappa(l - x)]/\sin(\kappa l), & 0 \le x \le l \end{cases} \qquad (4.72)$$

(it is assumed that $\kappa^2 = \alpha^2 - k^2 > 0$), and it yields $\Delta' = -2\kappa \cot(\kappa l)$. Thus, the field becomes tearing unstable ($\Delta' > 0$) when $\kappa l = (\alpha^2 - k^2)^{1/2} l > \pi/2$ (a sketch of $\psi_1(x)$ for both cases is drawn in Figure 4.12). It means that perturbations with $k \to 0$ are most unstable, which leads to the instability threshold $\alpha > \alpha_{cr} = \pi/2l$.

The instability growth rate $\gamma$ is determined by the current sheet structure (the internal solution), which should match the magnitude of $\Delta'$ prescribed by the external solution (4.72). Formally, the whole procedure is very similar to the one explored in Problem 4.3.3 for the case of forced magnetic reconnection. Thus, an incompressible flow inside the current sheet is described by the stream function $\phi(x, y, t) = \phi(x) \sin(ky) \exp(\gamma t)$, while the perturbation flux function $\psi_1(x, y, t) = \psi_1(x) \cos(ky) \exp(\gamma t)$. Then, instead of equations (4.52) and (4.53) one now gets

$$\gamma \rho \phi''(x) + \frac{\alpha B_0}{4\pi} kx \psi_1''(x) = 0,$$

$$\gamma \psi_1(x) = \alpha B_0 kx \phi(x) + \eta \psi_1''(x),$$

which by using the non-dimensional units ($\phi$ is now scaled by $\gamma \alpha^{-2}$) translate into

$$(\gamma \tau_A)^2 \phi'' + kx \psi_1'' = 0, \qquad (4.73)$$

$$\psi_1 = kx\phi + (\gamma \tau_\eta)^{-1} \psi_1'' \qquad (4.74)$$

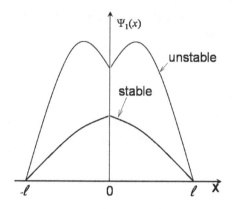

**FIGURE 4.12**
Sketch of the poloidal flux function perturbation for tearing stable and unstable magnetic configurations

After the re-scaling similar to that of (4.56):

$$x = (\gamma\tau_A/\mathcal{S}k^2)^{1/4}\xi, \ \phi = \psi_1(0)\left[\frac{\mathcal{S}}{(\gamma\tau_A)k^2}\right]^{1/4}\chi(\xi),$$

and the "*constant* $-\psi$" assumption (see below), one gets the following parameter-free equation for $\chi(\xi)$: $\chi'' + \xi^2\chi - \xi = 0$. Therefore, the order of magnitude estimates can now proceed as follows. The current sheet width is

$$(\Delta x) \sim (\gamma\tau_A/\mathcal{S}k^2)^{1/4}, \tag{4.75}$$

while the streamfunction is $\phi \sim \psi_1(0)[\mathcal{S}/(\gamma\tau_A k^2)]^{1/4}$. This yields the current density

$$\psi_1'' \sim \frac{(\gamma\tau_A)^2}{k(\Delta x)}\phi'' \sim \frac{(\gamma\tau_A)^2}{k(\Delta x)^3}\phi \sim (\gamma\tau_A)\mathcal{S}\psi_1(0).$$

Hence, the total current inside the current sheet is

$$i_z \sim \int \psi_1'' dx \sim \psi_1''(\Delta x) \sim \psi_1(0)\mathcal{S}^{3/4}(\gamma\tau_A)^{5/4}k^{-1/2}.$$

Then, the matching condition, which reads

$$\Delta' = \int \psi_1'' dx/\psi_1(0) \sim \mathcal{S}^{3/4}(\gamma\tau_A)^{5/4}k^{-1/2},$$

yields the instability growth rate

$$\gamma \sim \tau_A^{-1}\mathcal{S}^{-3/5}(\Delta')^{4/5}k^{2/5}, \tag{4.76}$$

with the parameter $\Delta'$ that follows from equations (4.71-4.72). Thus, for

$k \sim \alpha \sim l^{-1}$ the characteristic timescale of **spontaneous magnetic re-connection** via tearing instability is $\gamma^{-1} \sim \tau_A \mathcal{S}^{3/5}$, which is of the same order of magnitude as the reconnection time of forced magnetic reconnection, defined by Equation (4.58) in Problem 4.3.3.

Finally, consider now justification of the "*constant* $- \psi$" approximation, that has been already explored for both types of magnetic reconnection. As seen from equations (4.73-4.74), $\psi_1(x)$ is an even function of $x$; therefore, its variation inside the current sheet can be estimated as $\Delta \psi_1 \sim \psi_1''(\Delta x)^2 \sim \psi_1(0)\Delta'(\Delta x)$. Since in our case $\Delta' \sim 1$ and, according to expressions (4.75) and (4.76), $(\Delta x) \sim (\gamma \tau_A)^{1/4} \mathcal{S}^{-1/4} \sim \mathcal{S}^{-2/5}$, one gets $\Delta \psi_1 / \psi_1(0) \sim \mathcal{S}^{-2/5} \ll 1$.

More details about the tearing instability can be found, for example, in R. B. White, Resistive instabilities and field line reconnection, in *Basic Plasma Physics*, vol.1, North-Holland, 1984.

## Problem 4.3.6

Investigate the condition for the current sheet formation at a magnetic neutral $X$-point in a perfectly conducting zero-$\beta$ plasma.

Consider a planar magnetic field invariant along the $z$-coordinate:

$$\vec{B} = (\vec{\nabla}\psi(x,y) \times \vec{e}_z)$$

If this field is potential with a neutral $X$-point at the origin of the polar coordinate system $(r, \theta)$ as shown in Figure 4.13, its respective flux function is

$$\psi_0(r, \theta) = \frac{B_0 r^2}{2R} \cos 2\theta, \tag{4.77}$$

where $R$ is a radius of the circular domain under consideration. Assume now that this field, while being embedded into a perfectly conducting fluid (ideal MHD), becomes deformed by some displacement of its field line footpoints at the boundary surface $r = R$, which can be described by the relation $\theta_1(\theta)$, where $\theta$ and $\theta_1$ are, respectively, the initial and the displaced azimuthal coordinates of the footpoints. Since the magnetic field is frozen-in, each field line preserves its flux function magnitude; thus, the latter remains unchanged for the magnetic footpoints at the boundary:

$$\psi[R, \theta_1(\theta)] = \psi_0(R, \theta) \tag{4.78}$$

(for instance, $\psi(R, \theta_s) = 0$, where $\theta = \theta_s$ are locations of the separatrix footpoints $S_{1,2,3,4}$ shown in Figure 4.13(a)). Our interest is in the resulting deformed magnetostatic equilibrium, which in the case of zero-$\beta$ plasma should be a force-free one.

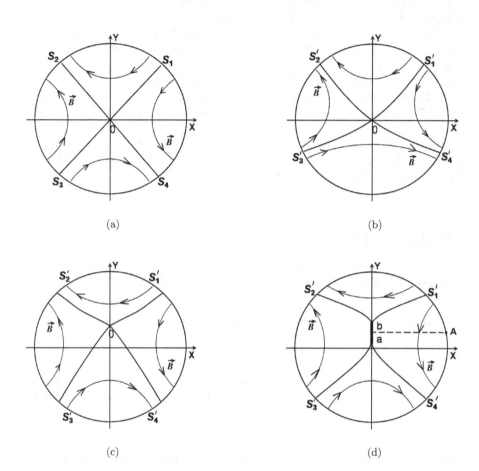

## FIGURE 4.13

(a) initial quadrupole magnetic field with the $m = 2$ flux distribution at the boundary $r = R$; (b) magnetic field under the boundary flux perturbation with $m = 3$: there is no change in the field lines connectivity compared to (a); (c) magnetic field perturbed with the $m = 1$ (dipole) mode: the null point is moved up, and the field lines connectivity is different from that in (a); (d) ideal MHD solution with the current sheet: the initial field lines connectivity of (a) is preserved

For a planar magnetic field with $B_z = 0$, the flux function of the deformed force-free equilibrium is, according to the Grad-Shafranov equation (4.10), a solution of the Laplace equation

$$\nabla^2 \psi(r, \theta) = 0, \qquad (4.79)$$

with the boundary condition specified by Equation (4.78) (which means that the deformed magnetic field remains a potential one). From the mathematical viewpoint this is a well-known Dirichlet problem (see, e.g., R. Courant and D. Hilbert, *Methods of Mathematical Physics*, v.2, Ch.4, Interscience, 1989) that always has a regular solution. However, it does not mean that such a solution can be realized under the ideal MHD frozen-in constraint on the magnetic field evolution. The reason lies in the particular structure of the initial magnetic field (4.77), where separatrix lines $OS_i$ (i=1,2,3,4) divide the whole domain into four quadrants, with the magnetic flux within each of them equal to $\Delta\psi_0 = \psi_0(R, 0) - \psi_s = B_0 R/2$ (note that $\psi_s = \psi_0(R, \pi/4) = 0$). Therefore, the appropriate solution of equation (4.79) must not only satisfy the boundary condition (4.78), but it should also preserve the magnetic footpoints' connectivity inside each of the quadrants and, hence, the given above amount of the magnetic flux $\Delta\psi$.

Consider, for example, the azimuthal deformation

$$\theta_1(\theta) = \theta + \frac{\delta_0}{R} \cos\theta,$$

with a small displacement amplitude $\delta_0 \ll R$. According to relation (4.78), it yields the following re-distribution of the magnetic flux at the boundary $r = R$:

$$\psi^{(R)}(\theta) = \psi_0^{(R)}(\theta - \delta\theta) \approx \psi_0^{(R)}(\theta) - \frac{d\psi_0^{(R)}}{d\theta} \delta\theta =$$

$$\psi_0^{(R)}(\theta) + B_0 \delta_0 \sin 2\theta \cos\theta =$$

$$\psi_0^{(R)}(\theta) + \frac{B_0 \delta_0}{2} \sin\theta + \frac{B_0 \delta_0}{2} \sin 3\theta \qquad (4.80)$$

As seen from expression (4.80), such a perturbation results in appearance of a dipole ($m = 1$) and $m = 3$ azimuthal components in the boundary flux distribution. As far as the latter is concerned, a regular solution of the Laplace equation (4.79), which accounts for the respective contribution in (4.80), reads

$$\psi_3(r, \theta) = \frac{B_0 \delta_0}{2R^3} r^3 \sin 3\theta \qquad (4.81)$$

By adding perturbation (4.81) to the initial flux function $\psi_0$ given by equation (4.77), one concludes that $\psi_0$ dominates in the vicinity of the origin $r = 0$. Therefore, the neutral $X$-point is not moved from the center, and the separatrix lines, that originate from the $X$-point, correspond to $\psi = \psi_s = 0$.

Thus, they end up at the displaced footpoints $(S_1', S_2', S_3', S_4')$ of the initial separatrix as shown in Figure 4.13(b). Therefore, the deformed field has the same field line connectivity as the initial one of Figure 4.13(a), since the amount of magnetic flux inside each quadrant remains unchanged and equal to $\Delta\psi_0 = B_0 R/2$. It also follows from this consideration that the same is the case for any harmonic of the boundary flux perturbation with the azimuthal number $m \geq 2$.

All is different for the dipole perturbation with $\psi_1^{(R)}(\theta) = \frac{B_0\delta_0}{2}\sin\theta$. Indeed, the respective regular solution inside the circle $r = R$ is

$$\psi_1(r,\theta) = \frac{B_0\delta_0}{2R}r\sin\theta = \frac{B_0\delta_0}{2R}y,$$

which corresponds to a uniform magnetic field $B_{1x} = \partial\psi_1/\partial y = B_0\delta_0/2R$. Being combined with the initial magnetic field (4.77), it moves the neutral $X$-point up from the center to the location $(x_1 = 0, y_1 = \delta_0/2)$. Moreover, this displaced $X$-point acquires now a magnitude of $\psi$ that is different from the initial one (which is equal to zero). Indeed,

$$\psi(x_1, y_1) = \psi_0(x_1, y_1) + \psi_1(x_1, y_1) = -\frac{B_0\delta_0^2}{8R} + \frac{B_0\delta_0^2}{4R} = \frac{B_0\delta_0^2}{8R} \qquad (4.82)$$

Since the separatrix originates from the $X$-point, its flux function magnitude $\psi_s = \psi(x_1, y_1) \neq 0$; therefore, all separatrix footpoints at the boundary $r = R$ cannot coincide with their displaced initial counterparts $(S_1', S_2', S_3', S_4'$, where $\psi^{(R)} = 0$ (see Figure 4.13(c)). Consequently, the amount of magnetic flux confined now in each of the four quadrants becomes different from the initial value of $\Delta\psi_0 = B_0 R/2$. Indeed, as seen from Figure 4.13, it is increased in the upper and lower quadrants by $\Delta\psi_s = B_0\delta_0^2/8R$, and, respectively, it is reduced by the same amount in the left-hand side and the right-hand side ones. Thus, the regular solution of Figure 4.13(c) is not compatible with the initial one of Figure 4.13(a). Hence, in the framework of ideal MHD such external perturbation leads to a singular deformed magnetic equilibrium with a current sheet (see Problem 4.3.7).

## Problem 4.3.7

Derive the singular ideal MHD equilibrium, which is formed under the dipole-type deformation of the magnetic field with the neutral X-point (see Problem 4.3.6).

The respective solution containing the current sheet (as shown in Figure 4.13(d)) can be constructed in the following way. Any potential planar magnetic field $[B_x(x,y), B_y(x,y)]$ can be represented as $B_x - iB_y = f(z) = $

$u(x, y) + iv(x, y)$, where $f(z)$ is an analytical function of the complex variable $z = x + iy$. Indeed, the divergence-free and curl-free requirements for $\vec{B}$, namely $\frac{\partial B_x}{\partial x} = -\frac{\partial B_y}{\partial y}$ and $\frac{\partial B_x}{\partial y} = \frac{\partial B_y}{\partial x}$, are equivalent to the Cauchy-Riemann equations for the real and imaginary parts of $f(z)$: $\frac{\partial u}{\partial x} = \frac{\partial v}{\partial y}, \frac{\partial v}{\partial x} = -\frac{\partial u}{\partial y}$. For example, the initial field, specified by the flux function (4.77), corresponds to $f(z) = iB_0z/R$. Then, the deformed magnetic field with the current sheet shown in Figure 4.13(d) can be described as

$$B_x - iB_y = \frac{iB_0}{R}(z - ia)^{1/2}(z - ib)^{1/2}, \tag{4.83}$$

where the current sheet corresponds to the branch cut in the complex plane between $z = ia$ and $z = ib$, where two neutral $Y$-points are formed. In order to find the values of $a$ and $b$, consider the magnetic field of Equation (4.83) far enough from the current sheet, i.e., for $|z| \gg a, b$. Thus,

$$B_x - iB_y = \frac{iB_0}{R}z\left(1 - \frac{ia}{z}\right)^{1/2}\left(1 - \frac{ib}{z}\right)^{1/2} \approx \frac{iB_0}{R}z + \frac{B_0}{2R}(a + b),$$

which means that this current sheet generates there an additional uniform magnetic field $B_x = B_0(a + b)/2R$. Therefore, in order to comply with the dipole contribution to the boundary flux perturbation, this extra field should be equal to $B_{1x} = B_0\delta_0/2R$, i.e., $(a + b) = \delta_0$. In other words, the current sheet is centered at $y = y_1 = \delta_0/2$, the location of the $X$-point of the regular solution of Figure 4.13(c). Then, if $\epsilon$ is a half-length of the current sheet, i.e., $a = \delta_0/2 - \epsilon$, $b = \delta_0/2 + \epsilon$, one can re-write (4.83) as

$$B_x - iB_y = \frac{iB_0}{R}\left[z - i\left(\frac{\delta_0}{2} - \epsilon\right)\right]^{1/2}\left[z - i\left(\frac{\delta_0}{2} + \epsilon\right)\right]^{1/2} \tag{4.84}$$

Another requirement applied to this singular solution, which allows to derive $\epsilon$, is that all separatrix lines and, therefore, the current sheet from which they originate should correspond to $\psi = \psi_s = 0$. This ensures that the separatrix lines end up at the boundary $r = R$ at the points $(S_1', S_2', S_3', S_4')$, preserving in this way the field lines' connectivity of the initial magnetic field (see Figure 4.13(d)). Thus, consider magnetic field (4.84) on the horizontal line $y = \delta_0/2$, which stretches from the center of the current sheet to the point $A$ at the boundary circle as shown in Figure 4.13(d). It follows from expression (4.84) that along this line, where $z = x + i\delta_0/2$, $B_y(x) = -\frac{B_0}{R}(x^2 + \epsilon^2)^{1/2}$, hence

$$\psi_A - \psi_s = -\int_0^{x_A} B_y dx = \frac{B_0}{R}\int_0^{x_A}(x^2 + \epsilon^2)^{1/2}dx =$$

$$\psi^{(R)}(\theta_A) = \psi_0^{(R)}(\theta_A) + \psi_1^{(R)}(\theta_A) \approx \frac{B_0R}{2}, \tag{4.85}$$

where $x_A = (R^2 - \delta_0^2/4)^{1/2}$. A straightforward derivation of the integral in (4.85) yields

$$\int_0^{x_A} (x^2 + \epsilon^2)^{1/2} dx \approx \frac{1}{2}[R^2 - \frac{\delta_0^2}{4} + \epsilon^2 \sinh^{-1}(R/\epsilon)],$$

which being inserted into equation (4.85) results in

$$\epsilon \approx \frac{\delta_0}{2\mathcal{L}^{1/2}}, \tag{4.86}$$

where $\mathcal{L} = \ln(R/\delta_0) \gg 1$.

Thus, two different magnetostatic equilibria are associated with the dipole-type deformation of the initial $X$-point magnetic configuration (4.77). There is the regular equilibrium shown in Figure 4.13(c), and the singular current sheet equilibrium of Figure 4.13(d). The latter corresponds to a perfectly conducting fluid, where connectivity of the magnetic field-line footpoints is preserved. However, presence of a very small but finite resistivity leads to magnetic reconnection, which eliminates the magnetic field singularity (the current sheet), and enables transition to the regular equilibrium with a lower magnetic energy. Therefore, this process is similar to forced magnetic reconnection discussed in Problems 4.3.1 and 4.3.2.

# 5

# Theory of elasticity

## 5.1 Mechanics of solid bodies

Small deformations of a solid body are described by the **strain tensor**

$$u_{ik} = \frac{1}{2}\left(\frac{\partial u_i}{\partial x_k} + \frac{\partial u_k}{\partial x_i}\right),$$

where $\vec{u}$ is the **displacement vector**. The deformations result in the appearance of internal forces, which are characterized by the **stress tensor** $\sigma_{ik}$, so that the force $\vec{F}$ exerted on a portion of the body $V$ is equal to

$$F_i = \int \sigma_{ik} dS_k, \tag{5.1}$$

where integration in (5.1) is carried out over the boundary surface of $V$, with the vector $d\vec{S}$ directed along the external normal. Thus, the force per unit volume, $f_i$, can be defined as

$$f_i = \frac{\partial \sigma_{ik}}{\partial x_k}, \tag{5.2}$$

so that $F_i = \int \sigma_{ik} dS_k = \int dV (\partial \sigma_{ik}/\partial x_k) = \int f_i dV$.

For a small deformation the stress tensor $\sigma_{ik}$ is proportional to the strain tensor (Hooke's law). In a macroscopically isotropic medium (such as, for example, a polycrystalline) this relation takes the form

$$\sigma_{ik} = K u_{ll}\delta_{ik} + 2\mu\left(u_{ik} - \frac{1}{3}u_{ll}\delta_{ik}\right), \tag{5.3}$$

$$u_{ik} = \frac{1}{9K}\sigma_{ll}\delta_{ik} + \frac{1}{2\mu}\left(\sigma_{ik} - \frac{1}{3}\sigma_{ll}\delta_{ik}\right), \tag{5.4}$$

which involves two elastic coefficients: the **modulus of compression**, $K$, and the **modulus of rigidity**, $\mu$. Then the free elastic energy per unit volume of a deformed medium reads

$$\mathcal{F} = \frac{1}{2}\sigma_{ik}u_{ik} = \frac{1}{2}K(u_{ll})^2 + \mu\left(u_{ik} - \frac{1}{3}u_{ll}\delta_{ik}\right)^2 \tag{5.5}$$

In the absence of external forces a state of thermodynamic equilibrium with no deformation should correspond to a minimum of the free energy. Therefore, it follows from expression (5.5) that the moduli of compression and rigidity are always non-negative: $K \geq 0$, $\mu \geq 0$.

Another set of the commonly used elastic coefficients is the **modulus of extension**, also called the **Young's modulus**, $E$, and **Poisson's ratio**, $\sigma$, which are related to $K$ and $\mu$ as follows:

$$E = 9K\mu/(3K + \mu), \quad \sigma = \frac{1}{2}(3K - 2\mu)/(3K + \mu) \tag{5.6}$$

As seen from relations (5.6), $E \geq 0$, while $-1 \leq \sigma \leq 1/2$. Although a negative Poisson's ratio is permitted by the laws of thermodynamics, such a material is unknown in the natural world (there have been, however, numerous efforts to create it artificially (see, e.g., M. C. Rechtsman et al., Physical Review Letters, vol. 101, 085501, 2008, and references therein). Therefore, in what follows it is assumed that $0 \leq \sigma \leq 1/2$. In terms of $E$ and $\sigma$ one gets, instead of expressions (5.3) and (5.4), that

$$\sigma_{ik} = \frac{E}{(1+\sigma)}\left(u_{ik} + \frac{\sigma}{(1-2\sigma)}u_{ll}\delta_{ik}\right), \tag{5.7}$$

$$u_{ik} = \frac{1}{E}[(1+\sigma)\sigma_{ik} - \sigma\sigma_{ll}\delta_{ik}] \tag{5.8}$$

The equation of equilibrium, according to (5.2), takes the form

$$\frac{\partial \sigma_{ik}}{\partial x_k} + \rho g_i = 0, \tag{5.9}$$

where $\rho$ is the density of the medium, and $\vec{g}$ is the external non-elastic force per unit mass (for example, gravity). In terms of the displacement vector $\vec{u}$, the equilibrium equation (5.9) in an isotropic medium reads

$$\vec{\nabla}^2 \vec{u} + \frac{1}{(1-2\sigma)}\vec{\nabla}(\vec{\nabla} \cdot \vec{u}) = -\rho\vec{g}\frac{2(1+\sigma)}{E} \tag{5.10}$$

## Problem 5.1.1

A cube of edge $l$, made from a material with given elastic coefficients, $E$ and $\sigma$, is immersed into an absolutely rigid cavity of the same size. Derive the deformation of the cube due to an external pressure $p$ acting on its side, as shown in Figure 5.1.

By using the coordinate system depicted in Figure 5.1, one can conclude that the displacement vector $\vec{u}$ has in this case only one non-zero component,

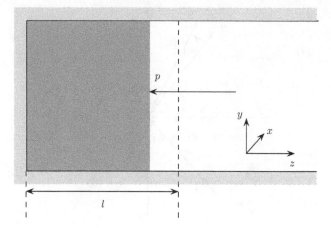

**FIGURE 5.1**
Compressed elastic cube in a rigid cavity

$u_z$, which depends only on the $z$-coordinate: $u_z(z) \equiv u(z)$. Consequently, in the strain tensor $u_{ik}$ defined in equation (5.1), only one non-zezo component $u_{zz} = du_z/dz$ is present. Therefore, the only significant components of the stress tensor that follow from the Hooke's law (5.7) are:

$$\sigma_{xx} = \sigma_{yy} = \frac{E\sigma}{(1+\sigma)(1-2\sigma)}u_{zz}, \ \sigma_{zz} = \frac{E(1-\sigma)}{(1+\sigma)(1-2\sigma)}u_{zz} \qquad (5.11)$$

The equilibrium condition, $\partial\sigma_{ik}/\partial x_k = 0$, yields that $u_{zz} = const$, the value of which is determined by the boundary condition at the side $z = l$: $\sigma_{zz}(l) = -p$. Therefore, it follows from expressions (5.11) that

$$u_{zz} = -\frac{p(1+\sigma)(1-2\sigma)}{E(1-\sigma)}$$

## Problem 5.1.2

An isotropic elastic material occupies the half-space $z > 0$, with its boundary surface $z = 0$ being "glued" to an absolutely rigid medium lying beneath this surface. Find the reflection coefficient for a longitudinal wave with the angle of incidence $\theta_l$ at the boundary plane (Figure 5.2).

In a general case, both longitudinal (wave 2) and transverse (wave 3) reflected waves will be present. Let us denote their displacement amplitudes as $\vec{u}_2$ and $\vec{u}_3$, when $\vec{u}_1$ is the amplitude of the incident wave (wave 1) (the directions of these vectors is shown in Figure 5.2). For a given $\vec{u}_1$, the vectors $\vec{u}_2$ and $\vec{u}_3$ have to be found from the boundary condition at the plane

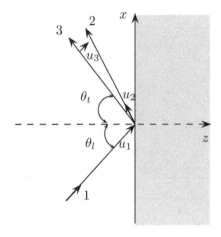

**FIGURE 5.2**
Reflection of the elastic wave

$z = 0$. In the present case of an absolutely rigid attachment it means zero total displacement there. Thus, one comes to the following relations:

$$u_z = u_1 \cos\theta_l - u_2 \cos\theta_l + u_3 \sin\theta_t = 0,$$
$$u_x = u_1 \sin\theta_l + u_2 \sin\theta_l + u_3 \cos\theta_t = 0,$$

which yield $u_2 = u_1 \cos(\theta_l + \theta_t)/\cos(\theta_l - \theta_t)$ and, hence, the reflection coefficient is equal to

$$R_l = (u_2/u_1)^2 = \cos^2(\theta_l + \theta_t)/\cos^2(\theta_l - \theta_t) \tag{5.12}$$

The angle, $\theta_t$, of the reflected transverse wave can be expressed in terms of the angle $\theta_l$, the longidutinal sound speed, $c_l$, and the transverse sound speed, $c_t$, in the following way. Since all three waves have the same frequency and the same $k_x$ ($x$-component of their wave vectors $\vec{k}_{l,t}$), one gets that $c_l k_l = c_t k_t$ and $k_l \sin\theta_l = k_t \sin\theta_t$. These yield $\sin\theta_t = (c_t/c_l)\sin\theta_l$, which together with expression (5.12) completes the answer. Finally, one may notice from (5.12) that the longitudinal reflection coefficient, $R_l$, becomes equal to zero when $\theta_l + \theta_t = \pi/2$, which has a simple geometrical explanation. Indeed, in this case the vectors $\vec{u}_1$ and $\vec{u}_3$ are collinear, and, thus, the required boundary condition $\vec{u}_{z=0} = 0$ is satisfied with $\vec{u}_3 = -\vec{u}_1$ and $\vec{u}_2 = 0$.

## Problem 5.1.3

A plug of length $L$ and radius $a$, made from elastic material with Young's modulus $E$ and Poisson's ratio $\sigma$, is squeezed into an absolutely rigid cylindrical channel so that the pressure $p$ acts on the lateral surface of the plug. What minimum pushing force, $F_{min}$, is required to move the plug, if the friction coefficient at the surface of the channel is equal to $\kappa$ such that $\kappa \ll 1$?

The problem becomes non-trivial because of the non-zero Poisson's ratio of the plug material. Indeed, in the case of $\sigma = 0$ the longitudinal stresses caused by the pushing force have no effect on the stresses in the plane transverse to the $z$-axis. Thus, the pressure acting on the lateral surface of the plug would remain equal to $p$, making the required pushing force simply equal to $F_{min} = 2\pi a L \kappa p$. However, for $\sigma > 0$ the pushing force results in an increased lateral pressure, which makes $F_{min}$ much bigger for a long plug with $L \gg a$. By using the cylindrical coordinates $(r, \theta, z)$, consider now the $r$ and $z$ components of the equilibium equation (5.9):

$$\frac{\partial(r\sigma_{rr})}{r\partial r} - \frac{\sigma_{\theta\theta}}{r} + \frac{\partial\sigma_{zr}}{\partial z} = 0 \qquad (5.13)$$

$$\frac{\partial(r\sigma_{zr})}{r\partial r} + \frac{\partial\sigma_{zz}}{\partial z} = 0 \qquad (5.14)$$

(in this case $\vec{g} = 0$, and the azimuthal equilibrium is satisfied identically due to the azimuthal symmetry). In the absence of the pushing force there are only two non-zero components of the stress tensor:

$$\sigma_{\theta\theta} = \sigma_{rr} = -p,$$

where the first equality follows from the radial equilibrium condition (5.13), and the second one comes from the boundary condition at the plug's lateral surface. Thus, the plug is uniformly deformed and, according to the Hooke's law (5.8),

$$u_{rr} = u_{\theta\theta} = -p(1 - \sigma)/E, \; u_{zz} = 2\sigma p/E \qquad (5.15)$$

Under the action of the pushing force the stress tensor components $\sigma_{zz}$ and $\sigma_{zr}$ also arise. However, if $\kappa \ll 1$, one may expect (and it is confirmed by what follows) that the component $\sigma_{zr}$ remains small. Therefore, the respective term in equation (5.13) can be neglected and, hence, as before, $\sigma_{rr} \approx \sigma_{\theta\theta}$, but now they are $z$-dependent. Since the channel is absolutely rigid, the radial deformation of the plug remains unchanged and equal to the one given in (5.15): $u_{rr} = -p(1 - \sigma)/E$. On the other hand, from Hooke's law (5.8),

$$u_{rr} = \frac{1}{E}[(1 + \sigma)\sigma_{rr} - \sigma(\sigma_{rr} + \sigma_{\theta\theta} + \sigma_{zz})] =$$

$$\frac{1}{E}[(1 - \sigma)\sigma_{rr} - \sigma\sigma_{zz}],$$

which yields the following relation between $\sigma_{rr}$ and $\sigma_{zz}$:

$$\sigma_{rr} = -p + \frac{\sigma}{(1 - \sigma)}\sigma_{zz} \qquad (5.16)$$

Then, integration of equation (5.14) over $r$ with the requirement that $\sigma_{zr}$ is finite at $r = 0$ gives

$$\sigma_{zr} = -\frac{1}{2}r\frac{\partial \sigma_{zz}}{\partial z},$$

and hence, by using the boundary condition $\sigma_{zr}(a) = \kappa \sigma_{rr}$ and expression (5.16), one gets the following equation for $\sigma_{zz}(z)$:

$$-\frac{a}{2}\frac{\partial \sigma_{zz}}{\partial z} = \kappa\left(-p + \frac{\sigma}{(1-\sigma)}\sigma_{zz}\right)$$

The solution with the required boundary condition $\sigma_{zz}(0) = -F/\pi a^2$ reads:

$$\sigma_{zz}(z) = \frac{(1-\sigma)}{\sigma}p\left[\exp\left(\frac{2\kappa\sigma z}{a(1-\sigma)}\right) - 1\right] - \frac{F}{\pi a^2} \tag{5.17}$$

Consider now the variation of $\sigma_{zz}$ along the plug under different magnitudes of the pushing force $F$. If the latter is not too big, the absolute value of $\sigma_{zz}$ decreases monotonically from the initial one at $z = 0$ (which is equal to $F/\pi a^2$), down to zero at some point $z = z_0 < L$. In this case Equation (5.17) becomes not applicable at $z_0 < z < L$, where one can put $\sigma_{zz} = 0$. In physical terms it means that the friction force exerted on the $0 \leq z \leq z_0$ portion of the plug is sufficient to withstand the pushing force; hence, the rest of the plug plays no role and can be cut off and removed without making any difference. When the force $F$ is increasing, so is the value of $z_0$, and the critical point is reached when $z_0 = L$. This, according to expression (5.17), occurs when

$$F = F_0 = \pi a^2 p\frac{(1-\sigma)}{\sigma}\left[\exp\left(\frac{2\kappa\sigma L}{a(1-\sigma)}\right) - 1\right] \tag{5.18}$$

Clearly, there is no equilibrium if the pushing force exceeds $F_0$, which, therefore, is the sought after minimum force required to move the plug. As seen from (5.18), this force becomes exponentially large when the plug's aspect ratio $L/a \gg 1$. The reason lies in the non-zero Poisson's ratio. Indeed, in the limit $\sigma \to 0$ expression (5.18) reduces to the expected one: $F_{min}(\sigma \to 0) = 2\pi La\kappa p$.

## Problem 5.1.4

A long elastic cylinder of radius $R$ is slit along the plane formed by the axis of the cylinder and its generatrix. This cut is then moved apart by an external force to a slice of angle $\alpha \ll 2\pi$ (see Figure 5.3), and a wedge of the same angle, which is made from the same unstressed material, is put in to fill the cavity. After that, the cut is set free by removing the external force. Find the resulting stresses present in the cylinder, and its elastic energy per unit length.

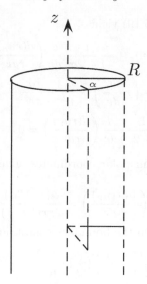

**FIGURE 5.3**
Elastic cylinder with a cut

The cylinder can be considered now as a continuous elastic body with the imposed azimuthal displacement equal to $u_\theta = \alpha r\phi/2\pi$, which also leads to the radial displacement being present (the cylinder becomes inflated). Note that, under this idealization, $u_\theta$ becomes discontinuous at the cut, resulting in logarithmic singularities in the deformations and stresses at the origin $r = 0$ (see below). Due to the azimuthal symmetry, the radial displacement varies only with the radius, $u_r(r)$, and it has to be determined from the equilibrium equation (5.10), which now yields

$$\vec{\nabla}^2 \vec{u} + \frac{1}{(1 - 2\sigma)} \vec{\nabla}(\vec{\nabla} \cdot \vec{u}) = 0 \qquad (5.19)$$

Since

$$(\vec{\nabla} \cdot \vec{u}) = \frac{1}{r}\left[\frac{\partial(ru_r)}{\partial r} + \frac{\partial u_\theta}{\partial \theta}\right] = \frac{\partial(ru_r)}{r\partial r} + \frac{\alpha}{2\pi}$$

and does not depend on $\theta$, the azimuthal component of equation (5.19) is satisfied if

$$(\vec{\nabla}^2 \vec{u})_\theta = \vec{\nabla}^2 u_\theta - \frac{u_\theta}{r^2} = \frac{1}{r}\frac{\partial}{\partial r}\left(r\frac{\partial u_\theta}{\partial r}\right) - \frac{u_\theta}{r^2} = 0$$

The radial component of (5.19) yields

$$(\vec{\nabla}^2 \vec{u})_r + \frac{1}{(1 - 2\sigma)} \frac{d}{dr}\left(\frac{d(ru_r)}{rdr}\right) =$$

$$\frac{1}{r}\frac{d}{dr}\left(r\frac{du_r}{dr}\right) - \frac{u_r}{r^2} - \frac{\alpha r}{\pi r^2} +$$

$$\frac{1}{(1 - 2\sigma)}\frac{d}{dr}\left(\frac{d(ru_r)}{rdr}\right) = 0$$

(the expression for $(\vec{\nabla}^2 \vec{u})$ in polar coordinates, given in Chapter 1, is used here). Hence,

$$\frac{d}{dr}\left(\frac{1}{r}\frac{d(ru_r)}{dr}\right) = \frac{\alpha(1 - 2\sigma)}{2\pi r(1 - \sigma)},$$

which, after integration with the boundary condition $u_r(0) = 0$, yields

$$u_r = \frac{\alpha(1 - 2\sigma)}{2\pi(1 - \sigma)}\left(\frac{r}{2}\ln\frac{r}{R} - \frac{r}{4} + A\frac{r}{2}\right)$$

Here $A$ is a yet unknown constant, that has to be found from the boundary condition at the lateral surface of the cylinder: $(\sigma_{ik}n_k)_{r=R} = 0$, i.e.,

$$\sigma_{rr}(R) = \sigma_{\theta r}(R) = \sigma_{zr}(R) = 0$$

In the present case there are only two non-zero components of the strain tensor:

$$u_{rr} = \frac{du_r}{dr}, \quad u_{\theta\theta} = \frac{1}{r}\frac{\partial u_\theta}{\partial \theta} + \frac{u_\theta}{r}$$

Therefore, only the diagonal components of the stress tensor are present, and the value of $A$ follows from the boundary condition for $\sigma_{rr}$. Thus, according to equation (5.7),

$$\sigma_{rr} = \frac{E}{(1 + \sigma)}\left(u_{rr} + \frac{\sigma}{(1 - 2\sigma)}(u_{rr} + u_{\theta\theta})\right),$$

where

$$u_{rr} = \frac{\alpha(1 - 2\sigma)}{4\pi(1 - \sigma)}\left(\ln\frac{r}{R} + \frac{1}{2} + A\right),$$

$$u_{\theta\theta} == \frac{\alpha(1 - 2\sigma)}{4\pi(1 - \sigma)}\left(\ln\frac{r}{R} - \frac{1}{2} + A\right) + \frac{\alpha}{2\pi}$$

Then, a straightforward calculation yields $A = -1/2(1-2\sigma)$, so that the stress tensor components are equal to

$$\sigma_{rr} = \frac{\alpha E}{4\pi(1 - \sigma^2)}\ln\frac{r}{R},$$

$$\sigma_{\theta\theta} = \frac{\alpha E}{4\pi(1 - \sigma^2)}\left(\ln\frac{r}{R} + 1\right),$$

$$\sigma_{zz} = \frac{\alpha\sigma E}{2\pi(1 - \sigma^2)}\left(\ln\frac{r}{R} + \frac{1}{2}\right)$$

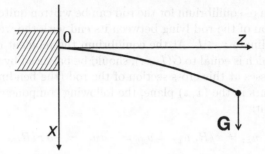

**FIGURE 5.4**
Bending of a loaded rod

Thus, the elastic energy density reads

$$\mathcal{F} = \frac{1}{2}\sigma_{ik}u_{ik} = \frac{1}{2}(\sigma_{rr}u_{rr} + \sigma_{\theta\theta}u_{\theta\theta}) =$$

$$\frac{\alpha^2 E(1-2\sigma)}{16\pi^2(1-\sigma^2)(1-\sigma)}\left[\left(\ln\frac{r}{R}\right)^2 + \ln\frac{r}{R}\right] + \frac{\alpha^2 E}{32\pi^2(1-\sigma^2)} \qquad (5.20)$$

The total elastic energy per unit length of the cylinder is equal to

$$\mathcal{W} = \int \mathcal{F}dS = 2\pi \int_0^R \mathcal{F}rdr$$

In turns out that the two logarithmic terms in (5.20) cancel each other after the integration, so that finally

$$\mathcal{W} = \frac{\alpha^2 E R^2}{32\pi(1-\sigma^2)}$$

## Problem 5.1.5

One end of a thin rod of length $L$ and radius $a$, $(L \gg a)$, is clamped, while the other one is loaded with a point weight $G$ (see Figure 5.4). The material of the rod has Young's modulus equal to $E$, and it breaks down under an internal stress exceeding $\sigma_c = \alpha E$, $\alpha \ll a/L \ll 1$. What is the maximum weight $G_{max}$ that the rod can sustain, if the weight of the rod itself is negligible?

If $\alpha \ll a/L$ the rod remains bent only slightly even under the maximum load allowed, and its deflection from the horizontal position is small. In this case the equation of equilibrium for the rod can be written quite simply. Thus, consider a portion of the rod lying between its end with the weight and some point with coordinate $z < L$. At the equilibrium the moment of force due to the weight $G$, which is equal to $G(L-z)$, should be balanced by the moment of the internal stresses at this cross section of the rod (the bending moment). If the bending occurs in the $(x, z)$ plane, the following components of the stress tensor are present:

$$u_{zz} = x/R, \ u_{xx} = u_{yy} = -\sigma u_{zz} = -\sigma x/R,$$

where $R(z)$ is the local radius of curvature of the rod axis, with $x$ being measured from the neutral line in the rod cross section (in this case the neutral line is the diameter parallel to the $y$-axis). It follows then from equation (5.7) that the stress tensor has only one non-zero component, $\sigma_{zz} = Eu_{zz} = Ex/R$. Therefore, the force $\sigma_{zz}dS$, acting in the $z$-direction at the surface element of the rod cross section, yields the bending momentum equal to

$$M = \int x\sigma_{zz}dS = \frac{E}{R}\int x^2 dS = \frac{EI}{R},$$

where $I = \pi a^4/4$ is the moment of inertia of a rod with circular cross section. Thus, the required balance of moments takes the form:

$$G(L - z) = EI/R(z), \tag{5.21}$$

showing that the largest curvature $R^{-1}$ and, hence, the largest stresses arise at the clamped end of the rod, where $R^{-1}(0) = GL/EI$. Therefore, the maximum stress is reached at the edge of the rod where $x = a$: $\sigma_{zz}^{(max)} = Ea/R(0) = GLa/I = 4GL/\pi a^3$, which yields the following upper limit for the sustainable weight:

$$G_{max} = \alpha(\pi a^2)E(a/4L) \tag{5.22}$$

In order to verify the validity of the above solution, one has to check that the maximum deflection of the rod from the horizontal line, which is $X(z = L)$, remains small, i.e., $X(L) \ll L$. Thus, the shape of the rod is determined by the following equation:

$$R^{-1}(z) \approx \frac{d^2X}{d^2z} = G\frac{(L-z)}{EI},$$

which after integration with the boundary conditions for the clamped rod, $X(0) = 0$, $(dX/dz)_{z=0} = 0$, yields $x(z) = GLz^2(1 - z/3L)/2EI$. Therefore, $X(L)/L \sim GL^2/Ea^4$ is small indeed even under the maximum load (5.22), if $\alpha \ll a/L$.

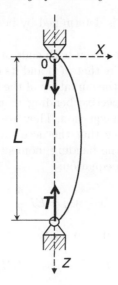

**FIGURE 5.5**
On the bending instability of a compressed rod

## Problem 5.1.6

Determine the critical compression force required to make a thin rod unstable with respect to bending (Euler's instability, Figure 5.5).

The magnitude of the sought after critical force, $T_{cr}$, depends on the boundary conditions at the ends of the rod. Consider first a rod with hinged ends, as shown in Figure 5.5. In this case both ends of the rod are free to turn; hence, there is no bending moment there and, therefore, the radius of curvature is equal to zero. The unperturbed straight form of the rod becomes unstable when another equilibrium, with a small bending $x(z)$, also exists. Thus, if the rod is under a given compressioin force $T$, the equilibrium requirement for its portion between the upper end, $z = 0$, and $z$ is as follows: the moment of the external force, which is equal to $-Tx(z)$, must be balanced by the bending moment there, i.e.,

$$-Tx = \frac{EI}{R(z)} \approx EI\frac{d^2x}{d^2z}$$

A non-trivial, $x(z) \neq 0$, solution of this equation with the boundary conditions $x(0) = x(L) = 0$ does exist when $(T/EI)^{1/2}L = \pi$, which yields the critical compression force $T_{cr} = \pi^2 EI/L^2$. If the cross section of the rod is not circular, its moment of inertia $I$ depends on the plane of bending. Therefore, in this case the instability threshold is determined by bending in the plane with the minimum moment of inertia.

Consider now another case, when both ends of the rod are clamped. Then the boundary conditions require that $x(z)$ and its derivative, $dx/dz$, are equal to zero at $z = 0, L$. However, the curvature of the rod differs from zero at the ends, and, therefore, the respective bending moments should be accounted for in deriving the equilibrium equation. They are not known beforehand and are established at such a value that the necessary boundary conditions are satisfied. Thus, by denoting the bending moment at $z = 0$ by $M_0$, one now gets the following equilibrium equation:

$$-Tx + M_0 = EI\frac{d^2x}{d^2z}$$

Its solution with $x(0) = x'(0) = 0$ reads

$$x(z) = M_0 T^{-1}\{1 - \cos[(T/EI)^{1/2}z]\},$$

and the boundary conditions at the other end, $z = L$, which are $x(L) = x'(L) = 0$, are satisfied when $(T/EI)^{1/2}L = 2\pi$. Therefore, the critical force in this case is equal to $T_{cr} = 4\pi^2 EI/L^2$, and, as before, the instability threshold is determined by bending in the plane with the minimum moment of inertia $I$.

## 5.2   Mechanics of liquid crystals

The state of **nematic liquid crystals**, or simply nematics, is defined by the distribution in space of $\vec{n}(\vec{r})$, where $\vec{n}$ is a unit vector called a **director**. This indicates the direction of preferred orientation for molucules in the vicinity of any point $\vec{r}$. The vectors $\vec{n}$ and $-\vec{n}$ are physically equivalent, so only a particular axis of the orientation can be specified. In the absence of any external action the equilibrium state of the nematic corresponds to a uniform field of directors $(\vec{n}(\vec{r}) = const)$. There are three independent types of deformation from this equilibrium, which are shown in Figure 5.6: (a) splay, (b) twist, and (c) bend. Splaying is characterized by $(\vec{\nabla} \cdot \vec{n}) \neq 0$, twisting by $\vec{n} \cdot (\vec{\nabla} \times \vec{n}) \neq 0$, and bending by $\vec{n} \times (\vec{\nabla} \times \vec{n}) \neq 0$. In a deformed state nematics possess excess elastic energy, and the volumetric density of this free energy $\mathcal{F}$ is formed by the additive contributions from all three types of deformation:

$$\mathcal{F} = K_1(\vec{\nabla} \cdot \vec{n})^2 + K_2(\vec{n} \cdot (\vec{\nabla} \times \vec{n}))^2 + K_3(\vec{n} \times (\vec{\nabla} \times \vec{n}))^2,$$

where $K_1$, $K_2$, $K_3$ are positive elastic moduli of the nematic: $K_1$ is the splay modulus, $K_2$ is the twist modulus, and $K_3$ is the bend modulus.

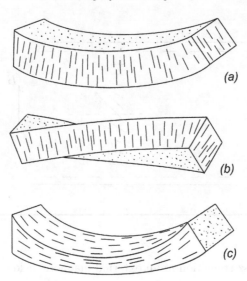

**FIGURE 5.6**
Different types of deformation in nematics: (a) splay, (b) twist, (c) bend.

## Problem 5.2.1

Determine the equilibrium field of directors for the ground state of a **cholesteric liquid crystal** (or simply cholesteric).

The difference between the cholesterics and nematics is that molecules of the former have no center of inversion. Therefore, they can be also characterized by pseudoscalar quantities, and their elastic free energy, $\mathcal{F}$, has an extra term of the form $\alpha(\vec{n} \cdot (\vec{\nabla} \times \vec{n}))$, where $\alpha$ is a pseudoscalar. Thus, the respective free energy takes the form

$$\mathcal{F} = K_1(\vec{\nabla} \cdot \vec{n})^2 + K_2(\vec{n} \cdot (\vec{\nabla} \times \vec{n}))^2$$
$$+ K_3(\vec{n} \times (\vec{\nabla} \times \vec{n}))^2 + \alpha(\vec{n} \cdot (\vec{\nabla} \times \vec{n})) \tag{5.23}$$

The equilibrium ground state corresponds to a minimum of $\mathcal{F}$, and, unlike the nematics, for the cholesterics this is not a uniform field of directors. Indeed, it follows from expression (5.23) that the extremum conditions for the free energy now read

$$\vec{\nabla} \cdot \vec{n} = 0, \quad \vec{n} \times (\vec{\nabla} \times \vec{n}) = 0,$$
$$(\vec{n} \cdot (\vec{\nabla} \times \vec{n})) = -\alpha/2K_2 \equiv -t_0 = const \tag{5.24}$$

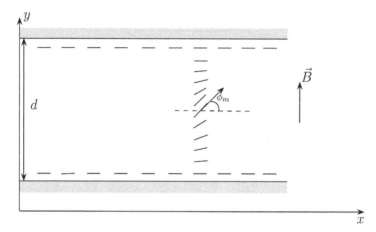

**FIGURE 5.7**

On the Fredericksz transition in an external magnetic field

It means the presence of a uniform twist deformation in the ground state of the cholesteric. For example, consider the solution of equations (5.24) that depends only on one coordinate, say $z$. Then, the condition $\vec{\nabla} \cdot \vec{n} = 0$ results in $n_z = const$, which, together with the $\vec{n} \times (\vec{\nabla} \times \vec{n}) = 0$, yields $n_z = 0$. This means a plane field of directors, $(n_x, n_y)$, which can be represented as: $n_x = \cos \phi(z)$, $n_y = \sin \phi(z)$. Since in this case $(\vec{\nabla} \times \vec{n})_x = -\cos \phi \, d\phi/dz$, and $(\vec{\nabla} \times \vec{n})_y = -\sin \phi \, d\phi/dz$, it follows from the last of equations (5.24) that $(\vec{n} \cdot (\vec{\nabla} \times \vec{n})) = -d\phi/dz = -t_0$, that is $\phi(z) = \phi_0 + t_0 z$. Thus, the ground state of the cholesteric liquid crystal has a helicoidal structure, in which $\vec{n}$ is uniform in each of the family of parallel planes, and it is uniformly twisted about the axis that is normal to these planes. This structure has a full pitch equal to $2\pi/t_0$, but since $\vec{n}$ and $-\vec{n}$ are equivalent, the physically significant period is equal to $\pi/t_0$.

### Problem 5.2.2

The domain between two parallel glass plates is filled with a nematic. Due to its relatively strong interaction with the glass, all directors are constrained to lie parallel to the glass surfaces at the boundaries (see Figure 5.7). Derive the magnitude of an external magnetic field which makes the uniform field of directors unstable (the Fredericksz transition). Consider, for simplicity, that the moduli $K_1$ and $K_3$ are equal: $K_1 = K_3 = K$.

The presence of an external magnetic field affects the free energy of the nematic. For a sufficiently weak field it can be accounted for with an additional term in the free energy:

$$\mathcal{F} = K_1(\vec{\nabla} \cdot \vec{n})^2 + K_2(\vec{n} \cdot (\vec{\nabla} \times \vec{n}))^2 +$$
$$K_3(\vec{n} \times (\vec{\nabla} \times \vec{n}))^2 - \beta(\vec{n} \cdot \vec{B})^2 \qquad (5.25)$$

In the absence of the field the directors are oriented uniformly between the plates (by using the coordinates chosen in Figure 5.7, one has $n_x = 1$, $n_y = 0$). However, if the constant $\beta$ in (5.25) is positive, the magnetic field tries to make $\vec{n}$ parallel to $\vec{B}$, which brings about a competition between the orientational actions of the field and of the glass. The result is as follows. If the strength of the external magnetic field is below some critical value $B_{cr}$, the orientation of the directors remains uniform (parallel to the glass surfaces) in the entire volume. Then, when $B$ exceeds $B_{cr}$, such a uniform orientation of directors becomes unstable, which yields a non-uniform stable equilibrium.

In order to determine the value of $B_{cr}$, one has to obtain the equilibrium equation for the field of directors $\vec{n}(\vec{r})$. In this particular case there are only $n_x(y)$, $n_y(y)$, which can be represented as $n_x = \cos\phi$, $n_y = \sin\phi$. Thus, there is no twist component of the deformation, and the total elastic energy, according to expression (5.25), takes the form:

$$W = \int\limits_0^d \mathcal{F} dy = \int\limits_0^d dy \left[ K_1 \cos^2\phi \left(\frac{d\phi}{dy}\right)^2 + K_3 \sin^2\phi \left(\frac{d\phi}{dy}\right)^2 - \beta B^2 \sin^2\phi \right]$$

$$= \int\limits_0^d dy \left[ K \left(\frac{d\phi}{dy}\right)^2 - \beta B^2 \sin^2\phi \right]$$

The equilibrium state corresponds to a minimum of $W$ as a functional of $\phi(y)$. Therefore, the respective Euler-Lagrange equation for the $\phi(y)$ reads

$$\frac{d^2\phi}{d^2y} + \frac{\beta B^2}{K} \sin\phi \cos\phi = 0 \qquad (5.26)$$

Whatever the value of $B$, it has the solution $\phi(y) = 0$, which corresponds to the uniform orientation of directors parallel to the glass surface. However, this solution becomes unstable at such a field $B = B_{cr}$, when another solution, with a small deformation of the nematic, $\phi \ll 1$, does exist (note the analogy with the Euler's instability of a compressed rod considered in Problem 5.1.6). Then, for small $\phi$ equation (5.26) reduces to

$$\frac{d^2\phi}{d^2y} + \frac{\beta B^2}{K}\phi = 0$$

It has the solution with the required boundary conditions $\phi(0) = \phi(d) = 0$, when $(\beta B^2/K)^{1/2}d = \pi$, therefore

$$B_{cr} = \frac{\pi}{d} \left(\frac{K}{\beta}\right)^{1/2}$$

## Problem 5.2.3

Determine the equilibrium field of directors for the setup of Problem 5.2.2, when the external magnetic field is only slightly above the critical one, i.e., $(B - B_{cr}) \ll B_{cr}$.

If $(B - B_{cr}) \ll B_{cr}$, the deformation of the otherwise uniform distribution of $\vec{n}$ is weak; therefore, the equilibrium equation (5.26) can be approximated as

$$\frac{d^2\phi}{d^2y} \approx -\frac{\beta B^2}{K}\left(\phi - \frac{2}{3}\phi^3\right)$$

Its first integral (which can be easily found with the help of the mechanical analogy, as in Problem 3.5.7) reads:

$$\frac{1}{2}\left(\frac{d\phi}{dy}\right)^2 + \frac{\beta B^2}{2K}\left(\phi^2 - \frac{\phi^4}{3}\right) = const = \frac{\beta B^2}{2K}\left(\phi_m^2 - \frac{\phi_m^4}{3}\right),$$

where $\phi_m$ is the maximum tilting angle of directors, corresponding to $y = d/2$ as shown in Figure 5.7. Another integration with the boundary condition $\phi(0) = 0$ yields

$$B(\beta/K)^{1/2}y = \int\limits_0^\phi d\xi \left(\phi_m^2 - \xi^2 - \frac{1}{3}\phi_m^4 + \frac{1}{3}\xi^4\right)^{-1/2}$$

Then, the maximum tilt angle $\phi_m$ is determined by the following relation:

$$B\left(\frac{\beta}{K}\right)^{1/2}\frac{d}{2} = \int\limits_0^{\phi_m} d\xi \left(\phi_m^2 - \xi^2 - \frac{1}{3}\phi_m^4 + \frac{1}{3}\xi^4\right)^{-1/2} =$$

$$\int\limits_0^1 dx \left[1 - x^2 - \frac{1}{3}\phi_m^2(1 - x^4)\right]^{-1/2} \tag{5.27}$$

Under $\phi_m \ll 1$, the integral in equation (5.27) can be calculated approximately as

$$\int\limits_0^1 dx \left[1 - x^2 - \frac{1}{3}\phi_m^2(1 - x^4)\right]^{-1/2} \approx$$

$$\int\limits_0^1 \frac{dx}{(1 - x^2)^{1/2}}\left[1 + \frac{\phi_m^2}{6}(1 + x^2)\right] = \frac{\pi}{2} + \frac{\pi}{8}\phi_m^2$$

Therefore,

$$\phi_m \approx 2 \left[ \frac{B}{B_{cr}} - 1 \right]^{1/2},$$

so that when the external magnetic field slightly exceeds the critical one, the respective deformation in the distribution of directors is proportional to a square root of the supercriticality.

## Problem 5.2.4

If the external magnetic field applied to a cholesteric is directed perpendicular to the twist axis of the latter, it has the unwinding effect on the spatial orientation of its molecules. Determine the critical magnetic field, under which the ground state of the cholesteric corresponds to a uniform distribution of the directors $\vec{n}$.

If the twist axis of the cholesteric is directed along the $z$-axis, the distribution of the directors can be written as:

$$n_x(z) = \cos \phi(z), \ n_y(z) = \sin \phi(z), \ n_z = 0$$

Thus, if the external magnetic field is directed along the $y$-axis, the volumetric density of the elastic energy is equal, according to expressions (5.23)-(5.25), to

$$\mathcal{F} = K_2 \left( \frac{d\phi}{dz} - t_0 \right)^2 - \beta B^2 \sin^2 \phi, \ t_0 = \frac{\alpha}{2K_2} \qquad (5.28)$$

As seen from (5.28), if $\beta > 0$, the magnetic field tries to orient directors along the $y$-axis; hence, its effect is unwinding. In order to find out how it competes with the originally twisted structure of the cholesteric, one has to obtain the equation of equilibrium, which follows from the minimization of the total elastic energy

$$W = \int dz \mathcal{F} = \int dz \left[ K_2 \left( \frac{d\phi}{dz} - t_0 \right)^2 - \beta B^2 \sin^2 \phi \right] \qquad (5.29)$$

Then, the respective Euler-Lagrange equation (the equlibrium equation) reads

$$\frac{d^2\phi}{d^2z} + \frac{\beta B^2}{K_2} \sin \phi \cos \phi = 0$$

Its first integral is

$$\gamma^2 \left( \frac{d\phi}{dz} \right)^2 + \sin^2 \phi = const = \kappa^{-2}, \qquad (5.30)$$

where $\gamma \equiv (K_2/\beta B^2)^{1/2}$, and the integration constant, $\kappa^{-2}$, must be greater than unity since $\phi$ is a periodic function of $z$. Then, the dependence of $\phi$ on $z$ can be written implicitly as

$$\frac{z}{\gamma\kappa} = \int_0^\phi dx(1 - \kappa^2 \sin^2 x)^{-1/2} \equiv F(\kappa, \phi),$$

where $F(\kappa, \phi)$ is the elliptic integral of the first kind (see, e.g., M. Abramowitz and I. Stegun, *Handbook of Mathematical Functions*, §17, Dover, New York). The spatial period of the structure, $L$, which corresponds to the variation of $\phi$ by $2\pi$, is equal to $L = 4\gamma\kappa K(\kappa)$, where

$$K(\kappa) = \int_0^{\pi/2} dx(1 - \kappa^2 \sin^2 x)^{-1/2} \equiv F(\kappa, \pi/2)$$

is the complete elliptic integral of the first kind. Hence, the elastic energy per unit volume, averaged over the period, is, according to expression (5.29), equal to

$$\tilde{W} = \frac{1}{L} \int_0^L dz \left[ K_2 \left( \frac{d\phi}{dz} - t_0 \right)^2 - \beta B^2 \sin^2 \phi \right] =$$

$$\frac{4\beta B^2}{L} \int_0^{\pi/2} \frac{d\phi}{d\phi/dz} \left[ \gamma^2 \left( \frac{d\phi}{dz} - t_0 \right)^2 - \sin^2 \phi \right] \qquad (5.31)$$

By substitution of $d\phi/dz$ from (5.30), the expression (5.31) for $\tilde{W}$ can be transformed into

$$\tilde{W} = \beta B^2 \left( 2\kappa^{-2} \frac{E(\kappa)}{K(\kappa)} - \frac{\gamma\pi t_0}{\kappa K(\kappa)} - \kappa^{-2} + \gamma^2 t_0^2 \right) \qquad (5.32)$$

where

$$E(\kappa) \equiv \int_0^{\pi/2} dx(1 - \kappa^2 \sin^2 x)^{1/2}$$

is the complete elliptic integral of the second kind (the identity

$$\int_0^{\pi/2} \frac{\sin^2 x \, dx}{(1 - \kappa^2 \sin^2 x)^{1/2}} = \frac{K(\kappa) - E(\kappa)}{\kappa^2}$$

has been used here). The yet unknown value of $\kappa$ has to be obtained by seeking the minimum of the energy (5.32) with respect to this parameter. By taking into account the following differentiation rules for the complete elliptic integrals:

$$\frac{dK}{d\kappa} = \frac{E}{\kappa(1 - \kappa^2)} - \frac{K}{\kappa}, \quad \frac{dE}{d\kappa} = \frac{E - K}{\kappa},$$

the required condition $d\tilde{\mathcal{W}}/d\kappa = 0$ yields

$$\frac{E(\kappa)}{\kappa} = \frac{\pi}{2}\gamma t_0 \tag{5.33}$$

This relation determines $\kappa$ as a function of $\gamma$ and, hence, its dependence on the applied magnetic field $B$. By using (5.33), one can get the following expression for the spatial period of the structure:

$$\frac{L}{L_0} = \frac{4}{\pi^2} K(\kappa)E(\kappa), \tag{5.34}$$

where $L_0 = 2\pi/t_0$ is the pitch length of the helical structure in the absence of the magnetic field. It follows from relations (5.33) and (5.34), that as the magnetic field is increasing, the cholesteric unwinds (the pitch length grows). For a weak field, when $(\gamma t_0)^{-1} = Bt_0(\beta/K_2)^{1/2} \ll 1$, the parameter $\kappa$ is small: $\kappa \approx (\gamma t_0)^{-1}$, and, according to relation (5.34), $L/L_0 \approx 1 + (\gamma t_0)^{-4}/32$, so the pitch grows slowly with the applied magnetic field. The sought after transformation from the cholesteric helicoidal structure to the uniform field of directors corresponds to the infinite pitch length, which occurs at $\kappa = 1$. As seen from relation (5.33), at this point $\gamma_{cr}t_0 = 2/\pi$, thus the critical magnetic field is equal to $B_{cr} = (\pi t_0/2)(K_2/\beta)^{1/2}$. If the magnitude of the external magnetic field exceeds this value, the cholesteric liquid crystal acquires a uniform structure.

## Problem 5.2.5

Determine the field of directors $\vec{n}(\vec{r})$ for a straight **disclination** with the **Frank index** $m$ in a nematic, when the moduli $K_1$ and $K_3$ are equal to each other: $K_1 = K_3 = K$.

If the disclination line is directed along the $z$-axis, the field of directors is plane, with the non-zero components $n_x = \cos\phi(x, y)$, $n_y = \sin\phi(x, y)$. Clearly, for a given field of directors $\vec{n}(x, y)$ the value of the angle $\phi(x, y)$ is not specified uniquely: any quantity of the form $l\pi$, where $l$ is either a positive or a negative integer, can be added to $\phi$ without making any difference in $\vec{n}(x, y)$. The very presence of the disclination means that while passing along any closed path in the $(x, y)$ plane around the origin of the coordinate system (which is the projection of the disclination line), the angle $\phi$ acquires the increment equal to $m\pi$, with the integer $m$ called the Frank index of the disclination: $\int \vec{\nabla}\phi \cdot \vec{dl} = m\pi$.

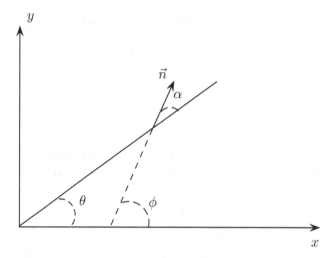

**FIGURE 5.8**
On the derivation of the field of directors for a straight disclination

The equilibrium equation for the function $\phi(x, y)$ can be obtained in the same way as it has been done in Problem 5.2.2. In this case there is no twisting deformation, so the free energy density takes the simple form: $\mathcal{F} = K(\vec{\nabla}\phi)^2$. Therefore, the total energy per unit length of the disclination is equal to $\mathcal{W} = \int K(\vec{\nabla}\phi)^2 dxdy$, and the required minimum of $\mathcal{W}$ yields the equilibrium equation $\vec{\nabla}^2\phi = 0$. By introducing the polar coordinates $(r, \theta)$ in the $(x, y)$ plane, one concludes that its relevant solution must not depend on $r$ (otherwise the angle increment $\int \vec{\nabla} \cdot d\vec{l}$ would be different for the circles with the different radii). Thus, the equilibrium equation reduces to $d^2\phi/d^2\theta = 0$, so that $\phi(\theta) = A\theta + \phi_0$. The constant $A$ is related to the Frank index as $\int \vec{\nabla}\phi \cdot d\vec{l} = m\pi$, hence

$$\phi(\theta) = \frac{m}{2}\theta + \phi_0$$

The resulting field of directors $\vec{n}$ can be sketched by plotting the "streamlines" of the directors, which are tangential to $\vec{n}$ at each point in the $(x, y)$ plane. Thus, their differential equation is (see Figure 5.8):

$$\frac{1}{r}\frac{dr}{d\theta} = \cot\alpha = \cot\left[\left(\frac{m}{2} - 1\right)\theta + \phi_0\right]$$

If the Frank index of the disclination is such that $m \neq 2$, the constant $\phi_0$ can be eliminated by the appropriate choice of the line $\theta = 0$. The field of directors for the cases $m = 1$ and $m = -1$ are shown in Figure 5.9. When $m = 2$, the result depends on $\phi_0$; hence some of these possibilities are also depicted there.

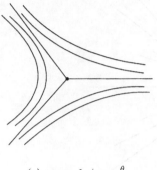

(a) $m = -1, \phi = -\frac{\theta}{2}$

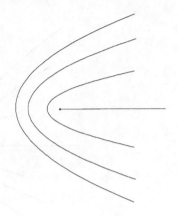

(b) $m = 1, \phi = \frac{\theta}{2}$

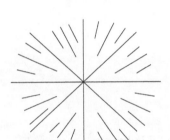

(c) $m = 2, \phi_0 = 0, \phi = \theta$

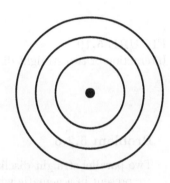

(d) $m = 2, \phi_0 = \frac{\pi}{2}, \phi = \theta + \frac{\pi}{2}$

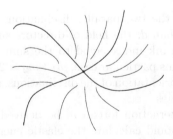

(e) $m = 2, \phi_0 = \frac{\pi}{4}, \phi = \theta + \frac{\pi}{4}$

**FIGURE 5.9**
Disclinations with different Frank indices

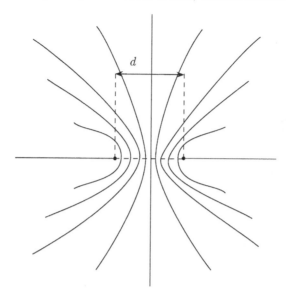

**FIGURE 5.10**
Interaction of two straight disclinations

## Problem 5.2.6

Two parallel straight disclinations with the Frank indices $m_1 = m_2 = 1$
are present in a nematic with equal elastic moduli ($K_1 = K_2 = K$). Plot
the resulting field of directors and derive the interaction force per unit
length of the disclinations.

If the separation of the two parallel disclinations is equal to $d$, then at
distances much larger than $d$, the field of directors should resemble a single
disclination with the Frank index equal to the sum of the indices of both
disclinations. Since, in this particular case, $m_1 + m_2 = 2$, the resulting diagram
depends on the relative orientation of the disclinations, and Figure 5.10 depicts
just one of the possibilities.

The sought after interaction force can be derived from the energy consideration. First, one should calculate the elastic energy of a single straight
disclination with the Frank index $m$, which is equal to

$$W = K \int dx dy (\vec{\nabla}\phi)^2 = K \int \left( \frac{1}{r} \frac{d\phi}{d\theta} \right)^2 2\pi r dr = \frac{\pi m^2 K}{2} \int \frac{dr}{r} \quad (5.35)$$

As seen from expression (5.35), the resulting integral is logarithmically divergent both at small and large distances from the disclination line. Therefore, the global size, $L$, of the domain occupied by the liquid crystal should be used as the upper limit of integration in (5.35). As far as the lower limit is concerned, it has to be of the order of the molecular size $r_0$, because the macroscopic description used here is not applicable for $r \leq r_0$. Hence, equation (5.35) can be written as

$$W \approx \frac{1}{2} \pi m^2 K \ln \left( \frac{L}{r_0} \right)$$

For a macroscopic sample of a liquid crystal the ratio $L/r_0 \gg 1$, so the energy of the disclination is proportional to a large logarithm, the consequences of which are twofold. Firstly, it makes the disclination elastic energy practically insensitive to the particular value of $L$ and $r_0$. Secondly, this energy greatly exceeds the elastic energy concentrated at the core of the disclination (at $r \leq r_0$), and this very feature allows one to consider disclination as a macroscopic object.

One may notice from expression (5.35) that from the energy viewpoint such a disclination is similar to a straight uniformly charged line. Indeed, if the charge per unit length is $\kappa$, the respective energy of the electric field is

$$\mathcal{W}_{el} = \int \frac{E^2}{8\pi} 2\pi r dr = \kappa^2 \int \frac{dr}{r}$$

Thus, by comparing this with (5.35), one concludes that the "effective" charge of the disclination is equal to $\tilde{\kappa} = m(\pi K/2)^{1/2}$, and, therefore, the interaction force of the two parallel disclinations is equal to

$$T = \tilde{\kappa}_1 (2\tilde{\kappa}_2/d) = \pi K m_1 m_2 / d$$

If $m_1 = m_2 = 1$, the disclinations repel each other with the force $T = \pi K/d$.

## Problem 5.2.7

Determine the force acting on a straight disclination with the Frank index $m$, if it is located at a distance $l$ from the boundary plane of the nematic with the fixed orientation of directors (see Figure 5.11).

This problem can be solved by exploring the above mentioned electrostatic analogy. As seen from Figure 5.11, the equilibrium distribution of directors, which satisfies the required boundary condition, can be formed by two disclinations: the original one, and the "reflected" one with the same Frank index. Indeed, the angles $\theta_1$ and $\theta_2$ are equal to each other; therefore, while moving along the boundary surface, the one discinlation provides the clockwise rotation of the director there, while the other one makes it counterclockwise with

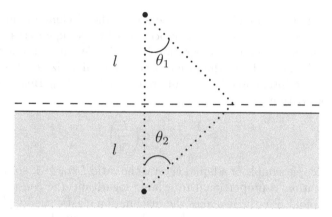

**FIGURE 5.11**
Repulsion of disclination from a boundary

the same pace. As a result, the orientation of directors at the boundary does not change. Then, it follows from the previous problem, that the disclination under consideration is repelled from the boundary with the force per unit length equal to $\pi K m^2 / 2l$.

## Problem 5.2.8

A nematic liquid crystal with the equal elastic moduli ($K_1 = K_3 = K$) occupies the domain inside a cylindrical tube of radius $R$. At the tube surface the directors are oriented parallel to the boundary and lie in the plane perpendicular to the axis of the tube (see Figure 5.12). Determine the ground energy equilibrium state of directors.

It follows from the definition of the problem, that the distribution of directors should be a plane one, with $n_z = 0$ (the $z$-axis is directed along the axis of the tube). Furthermore, since at the boundary circle (at $r = R$) the directors are parallel to it, the increment of the angle $\phi$ while moving along the boundary circle is equal to $2\pi$. It means that the disclinations must be present in the nematic, with the sum of their Frank indices equal to 2. An obvious solution is just one disclination with $m = 2$, located at the center, with the circular pattern of directors orientation (see Figure 5.9). However, such a configuration does not provide a minimum of the elastic energy, and, therefore, it is unstable. Indeed, suppose that this single disclination with $m = 2$ splits into two very close to each other disclinations with $m_1 = m_2 = 1$. Then, according to Problem 5.2.6, they would repel each other trying to increase their separation. On the other hand, as follows from Problem 5.2.7, they can-

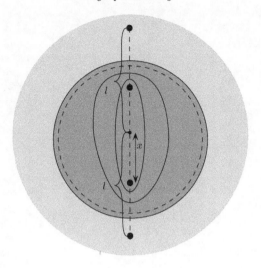

**FIGURE 5.12**
Equilibrium with the minimum of elastic energy

not become too close to the boundary with a fixed orientation of directors due to another respective repulsion. Therefore, in the ground energy equilibrium state their separation $x$ from the center should be such that the two repulsive actions balance each other. The respective solution can be easily found by using again the electrostatic analogy, so that the two "reflected" disclinations with $m = 1$, located outside the tube at a distance $l = R^2/x$ from the center, are added. Then, the balance of forces reads:

$$\frac{1}{(l-x)} = \frac{1}{2x} + \frac{1}{(l+x)},$$

yielding $x = R/5^{1/4}$. The resulting pattern of directors is shown in Figure 5.12.

# Index